SO-AHG-463

Sensory
Transduction

NATO ASI Series

Advanced Science Institutes Series

A series presenting the results of activities sponsored by the NATO Science Committee, which aims at the dissemination of advanced scientific and technological knowledge, with a view to strengthening links between scientific communities.

The series is published by an international board of publishers in conjunction with the NATO Scientific Affairs Division

A	**Life Sciences**	Plenum Publishing Corporation
B	**Physics**	New York and London
C	**Mathematical**	Kluwer Academic Publishers
	and Physical Sciences	Dordrecht, Boston, and London
D	**Behavioral and Social Sciences**	
E	**Applied Sciences**	
F	**Computer and Systems Sciences**	Springer-Verlag
G	**Ecological Sciences**	Berlin, Heidelberg, New York, London,
H	**Cell Biology**	Paris, and Tokyo

Recent Volumes in this Series

Volume 189—Free Radicals, Lipoproteins, and Membrane Lipids
edited by A. Crastes de Paulet, L. Douste-Blazy,
and R. Paoletti

Volume 190—Control of Metabolic Processes
edited by Athel Cornish-Bowden and María Luz Cárdenas

Volume 191—Serine Proteases and Their Serpin Inhibitors
in the Nervous System:
Regulation in Development and in Degenerative and
Malignant Disease
edited by Barry W. Festoff

Volume 192—Systems Approaches to Developmental Neurobiology
edited by Pamela A. Raymond, Stephen S. Easter, Jr.,
and Giorgio M. Innocenti

Volume 193—Biomechanical Transport Processes
edited by Florentina Mosora, Colin G. Caro, Egon Krause,
Holger Schmid-Schönbein, Charles Baquey, and Robert Pelissier

Volume 194—Sensory Transduction
edited by Antonio Borsellino, Luigi Cervetto, and Vincent Torre

Volume 195—Experimental Embryology in Aquatic Plants and Animals
edited by Hans-Jürg Marthy

Series A: Life Sciences

Sensory Transduction

Edited by

Antonio Borsellino

International School for Advanced Studies (SISSA)
Trieste, Italy

Luigi Cervetto

Institute of Neurophysiology, CNR
Pisa, Italy

and

Vincent Torre

University of Genoa
Genoa, Italy

Plenum Press
New York and London
Published in cooperation with NATO Scientific Affairs Division

Proceedings of a NATO Advanced Study Institute/
18th Course of the International School of
Biophysics on Sensory Transduction,
held June 9–19, 1988,
in Erice, Italy

Library of Congress Cataloging in Publication Data

(Revised for vol. 194)

Sensory transduction.

(NATO ASI series. Series A, Life sciences; vol. 194)
"Proceedings of a NATO Advanced Study Institute/18th Course of the
International School of Biophysics on Sensory Transduction, held June 9–19,
1988, in Erice, Italy"—T.p. verso.
"Published in cooperation with NATO Scientific Affairs Division."
Includes bibliographical references and index.
1. Senses and sensation Congresses. 2. Cellular signal transduction
Congresses. I. Borsellino, Antonio. II. Cervetto, L. III. Torre, Vincent,
1950–   . IV. North Atlantic Treaty Organization. Scientific Affairs Division.
V. Series: NATO ASI series. Series A, Life science; v. 194.
QP431.N27   1988   591.1′82   90-42555
ISBN 0-306-43677-9

© 1990 Plenum Press, New York
A Division of Plenum Publishing Corporation
233 Spring Street, New York, N.Y. 10013

All rights reserved

No part of this book may be reproduced, stored in a retrieval system, or transmitted
in any form or by any means, electronic, mechanical, photocopying, microfilming,
recording, or otherwise, without written permission from the Publisher

Printed in the United States of America

PREFACE

Present knowledge of the mechanisms underlying any single sensory modality is so massive as to discourage effort directed towards completeness. The idea underlying the structure of this volume on "Sensory transduction" was to select just a few topics of general interest, which are currently being investigated and for which a reasonably clear picture is now available.

During the last five years there has been a revolution in the way sensory physiologists think about transduction, and a series of exciting advances have been made in understanding the basic processes of phototransduction, chemotransduction and mechanotransduction. It is clear that in many cases the fundamental processes by which nature attains optimization of performance are similar, and that they have much in common with more general processes of signal recognition by living structures. The molecular events underlying the detection of photons by visual cells, the recognition of a given molecule by a chemoreceptor, or the level of a hormone in the extracellular fluid by a target cell, are all very similar, and involve the activation of a sequence of events leading to a second messenger.

The 20 papers that form the present volume cover various topics in the field of sensory transduction. They originate from the lectures, seminars and discussions which made up the XVIII Course of the International School of Biophysics held in Erice, 9th - 19th June 1988.

The contributors to this volume wish to dedicate their papers to the memory of Brian Nunn, who for many of them has been an inspiring colleague and friend.

A. Borsellino

L. Cervetto

V. Torre

BNA 2-14-91

AN APPRECIATION OF THE SCIENTIFIC WORK OF BRIAN J. NUNN

P.A. McNaughton

The 1988 School on Sensory Transduction, held in Erice, Sicily, was dedicated to the memory of Brian Nunn, who was a close friend and collaborator of several of the organisers and participants. Brian had spent his scientific life - which was all too brief - in the study of visual transduction, so this dedication was especially appropriate.

Brian was born in Bristol, England, on 15th November, 1951, but his parents left soon after for South Africa, where he grew up. He suffered from a congenital heart defect which prevented him from participating in the sporting activities which are such an important part of life in South Africa. Brian never mentioned this period of his life to me, but the enforced solitude and the consequent freedom from peer pressure may have contributed to his total rejection of the system of racial segregation in South Africa. The same isolation, with its time to think things through to their logical conclusion, may also have explained his extraordinary tenacity in pursuing meaningful experiments and clearcut answers in science.

Brian returned to England in 1974 to study for his Ph.D. degree at Imperial College, London, where he carried out psychophysical experiments on colour blindness and normal colour vision in man under the supervision of Dr. Keith Ruddock. At Imperial, Brian showed no sign of his illness and soon after his arrival acquired a bicycle for getting around. He fully enjoyed the three years in London, sharing a house with fellow post-graduate students. During this period, he demonstrated a natural ability to make friendships, including with scientific collegues and enjoyed having regular dinner parties and social discussions with his friends. His research into the cellular basis of photo-transduction began when he moved in 1978 to Denis Baylor's laboratory in Stanford. In collaboration with Denis Baylor, and later with Gary Matthews and Julie Schnapf, he began work on two separate but related questions: the relation between the outer segment membrane current and the transmembrane voltage in rods, and the response properties of primate photoreceptors. For the first study Brian and Denis (in common with other researchers) chose the salamander rod as a model. Salamanders have large, robust photoreceptors, which makes them a useful system when technically difficult experiments such as those they proposed are to be carried out. The main problem is that salamander vision would be of limited interest if the findings applied only to salamanders. This problem was tackled by the second approach: from direct recordings of the responses of monkey rods Brian and Denis hoped to find out to what extent the recordings from lower vertebrates were typical of the responses of the photoreceptors of higher primates such as ourselves. To anticipate somewhat, they found that in general the responses of salamander and monkey photoreceptors are remarkably similar, in spite of the evolutionary separation of many millions of years, but that there was one surprising exception: light adaptation, which is well developed in salamander rods, is weak or absent in primate rods.

Recording membrane current and voltage simultaneously was no mean feat. The outer segment of an isolated salamander rod had to be drawn into a suction pipette to record the current, and the inner segment then penetrated with a fine electrode to record voltage. Anyone who has attempted similar experiments knows how difficult this work must have been, but there is no trace of this in the published paper (Baylor, Matthews and Nunn, 1984); the precautions taken to avoid artefacts are clearly explained, and every trace is crisp and noise-free. This elegance is more than just a matter of aesthetics, because with clear data the conclusions are correspondingly clear. The experiments showed that normal voltage responses, with the peak-plateau relaxation which workers recording from intact retina had come to expect, were observed in these isolated rods without any trace of a similar relaxation in the recording of the outer segment current. The conclusion was simple: the ionic channels responsible for the relaxation are present only in the inner segment, and the outer segment acts as a high-impedance current source which drives the light-sensitive current through the inner segment almost irrespective of the voltage-sensitive conductance changes in the inner segment. A simple and sensible system, as the two functions of the ionic currents - transduction and response modulation - are segregated spatially, and the internal environment of the outer segment is shielded from changes in internal ionic concentrations which might result from the operation of voltage-sensitive channels.

In a further series of experiments (Baylor and Nunn, 1986) the properties of the light-sensitive channel were examined in more detail. To do this the cell membrane potential was voltage- clamped, which required the further experimental complication of a second current-passing microelectrode inserted into the inner segment. From these experiments the current-voltage relation of the light-sensitive channel was established reliably for the first time. Baylor and Nunn found that the steep outward rectification could be explained by blockage of the channel by a divalent ion. Subsequent experiments in Baylor's laboratory and in others have confirmed this idea, and have shown that the light-sensitive channel is blocked most of the time by divalent ions (in physiological solution calcium is normally the chief blocker). This apparently pointless exercise - channels are held open by cyclic GMP only to be blocked by calcium - in fact serves the useful function of reducing the single-channel conductance, because the block by calcium is both rapid and incomplete. The noisiness of the dark current and of the single-photon response are thereby minimised.

The experiments on primate rods were more straightforward from the point of view of interpretation, although the experimental difficulties involved in developing the technique of recording from the tiny rods of the monkey's retina can scarcely have been easier than inserting two microelectrodes into a salamander rod. Baylor, Nunn and Schnapf (1984) found that the dark current and single-photon response of monkey rods were surprisingly similar to those recorded from the rods of lower vertebrates; even the faster response time course was explained by the higher operating temperature of the primate. From the frequency of single-photon events they were able to show that the "dark light" which is observed in psychophysical experiments is not just a mathematical fiction required to make Weber's law valid at very low light intensities, but instead is close to the real rate of thermal isomerizations of rhodopsin in the primate retina. The spectral sensitivity of rod vision was determined with greater accuracy than previously possible. Finally, the most surprising observation was that the light adaptation seen in lower vertebrates is not prominent in primate rods. The conclusion from this last observation must be that the adaptation which is such a striking property of our scotopic vision must have its origin in neural processing in the retina, as it does not appear to be located in the rods themselves.

This work was extended to the cones of the monkey on Brian's several visits back to Stanford. Baylor, Nunn and Schnapf (1987) measured the spectral sensitivities and response properties of the three cone types, and were able to correlate these findings with the human colour discrimination measured many years ago in psychophysical experiments by W. S. Stiles. Brian's interest in the biophysical basis of human vision dated back to the days of his doctoral work in London, and was carried through to fruition in these precise and elegant experiments which have added so much to our understanding of our own visual system.

In 1982 Brian moved to Cambridge to join Alan Hodgkin and myself in experiments on the ionic basis of the dark current. We found (Hodgkin, McNaughton and Nunn, 1985) that the ionic selectivity of the dark current for sodium was less good than had been supposed; in fact the light-sensitive channel discriminated rather poorly between the alkali metal cations. This finding tied in nicely with the observation by Baylor and Nunn (made during Brian's sojourn in Stanford, but not published until 1986) that the reversal potential of the light-sensitive current lay between the Na and K equilibrium potentials. We also found that a large influx of calcium through the light-sensitive channel suppressed the dark current too slowly for calcium to be the internal transmitter released by light. This observation, which was first made in Cambridge, provided the first clear evidence against the "calcium hypothesis"; it was, of course, soon followed by the experiments of Fesenko and collaborators which established that isolated patches of outer segment membrane responded to cyclic GMP, as expected if cGMP was the internal transmitter, but were insensitive to changes in calcium. In this paper we proposed that the basis of the action of internal calcium was not directly on the light-sensitive channel, as in the original calcium hypothesis, but was instead an inhibition of the guanylate cyclase which synthesizes cGMP from GTP. This was at the time a somewhat serendipitous guess, backed up by some earlier evidence from biochemical studies, but later evidence has borne it out, and at the time of writing it still seems to be the most likely mechanism of action of internal calcium.

From 1985 onwards Brian worked with Alan Hodgkin on the properties of the Na:Ca exchange mechanism, and on the control of the light-sensitive current by internal calcium. For a few months, though, he lent his considerable experimental skills to the effort by Luigi Cervetto and myself to measure calcium inside single rod outer segments. This project was in the end successful (McNaughton, Cervetto and Nunn, 1986), and established that calcium declines in the rod outer segment after a flash of light, rather than increasing as required by the calcium hypothesis. This study put the final nail in the coffin of the calcium hypothesis, and in addition established both that the light-sensitive channel was the only significant means by which calcium enters the outer segment, and that the Na:Ca exchange was the only significant means by which it is extruded.

The photoreceptor outer segment is a wonderful preparation in which to study the Na:Ca exchange. The cell can be loaded with a known amount of calcium through the light-sensitive channels, which can then be closed with a bright flash of light, and the efflux of calcium through the Na:Ca exchange can then be measured from the membrane current which is produced by the electrogenic nature of the exchange. In two papers

Brian and Alan - and to a limited extent myself - measured the exchange stoichiometry, which was found to be close to one charge flowing into the cell for every Ca^{2+} extruded, and analysed the interactions of Na^+, Ca^{2+} and H^+ ions with the exchange (Hodgkin, McNaughton and Nunn, 1987; Hodgkin and Nunn, 1987). This work is important, of course, for the Na:Ca exchange in photoreceptors, but it also has relevance for the function of this exchange process in many other cells.

The other strand of Brian's recent work in Cambridge concerned the mechanism of control of the light-sensitive current. With Alan Hodgkin he developed a method for measuring the rates of synthesis and breakdown of cyclic GMP in an intact cell. Instantaneous inhibition of the PDE which breaks down cGMP should produce a rate of increase of dark current proportional to the rate of synthesis of cGMP. On the other hand, instantaneous inhibition of the guanylate cyclase which synthesizes cGMP should produce a rate of reduction of dark current proportional to the rate of destruction of cGMP by the PDE. In their experiments inhibition of the PDE was produced by rapid application of the phosphodiesterase inhibitor IBMX, which readily permeates the cell membrane, and the production of cGMP was inhibited by the somewhat more indirect method of removing external Na, which inhibits the Na:Ca exchange, causes a rapid buildup of calcium inside the cell, and consequently inhibits the calcium-sensitive guanylate cyclase. They were able to show (Hodgkin and Nunn, 1988) that not only is the rate of destruction of cGMP speeded by a flash of light or a steady background, as might be expected, but that the rate of its synthesis is also accelerated. The acceleration of the synthesis appeared with a short delay, which could be very well accounted for by the delay which is known to exist between the suppression of the light- sensitive current and the fall in internal calcium in the outer segment (McNaughton, Cervetto and Nunn, 1986). The idea that the breakdown of cGMP is controlled by light and that its synthesis is controlled by the level of calcium in the cell (Hodgkin, McNaughton and Nunn, 1985) was entirely consistent with their results.

In the autumn of 1986 Brian moved to a tenure-track position at Duke University in Durham, North Carolina. He worked frantically hard to get his apparatus going, and he was beginning to get results on the ionic selectivity of the light-sensitive channel in isolated patches of outer segment membrane (Nunn, 1987a, b) when he picked up an acute form of pneumonia, from which he died in a few days, on 18th September, 1987. He leaves a widow, Rashmi, and young twins, Aneirin and Anjuli.

This brief review of Brian's scientific achievements has hardly touched on his personality and approach to life. Suffice to say that Brian was liked by almost everyone who met him, and that his open and generous approach to life was quite unmarred by that aggressiveness that one meets so often in successful scientists. I could not do better than conclude by quoting Denis Baylor's words at Brian's funeral:

"Part of Brian's success came from the fact that he was very bright, but even more important were his exceptional character and personality traits. Brian loved science in a unique way. Many people do science, but few love it in the way that he did. He had a deep and genuine curiosity about nature. He wanted to understand how things work, and he loved to do experiments to find out how they work. Brian also had wonderful self-discipline and the ability to get the very best out of himself at all times. Because of these personal qualities Brian was a wonderful man to work with. No one who has worked with him can forget the excitement and pleasure of discussing a new experiment, the twinkle in his eye, and his saying "Let's do it!". The loss we mourn today is not only the loss of a wonderful person, but also the loss of a wonderful scientist. Science has lost a champion, a young hero who carried the banner high."

REFERENCES

Baylor, D.A., Matthews, G. and Nunn, B.J., 1984 Location and function of voltage-sensitive conductances in retinal rods of the salamander, Ambystoma tigrinum, *J. Physiol.*, 354:203-223.

Baylor, D.A. and Nunn, B.J., 1986, Electrical properties of the light-sensitive conductance of rods of the salamander Ambystoma tigrinum, *J. Physiol.*, 371:115-145.

Baylor, D.A., Nunn, B.J. and Schnapf, J.L., 1984, The photocurrent, noise and spectral sensitivity of rods of the monkey, Macaca fascicularis, *J. Physiol.*, 357:575-607.

Baylor, D.A., Nunn, B.J. and Schnapf, J.L., 1987, Spectral sensitivity of cones of the monkey Macaca fascicularis, *J. Physiol.*, 390:145-160.

Hodgkin, A.L., McNaughton, P.A. and Nunn, B.J., 1985, The ionic selectivity and calcium dependence of the light-sensitive pathway in toad rods, *J. Physiol.*, 358:447-468.

Hodgkin, A.L., McNaughton, P.A. and Nunn, B.J., 1987, Measurement of sodium-calcium exchange in salamander rods, *J. Physiol.*, 391:347-370.

Hodgkin, A.L. and Nunn, B.J., 1987, The effects of ions on sodium- calcium exchange in salamander rods, *J. Physiol.*, 391:371-398.

Hodgkin, A.L. and Nunn, B.J., 1988, Control of light-sensitive current in salamander rods, *J. Physiol.*, 403:439-471.

McNaughton, P.A., Cervetto, L. and Nunn, B.J., 1986, Measurement of the intracellular free calcium concentration in salamander rods, *Nature*, 322:261-263.

Nunn, B.J., 1987a, Precise measurement of cyclic-GMP activated currents in membrane "patches" from salamander rod outer segments, *J. Physiol.*, 394:8P.

Nunn, B.J., 1987b, Ionic permeability ratios of the cyclic GMP-activated conductance in the outer segement membrane of salamander rods, *J. Physiol.*, 394:17P.

CONTENTS

IONS AND CHANNELS

MECHANOTRANSDUCTION

CHEMOTRANSDUCTION

PHOTOTRANSDUCTION

Ions and Channels

H. Horn
S. Masetto, M. Toselli and V. Taglietti

A PRIMER OF PERMEATION AND GATING

R. Horn

Neurosciences Department Roche
Institute of Molecular Biology, Nutley, U.S.A.

We are in the midst of an explosive interest in ion channels. This is due to three factors. First, ion channels have been discovered in virtually every type of cell, not just in nerve and muscle cells, the classical "excitable" cells. Second, patch recording has greatly simplified electrophysiology, especially for small cells. Third, molecular biology has produced primary sequences of channel proteins, making structure-function studies more than just a dream.

More specifically, ion channels in the plasma membrane of sensory cells play a fundamental role in sensory transduction. The sensory signal leads to a change in membrane conductance, either by direct or by indirect effects on ion channels. Experimental studies of the functional properties of ion channels fall into the domain of electrophysiology. In recent years single-channel recording has refined our concepts of channel function. For many of us "channelologists" the function of ion channels is separated conceptually into two broad categories, gating and permeation. Loosely speaking, gating is the process (e.g. conformational fluctuations of the channel protein) that opens and closes a pore, and permeation is the process of ion transport through the open pore. A little reflection shows that a gray area exists between these two categories. For example, a permeant ion may bind to an open channel long enough that its entry and exit from the channel produces current transitions that look, in single channel records, like gating transitions (e.g., see Lansman et al., 1986). In other words, one permeant ion may produce gating transitions for another species of ion that traverses the pore more rapidly. On the other end of the spectrum, fluctuations of the protein may produce "gating transitions" that are so fast (e.g. less than 10 μsec) that the measured current through an open pore, filtered by the recording amplifier, is decreased in magnitude. Thus gating events, measured and analysed by our imperfect methods, fall into the realm of permeation. There are, to be sure, experimental and theoretical studies of this gray area. However I will restrict most of my comments to the two extremes.

An extensive review on permeation and gating is far beyond the scope of this chapter. Books have been written on this subject (e.g. Sakmann and Neher, 1983; Hille, 1984; Miller, 1986). I will instead present an overview of some of the main concepts and the experimental/theoretical methods used for exploring permeation and gating. The emphasis of this chapter is, therefore, personal rather than comprehensive. For conciseness the organization is in outline form.

Sensory Transduction
Edited by A. Borsellino *et al.*
Plenum Press, New York, 1990

PERMEATION

Ion channels reduce the energy barrier for ions to traverse the low-dielectric environment of the lipid bilayer membrane. The process of ion movement through an open pore is complex, leading to a choice among several different perspectives. For example, it is possible to focus attention on which ions are able to traverse the channel. Alternatively, one may explore the mechanisms by which a channel reduces the energy barrier for ion movement. I will describe some of these perspectives here and the approaches used to characterize permeation through a particular type of ion channel. I assume, for simplicity, that it is possible to examine a single class of ion channel in isolation from other types. In practice this may not be so easy. Most cells, for example, have a variety of ion channels.

I. PERSPECTIVES OF PERMEATION STUDIES

A. Selectivity

One of the striking features of ion channels is their ability to choose which types of ions are allowed to pass through and which are excluded. In some cases ion channels are relatively non-selective, permitting a wide range of both cations and anions to permeate (e.g., see Hancock, 1987). Such channels are likely to be large water-filled pores with inner walls that interact poorly with the ions or nonelectrolytes that flow through. Most channels are more selective, however, and may discriminate very specifically in favour of one physiologically relevant ion, such as Cl^- or K^+. Two forms of selectivity are typically used to characterize ion channels, i.e. permeability and conductance.

1. Permeability selectivity. The relative permeabilities of two ion species are determined from the reversal potential (i.e. zero-current voltage) of a membrane containing open channels. The Goldman-Hodgkin-Katz equation (Hodgkin and Katz, 1949) is used to calculate the relative permeability between any two ions. For two monovalent cations, say A^+ and B^+, the relative permeability is given by the permeability ratio P_A/P_B and may be determined from the reversal potential, E_r, and the GHK equation:

$$E_r = RT/F \cdot \ln(P_A/P_B[A^+]_1 + [B^+]_1)/(P_A/P_B[A^+]_2 + [B^+]_2),$$

where the subscripts 1 and 2 refer to different sides of the membrane, and R, T, and F are familiar physical constants. The best-studied permeability sequences are those for the alkali metal cations Cs^+, Rb^+, K^+, Na^+, and Li^+. The number of possible sequences is the number of ordered ways of arranging the five ions, namely $5! = 120$. In practice, however, one commonly observes one of 11 sequences, the well-known Eisenman sequences (Fig. 1). Eisenman et al. (1957) were able to account for these sequences by a relatively simple model involving electrostatic interactions between the hydrated cation and a nonpolarizable anionic site. Eisenman sequences are easy to identify when permeability ratios are plotted as shown in Fig. 1. Such sequences are always *convex* upward and have a single maximum. In other words cations of intermediate size tend to be favoured. In contrast, when the anionic site is polarizable, such plots are *concave* upward with a single minimum (Eisenman and Horn, 1983). Permeability sequences have been described, both experimentally and theoretically, for other types of ions (e.g. anions and multivalents) as well (e.g., see Morf, 1981; Hille, 1984).

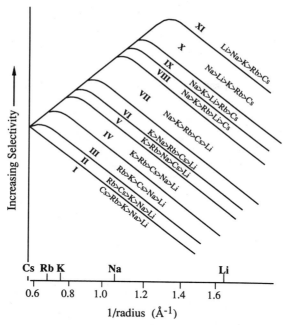

Fig. 1: The topology of Eisenman sequences, redrawn from Fig. 4 in Eisenman and Horn (1983). The relative selectivities, with respect to Cs^+, are plotted against reciprocal radius of the nonhydrated ion.

2. Conductance selectivity. The relative conductances between two ions is obtained by comparing the current carried by each species at a fixed driving force, e.g. 50 mV more depolarized than E_r. This is the method for obtaining conductance ratios, e.g. g_A/g_B, for any two ions. One peculiarity of selectivity among permeant ions is that the most permeant ion, by the above criterion of reversal potential, may be the least conductive ion, meaning that it has the lowest single-channel conductance at a fixed concentration and voltage. Qualitatively this may be explained by the fact that a highly permeant ion can easily enter an open channel where it gets stuck in a binding site (Lansman et al., 1986). Theoretical studies (see below) show that permeability is primarily determined by the height of energy barriers encountered by the ions of interest, whereas conductance is affected by many factors, including the height of barriers, the affinity of ions for binding sites within the channel, the concentrations of ions, and the membrane potential. One reason for the relative complexity of conductances is that the measurements of currents are, by definition, non-equilibrium, in contrast to measurements of reversal potentials.

B. I-V Shape

The shape of the current-voltage (I-V) relationship for an open channel provides insight into the energy profile encountered by an ion in its journey through the channel. The I-V shape is measured either at the single-channel level, using voltage ramps or holding potentials, or at the macroscopic level, using instantaneous I-V protocols. The latter are necessary at the macroscopic level to avoid a contribution of voltage-dependent gating to the I-V curves. If we assume, for simplicity, an equal concentration of a single species of permeant ion on both sides of the membrane, the I-V shapes are characterized by two features, rectification and linearity. An I-V relationship is said to show outward rectification if the outward (cytoplasmic to extracellular) current is larger than the inward current at equal and opposite voltages (Fig. 2A). Conversely, inward rectification signifies a larger inward current. The presence of rectification suggests that the channel is asymmetric with respect to the plane of the membrane. I-V shapes are sometimes linear. However they may be either sublinear or supralinear, either for inward or for outward current (Fig. 2B). A sublinear shape is flatter for greater voltages, meaning that the conductance decreases with increases of membrane potential away from the reversal potential. A supralinear shape is, by contrast, conductance-increasing with voltage. Fig 2C shows, for example, an I-V relationship which has inward rectification and in which the inward current limb is sublinear. The shape of the I-V curve may be different for different permeant ions, suggesting that the energy profile has a different shape (see section IIA, below).

C. Independence

The concept of independence of ion movements within a channel was an assumption of some of the earliest models of permeation (see discussion in Hille, 1984). These models were essentially electrodiffusive. Independence in this context means that each ion moves independently of other ions and independently of the interactions between other ions and the channel. It is, perhaps, not surprising that ion channels rarely have such properties, especially at high concentrations of permeant ions. The experimental symptoms of a channel obeying independence are given in section IIIA, below.

D. Single- and Multi-ion Channels

One way to violate independence is to restrict the number of ions that a channel may contain at one time. If, for example, a channel may contain at most five ions, then a fully-loaded channel cannot be entered by another ion. Thus the presence of the other ions affects the ability of the sixth ion to interact with the channel. For simplicity permeation models are often divided into two categories, namely single- and multi-ion channels. A single-ion channel may contain, at most, one ion. It therefore exists in at least two distinct configurations, empty or loaded with an ion. A loaded channel may have a number of distinct loaded states, depending on the position of the ion within the channel. Multi-ion channels may contain more than a single ion. Models of multi-ion channels are often very complicated, largely because of the large number of possible states.

E. Blockers

One of the most useful tools for understanding permeation is the open-channel blocker. Many blockers have the same charge and presumably interact with the channel at the same sites as permeant ions. However the kinetics of binding and dissociation are usually much slower and therefore appear, on single-channel records, as gating events (Neher and Steinbach, 1978). Analysis of the records allow the direct estimation of the rates of binding and dissociation of the blocker. Furthermore the voltage dependence of the rates provides information about the location within the membrane field of the blocking ion, and thus of its binding site (Neher and Steinbach, 1978; Miller, 1982). In some cases the voltage dependence of the block can provide evidence for multi-ion occupancy (Hille and Schwarz, 1978).

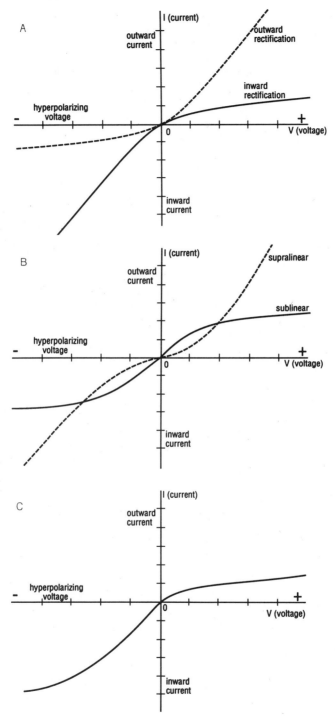

Fig. 2: I-V shapes. A shows rectification. B shows linearity. C shows an inwardly-rectifying I-V curve in which the inward current limb is sublinear.

Some blockers have more rapid kinetics that are obliterated by the low-pass filters in the recording system. These compounds appear to reduce the single-channel current amplitude and prolong open time (if a blocked channel cannot close normally). The analysis of these fast blockers, with some assumptions, can provide the same information as slow blockers; they may be used, therefore, as probes of the barriers and wells of an open channel (Coronado and Miller, 1982).

II. THE ENERGY PROFILE

Permeation may be completely characterized by the energy profile experienced by ions as they pass through an open channel. The lipid bilayer presents an essentially insurmountable free-energy barrier to ions. The ion channel reduces this barrier and leaves a bumpy pathway having energy maxima (peaks or barriers) and minima (wells or binding sites). Ions are often imagined to hop through the channel from site to site. An ion dwells in a site until it has enough kinetic energy to hop over an adjacent barrier. Although most models of permeation assume that the channel presents a static barrier to one-dimensional movement of an ion, more realistic models allow the channel to have a fluctuating energy profile and allow the ions three-dimensional freedom of movement through the channel (Polymeropoulos and Brickmann, 1985). Unfortunately molecular models of permeation require far too many assumptions to be realistic, even for such well-known structures as the gramicidin channel. However simplified models may provide insight into some of the global properties of permeation.

Perhaps the most useful simplified model is the rate-theory, or Eyring-barrier, model (Hille, 1984). I will describe some properties of one-dimensional, static rate-theory models. The one dimension refers not to physical location of the ion in the channel, but to position in the electrical field across the membrane. The reason for this distinction is that electrical measurements *per se* provide no information about the position of the ion or the energy profile in physical space (Miller, 1982). The simpler single-ion channel will be described first. The heights of barriers and depths of wells refer to the maximum and minimum free energies of an ion at barriers and wells, with respect to the free energy of the ion in solution.

A. Single-ion Channels

The properties of a single-ion channel are easily verified by rate-theory models of permeation (for more detailed discussions, see Hille and Schwarz, 1978; Eisenman and Horn, 1983 and Hille, 1984). They are:

1. Permeability selectivity is determined only by the energies and locations of barriers and not by well depths.

2. Conductance selectivity is determined by barriers, wells, and ion concentration.

3. I-V shape is mainly determined by the location and height of barriers.
    a. high barriers on the extracellular side of the membrane cause outward rectification.
    b. high barriers on the intracellular side of the membrane cause inward rectification.
    c. high barriers at the "mouths" of channels cause sublinear I-V's.
    d. high barriers in the middle of a channel cause supralinear I-V's.

4. Well depths, but not barrier heights, determine the saturation of conductance as a function of ion concentration.

5. Well depths determine the concentration dependence (i.e. the K_i) of blocker action. The barrier heights determine the kinetics of block. The location of the well (i.e. the blocking site) determines the voltage dependence of K_i.

III. FINGERPRINTS OF CLASSES OF ION CHANNELS

The various types of ion channels may be distinguished by experimental measurements. The distinction between single-ion and multi-ion channels uses information from rate-theory models. Further discussion of the results in section III may be found in Eisenman and Horn (1983) or in Hille (1984).

A. Independence

Two criteria may be used as characteristic fingerprints of channels obeying independence. For both criteria ion concentrations must be changed. Strictly this means changing the concentrations on both sides of the membrane by a constant factor, e.g. doubling the ion concentrations on both membrane faces.

1. The permeability ratio for any two permeant ions must equal the conductance ratio for the same two ions at all concentrations.

2. The I-V shape is independent of ion concentration.

Ion channels rarely, if ever, satisfy these criteria. Therefore ions significantly interact with the walls of known ion channels.

B. Single-ion Channels

The following criteria may be used to discriminate single-ion from multi-ion channels. Violation of any criterion is sufficient to reject a single-ion model of permeation.

1. The single-channel conductance increases monotonically with ion concentration.

2. P_A/P_B is independent of concentration (for precision, see the criteria in Eisenman and Horn, 1983).

3. P_A/P_B and the conductance ratio between any two species of ions depend monotonically on the mole fraction of a mixture between the two species. If either of these ratios shows a minimum with respect to mole fraction, the channel is said to have "anomalous mole fraction behavior", indicative of a multi-ion channel.

4. The voltage dependence of a blocking ion yields an estimate of the well location within the membrane field. If the blocking site is apparently outside the membrane field, the channel is a multi-ion pore.

5. The Ussing flux ratio exponent is less than or equal to one. The flux ratio exponent is determined from measurements of unidirectional flux of ions (e.g. Hodgkin and Keynes, 1955).

GATING

The process of gating is much slower, in general, than the flux of ions through an open channel. Consequently the molecular processes of gating are believed to be caused by fairly large conformational changes, meaning transitions with large free energy, of the channel protein.

I. GENERAL QUESTIONS

A. Types of Models

1. Diffusional. The process of gating may involve diffusional movements of parts of the channel protein. For example, small charged groups may diffuse into the vicinity of the pore, and therefore "close" the channel electrostatically (e.g., see Lauger, 1988).

2. Markov models. The models commonly used to describe gating are Markov models with a small number (say, less than a dozen) states. Each state has a fixed conductance and transitions, in general, are possible between any two states. In the most common models each state is either fully open or closed; these are two-conductance models. The characteristics of such models have been described in detail (e.g., see Colquhoun and Hawkes, 1983; Horn, 1984). Briefly, the transitions between states are time-homogeneous, meaning that the dwell time in each state is exponentially distributed and independent of the past history of the gating transitions. It is as if a channel always has amnesia. It knows which state it is in, but not how it got there or even how long it has been in that state. The transitions between states are very rapid, compared to the dwell times within states. Although the transition rates do not depend on time, they may be instantaneously dependent on membrane potential or on the concentration of some substance, for example an agonist. At steady state the single-channel dwell times of Markov models are multi-exponentially distributed (see below). The number of closed states, for example, corresponds to the number of time constants in the closed time distribution. The evidence in support of Markov models is quite good. The transitions between open and closed states are as rapid as one can measure (i.e., less than 10 μsec). The dwell time histograms are well-represented by sums of exponentials. Finally, the transition rates are often simple functions of voltage and concentrations of agonists or blockers. In some cases, however, the number of exponential components may be so large that one is suspicious of the accuracy of estimating and separating the components. In cases such as these it may be necessary to accumulate very large data sets and use fairly sophisticated statistics to verify the goodness and robustness of the fits. Models with many discrete states are not palatable to everyone, because they are complicated and possibly non-unique. However alternative types of models have not been successful in describing the details of single-channel gating.

3. Fractal models. Liebovitch et al. (1987) proposed an alternative to Markov models. It is a two-state (open and closed) model in which the transitions between states are dependent on the dwell time within the currently occupied state. In the "fractal model" the rate for leaving a state is

$$k = At^{(1-D)},$$

where A is a positive scaling parameter, D is the fractal dimension that may take on values between 1.0 and 2.0, and t is the time since arriving in the current state. By contrast the transition rates in a time-homogeneous Markov model are independent of time. This so-called fractal model produces dwell-time histograms that, in general, decay more gradually than a single-exponential. The difference between a multi-exponential Markov model and a fractal model is that the former model produces significant "humps" or "plateaus" on dwell-time distributions, depending on the method of plotting them. A recent statistical analysis (McManus et al., 1988; Korn and Horn, 1988) suggests that the bumpiness of dwell-time histograms from single-channel records cannot be explained as random noise superimposed on a smooth, fractal-type, distribution. Single-channel data are, therefore, inconsistent with the predictions of the above fractal model. It remains to be seen whether more elaborate non-Markovian models are useful in understanding gating mechanisms.

B. Single Channel Properties

For simplicity I will assume that ion channels obey Markovian kinetics of gating. Therefore questions about channel function are equivalent to questions about the Markov model underlying the observed behavior.

1. Number of states. The number of conformational states of a large protein, such as an ion channel, is certainly very big. Some conformational transitions are rapid, on the order of picoseconds, and involve motions such as the bending or rotation of single bonds. From the viewpoint of gating kinetics, as measured at the single-channel level, kinetics in the picosecond domain are lost. The slowest conformational changes may be on the order of minutes or longer in some channels. In fact, it is somewhat surprising, considering the vast time range of protein fluctuations, that dwell-time distributions are often fit by just a few exponential components. This fact suggests that certain, high-energy, transitions dominate the opening and closing of channels. If gating transitions have very high activation energies, compared with other possible transitions from a given conformational state, they may lead to the type of Markovian kinetics commonly observed in single-channel records. In practice the number of kinetic states is estimated by the number of time constants in the open- and closed-time histograms.

2. Number of conductance states. Although we have been speaking of open and closed channels, there are many reports of subconductance levels (e.g., see Hamill and Sakmann, 1981). In principle subconductance levels provide extra information about gating by allowing the direct measurement of the dwell times in a different type of state. In practice, however, subconductance states are often a nuisance, because they are harder to resolve than the main conductance level. A general problem in gating studies at the single-channel level is the possibility of missing brief events, either because of low-pass filters or because of the use of digital sampling intervals that are too long (Colquhoun and Sigworth, 1983). The missed-events problem is even more severe when the records are contaminated by currents at subconductance levels, because of the lower signal/noise ratio.

3. Rates of transition between states. When a model of gating is formulated, it consists of a number of discrete states with a conductance level assigned to each state. The model is fully defined by the rates of transitions between each of the states. For a Markov process these rates will not depend on time, but may depend on other relevant factors.

a. Voltage dependence. Each rate constant may be dependent on membrane potential. The form and extent of voltage dependence varies among ion channels, some (e.g.

sodium channels) being exquisitely sensitive to voltage. The form of the voltage dependence is often exponential. For example, a rate constant, k, may have the following form.

$$k = A\exp(\delta V F/RT),$$

where V is the membrane potential, and the voltage dependence is given by the dimensionless factor, δ. If $\delta=0$, the rate constant is voltage-independent. Its sign gives the direction of the voltage dependence. More complicated forms of voltage dependence are known (Stevens, 1978).

b. Agonist dependence. An agonist gated channel will bind each agonist molecule with a rate constant that depends, usually linearly, on the agonist concentration. The unbinding rate, in the same fashion as the reaction between a ligand and an enzyme, should be independent of agonist concentration.

c. S-dependence. I define S as everything else that may affect a transition rate. For example, S may be any sensory input (light, touch, stretch, temperature, etc.). For simple Markov models of gating all stimuli must modify rate constants. The form of modification could depend on the expected physical effects of a stimulus on particular conformational changes of the channel. For example, temperature will affect a given transition in accordance with the temperature dependence of the activation enthalpy and entropy of that transition. It is possible that some, but not all, rate constants in a gating process will depend on a sensory signal. For example, the gating of stretch-activated channels has been modelled by a four-state scheme in which three of the states are closed and one is open (Guharay and Sachs, 1984). Only one transition, between two of the closed states, was stretch sensitive.

C. Multiple Channels

The usual assumption in studies of gating is that the behavior of a large population of ion channels in a membrane is a simple consequence of the combined behavior of many identical channels that function independently of one another. This concept has a simplistic appeal which may be unrealistic. I will discuss two possible violations of this idea, namely heterogeneity and cooperativity. Section IID1, below, describes a test for these violations.

1. Heterogeneity. It is, perhaps, unreasonable to assume that all channels in a biological membrane are the same, even when examining a single class (e.g. voltage-activated sodium channels) of channel. First, several different genes may be transcribed in a given cell to produce a heterogeneous population of, for example, sodium channels. Even a single type of gene may produce a heterogeneous population of channels, either by alternative splicing (Papazian et al., 1987) or by post-translational modifications. Second, individual channels may exist in different forms due to reversible modifications, for example, by phosphorylation.

2. Cooperativity. Ion channels may be very densely packed. For example, acetylcholine receptors at endplates of skeletal muscle (Fambrough, 1979) and sodium channels in the node of Ranvier of myelinated nerves (Ritchie, 1986) are extremely dense. The close packing invites the possibility of interactions between channels. Nicotinic acetylcholine receptors may exist as dimers connected by a disulfide bond (Reynold and Karlin, 1978). In such a situation the conformational changes of one channel could affect its neighbour. One could also imagine electrostatic cross-talk between channels. Since

charge moves across the membrane during the gating of voltage-activated channels, the charge movement in one channel may electrostatically inhibit the charge movement in a neighbouring channel, depending on the screening of charges between the two channels. Cooperativity exists in two flavors, positive and negative. In terms of gating, positive (negative) cooperativity means that the opening of one channel increases (decreases) the probability that another channel will open.

II. METHODS OF STUDYING GATING

A. Macroscopic Currents

The macroscopic currents from a voltage-clamped membrane, with some simplifying assumptions, will obey the following equation.

$$I(V, S, t) = N \cdot i(V, S) \cdot P_o(V, S, t),$$

where I is the macroscopic current, V is the membrane potential, S is defined as above, N is the number of channels, i is the single-channel current, and P_o is the probability of a channel being open. This equation shows the dependence of the macroscopic current on both the single-channel current and the open probability. If a channel obeys Markovian kinetics, then the nonstationary kinetics of macroscopic current will be fit by a weighted sum of exponentials. The number of exponential components equals the total number of states minus one. These exponentials do not in general have the same time constants as those observed in the dwell-time distributions of single channels (see below). The weights and time constants of the exponentials may be determined exactly from a model in which one knows all of the rate constants and the initial conditions (Colquhoun and Hawkes, 1977).

B. Gating Currents

If channels are voltage dependent, the openings of channels must be accompanied by charge movement across the membrane. In the absence of ionic currents it may be experimentally feasible to measure this charge movement as a gating current. The kinetics of gating currents correspond closely with those of macroscopic currents in that they will be fit, for Markov models, by the same number of exponential components with the same time constants. Only the weights of the time constants are different (Horn, 1984).

C. Noise

Since the gating of ion channels is inherently stochastic, the number of open channels in a membrane with many channels is not constant, even under steady-state conditions. The macroscopic current therefore fluctuates. Analysis of these fluctuations is called noise analysis. In some cases (e.g. when single-channel currents are too small, or when the density of channels is either too high or too low for single-channel recording) it may be more convenient to do noise analysis than single-channel recording. Another advantage of noise measurements over single-channel recording is that the analysis of data is somewhat simpler. One estimate of gating kinetics from noise analysis is the autocovariance which, like gating current, has the form of a sum of exponentials (Colquhoun and Hawkes, 1977). Again, the autocovariance will have the same number of exponential components with the same time constants as macroscopic currents. Only the weights of the time constants are different (Colquhoun and Hawkes, 1977).

D. Single Channels

I will assume, for the remainder of the chapter, that channels obey Markovian kinetics and that they have only two conductance states, open and closed. It is straightforward to generalize the following discussion to the case of multiple conductance levels. There are many ways to record and analyze single-channel data. Three of the most common methods are listed here.

1. P_o. The stationary probability of a channel being open, P_o, may be estimated easily, in a patch containing one channel, as the fraction of time that the channel is open. For a membrane patch with multiple channels P_o is estimated by examining the fractions of time spent when exactly 0,1,2,... channels are open. These fractions will obey a binomial distribution if the channels are *identical* and *independent* of one another (Horn, 1984). In fact, a failure to fit this distribution is good evidence that one of these two assumptions is false. Unfortunately this test is neither reliable nor powerful. First, it is often difficult to estimate the number of channels in a patch. This is a critical parameter in the binomial distribution. Second, the binomial distribution is notoriously insensitive to violations of these two assumptions, which may be tested by more elaborate statistics (Yeramian et al., 1986).

2. Dwell-time distributions. For single-channel currents, the dwell times are usually measured from digitally sampled records by the half-maximal criterion (Colquhoun and Sigworth, 1983). Transitions between open and closed states are assumed to occur when the current crosses a threshold set at half the level of the open-channel current. An extensive evaluation of this method is given by Colquhoun and Sigworth (1983) and by McManus et al. (1987). The data from steady-state records are usually binned and plotted in the form of a probability density function. Theoretically this has the form of a sum of decaying exponentials. The histograms are fit statistically, optimally by maximum likelihood. This is often the first step in the kinetic analysis of single-channel gating. The most important information from such an analysis is the number of exponential components, which estimates the number of either closed or open states, for closed- and open-time histograms, respectively. The time constants and relative weights further characterize the data sets. The ultimate goal of such fits is an estimate of the rate constants for a given kinetic model. This may not be possible from this method alone, however, because the number of free parameters in a Markov model is usually greater than the number of parameters obtainable from fits to dwell-time histograms (Bauer and Kenyon, 1988). A new method of plotting histograms has appeared (Sigworth and Sine, 1987). This method is likely to replace the previous methods because it provides a simple, visual method for estimating the number of exponential components, their time constants, and their weights. On this plot each component appears as a bump with its peak aligned on the X-axis with its time constant. The relative height of each bump gives the relative amplitude of each component.

3. Maximum likelihood. Maximum likelihood is a statistical method for estimating parameters and testing models. As a method for estimation it has excellent properties for large data sets, as usually found in single-channel records. The method is, for most problems, guaranteed to produce the best estimates possible by a number of criteria

(Rao, 1973). Its main disadvantage is that it is often computationally tedious, requiring complicated, slow algorithms for data analysis. Its use in single-channel records was introduced by Colquhoun and Sigworth (1983) and by Horn and Lange (1983). At present the modified approach of Sigworth and Sine (1987) is probably the best method for fitting dwell-time distributions. The more elaborate method of Horn and Lange (1983) is used for estimating rate constants in Markov models and for statistical comparisons between models. This method may be used with nonstationary single-channel records, even with multi-channel patches.

ACKNOWLEDGEMENTS

I thank Dr. Stephen Korn for insightful comments on the manuscript.

REFERENCES

Bauer, R.J., Bowman, B.F. and Kenyon, J.L., 1987, Theory of the kinetic analysis of patch-clamp data, *Biophys. J.*, 52:961-978.

Colquhoun, D. and Hawkes, A.G., 1977, Relaxation and fluctuations of membrane cur-reents that flow through drug-operated channels, *Proc. R. Soc. Lond. B*, 199:231-262.

Colquhoun, D. and Hawkes, A.G., 1983, The principles of stochastic interpretation of ion-channel mechanisms, *In:* "Single Channel Recording," B. Sakmann and E. Neher, ed., Plenum Publishing Corp., New York, 135-175.

Colquhoun, D. and Sigworth, F., 1983, Fitting and statistical analysis of si ngle-channel records, *In:* "Single Channel Recording," B. Sakmann and E. Neher, ed., Plenum Publishing Corp., New York, 191-264.

Coronado, R. and Miller, C., 1982, Conduction and block by organic cations in a K^+-selective channel from the sarcoplasmic reticulum incorporated into planar phos-pholipid bilayers, *J. Gen. Physiol.*, 79:529-547.

Eisenman, G. and Horn, R., 1983, Ionic selectivity revisited: The role of ki netic and equilibrium processes in ion permeation through channels, *J. Membr. Biol.*, 76:197-225.

Eisenman, G., Rudin, D.O. and Casby, J.U., 1957, *Science*, 126:831-834.

Fambrough, D.M., 1979, Control of acetylcholine receptors in skeletal muscle, *Physiol. Rev.*, 59(1):165-227.

Guhary, F. and Sachs, F., 1984, Stretch-activated ion channel currents in tissue-cultured embryonic chick skeletal muscle, *J. Physiol.*, 352:685-701.

Hamill, O.P. and Sakmann, B., 1981, Multiple conductance state of single acetylcholine receptor channels in embryonic muscle cells, *Nature*, 294:462-464.

Hancock, R.E. 1987, Role of porins in outer membrane permeability, *J. Bacteriol,*, 169(3):929-933.

Hille, B. 1984, Ionic Channels of Excitable Membranes, Sinauer, Sunderland, MA.

Hille, B. and Schwarz, W., 1978, Potassium channels as multi-ion single-file pores, *J. Gen. Physiol.*, 72:409-442.

Hodgkin, A.L. and Katz, B., 1949, The effect of sodium ions on the electrical activity of the giant axon of the squid, *J. Physiol.*, 108:37-77.

Hodgkin, A.L. and Keynes, R.D., 1955, The potassium permeability of a giant nerve fibre, *J. Physiol.*, 128:61-88.

Horn, R. 1984, Gating of channels in nerve and muscle: a stochastic approach, *In:* "Ion Channels: Molecular and Physiological Aspects," W.D. Stein, ed., Academic Press,

Inc., New York, 53-97.

Horn, R. and Lange, K., 1983, Estimating kinetic constants from single channel data, *Biophys. J.*, 43:207-223.

Korn, S.J. and Horn, R., 1988, Statistical discrimination of fractal and Markov models of single channel gating, *Biophys. J.*, 54:871-877.

Lansman, J.B., Hess, P. and Tsien, R.W., 1986, Blockade of current through single calcium channels by Cd^{2+}, Mg^{2+}, and Ca^{2+}, Voltage and concentration dependence of calcium entry into the pore, *J. Gen. Physiol.*, 88:321-347.

Luger, P., 1988, Internal motions in proteins and gating kinetics of ionic channels, *Biophys. J.*, 53:877-884.

Liebovitch, L.S., Fischbarg, J. and Koniarek, J.P., 1987, Ion channel kinetics: a model based on fractal scaling rather than multistate Markov processes, *Math. Biosci.*, 84:37-68.

McManus, O.B., Blatz, A.L. and Magleby, K.L., 1987, Sampling, log binning, fitting, and plotting durations of open and shut intervals from single channels and the effects of noise, *Pfluegers Arch. Eur. J. Physiol.*, 410:530-553.

McManus, O.B., Weiss, D.S., Spivak, C.E., Blatz, A.L. and Magleby, K.L., 1988, Fractal models are inadequate for the kinetics of four different ion channels, *Biophys. J.*, 54:859-870.

Miller, C., 1982, Bis-quaternary ammonium blockers as structural probes of the sarcoplasmic reticulum K^+ channel, *J. Gen. Physiol.*, 79:869-891.

Miller, C., ed., 1986, "Ion Channel Reconstitution," Plenum Publishing Corp., New York.

Morf, W.E., 1981, "The Principles of Ion-Selective Electrodes and of Membrane Transport," Elsevier, Amsterdam.

Neher, E. and Steinbach, J.H., 1978, Local anaesthetics transiently block currents through single acetylcholine-receptor channels, *J. Physiol.*, 277:153-176.

Papazian, D.M., Schwarz, T.L., Tempel, B.L., Jan, Y.N. and Jan, L.Y., 1987, Cloning of genomic and complementary DNA from Shaker, a putative potassium channel gene from Drosophila, *Science*, 237:749-753.

Polymeropoulos, E.E. and Brickmann, J., 1985, Molecular dynamics of ion transport through transmembrane model channels, *Ann. Rev. Biophys. Chem.*, 14:315-330.

Rao, C.R., 1973, Linear Statistical Inference and Its Applications, 2nd ed., John Wiley and Sons, Inc., New York.

Reynold, J.A. and Karlin, A., 1978, Molecular weight in detergent solution of acetylcholine receptor from Torpedo californica, *Biochemistry,*, 17:2035-20 38.

Ritchie, J.M., 1986, Distribution of saxitoxin-binding sites in mammalian neural tissue, *In*, Tetrodotoxin, Saxitoxin, and the Molecular Biology of the Sodium Channel, C.Y. Kao and S.R. Levinson, editors, Annals of the New York Academy of Sciences, volume 479:385-401.

Sakmann, B. and Neher, E. (eds.), 1983, "Single Channel Recording," Plenum Publishing Corp., New York.

Sigworth, F.J. and Sine, S.M., 1987, Data transformations for improved display and fitting of single-channel dwell time histograms, *Biophys. J.*, 52:1047-1054.

Stevens, C.F., 1978, Interactions between intrinsic membrane protein and electric field; an approach to studying nerve excitability, *Biophys. J.*, 22:295-306

Yeramian, E., Trautmann, A. and Claverie, P., 1986, Acetylcholine receptors are not functionally independent, *Biophys. J.*, 50:253-263.

INTERACTIONS AMONG CATIONS IN CURRENT CONDUCTION THROUGH

THE STRETCH-ACTIVATED CHANNEL OF THE FROG OOCYTE

S. Masetto, M. Toselli and V. Taglietti

Istituto di Fisiologia Generale
Universita' di Pavia, Italy

INTRODUCTION

An increasing literature is being accumulating about stretch-activated (s.a.) channels. They were found first in chick skeletal muscle by the group of F. Sachs (Guharay and Sachs, 1984, 1985), and later on in many other structures (Hamill, 1983; Cooper et al., 1986; Methfessel et al., 1986; Yang et al., 1986; Falke et al., 1987; Gustin et al., 1987; Lansman et al., 1987; Ove, 1987). S.a. channels, unlike voltage- or chemically-activated channels, are primarily gated by mechanical distortion of the cell membrane. By using the improved patch-clamp technique in the cell-attached configuration, we are studying the s.a. channels present in the plasma membrane of the frog oocyte. An interesting property of this cation channel was evidenced in a previous report (Toselli et al., 1987): the current carried in the inward direction through the s.a. channel obviously increases when Ca^{2+} is removed from the standard Ringer. This was investigated thouroughly in the present experiments. Briefly, we find that the cations that can enter the channel do not move independently of one another: the Goldman-Hodgkin-Katz equations therefore do not adequately describe the modalities of ion permeation. Conversely, the experimental observations are in agreement with the expectations of a model representing the s.a. channel as a membrane pore with a single binding site accepting one ion at a time, and displaying different affinities for the various cations.

METHODS

All experiments were carried out with oocytes at maturation stages IV-V (Dumont, 1972). Pieces of ovary were removed from adult females of european frog (Rana esculenta, L.) and stored in sterile Barth's medium (in mM: NaCl 88, KCl 1, $MgSO_4$ 0.82, $Ca(NO_3)_2$ 0.33, $CaCl_2$ 0.41, $NaHCO_3$ 2.4, Tris-HCl 5; pH 7.4); Penicillin and Streptomycin were added at a concentration of 10 $\mu g/ml$. To obtain tight seals of high resistances, after mechanical removal of ovarian epithelium and connective layer the oocytes were incubated with Protease (1 mg/ml, Type XIV, Sigma) in Barth's medium for few minutes at room temperature to remove the follicular and vitelline envelopes. Finally the oocytes were transferred to a tissue culture dish containing standard Ringer solution (in mM: NaCl 107, KCl 2.5, $CaCl_2$ 2, Glucose 5, Hepes-KOH 10; pH 7.4). Single channel current mea-

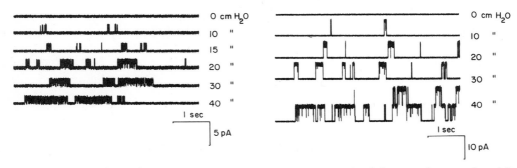

Fig. 1: Pressure/dependence of s.a. channel activity. On the left, sample records with standard Ringer solution in the pipette; on the right, sample records with "Ca-free" filling solution (in mM: NaCl 100, EGTA 10, HEPES-NaOH 10; pH 7.4). The patch membrane potential was kept in both cases at -80 mV.

surements were done at room temperature applying the improved patch-clamp technique in the cell-attached configuration (Hamill et al., 1981). After the formation of a high-resistance seal (10-100 GΩ), the membrane patch was stretched by applying suction (i.e. negative pressure) to the patch pipette. The resting potential of the oocytes was determined using a separate intracellular recording electrode filled with 3 M KCl of 60 to 100 MΩ resistance and inserted into the oocyte very close to the membrane patch explored. The membrane potentials had mean values around -40 mV, oocytes with membrane potentials lower than -15 mV being discarted. The actual potential difference across the membrane patch was calculated as the difference between the cell resting potential and the pipette voltage, clamped at different values during the experiments. The internal K^+, Na^+ and Cl^- concentrations of the frog oocyte were assumed to be 60, 20 and 50 mM respectively (Taglietti et al., 1984). Internal Ca^{2+} concentration was assumed to be negligible.

RESULTS AND DISCUSSION

Fig. 1 shows examples of single channel currents recorded in the absence and in the presence of Ca^{2+} respectively. Channel activation was obtained by applying different degrees of suction to the patch pipette. It's evident that:

1) Single channel activity increases with increasing suction at the patch pipette.
2) The amplitude of single channel current increases when no Ca^{2+} is present in the patch pipette filling solution.
3) Flickering decreases in absence of Ca^{2+}. This was not investigated further.

Fig. 2 shows the corresponding current/voltage relations and clearly indicates that the inward current rectifies when Ca^{2+} is removed from the physiological saline.

To investigate the mechanism underlying the particular interaction between Ca^{2+} and current conduction by other cations through the s.a. channel, we have recorded single channel currents at different membrane potentials and with different cations present separately or together in varying concentrations in the patch pipette filling solution.

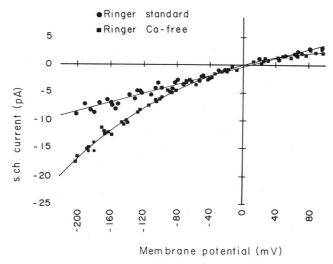

Fig. 2: Current/voltage relationship of the s.a. channel with standard Ringer (circles) or "Ca-free" solution (squares) used as patch pipette filling solutions.

In Fig. 3 single channel currents recorded at -150 mV patch membrane potential are plotted vs. the concentrations of K^+, Na^+ and Ca^{2+} present separately in the pipette filling solution. The unitary current amplitudes obviously saturated with increasing external cation concentration; data in fact could be fitted with very low standard error by a hyperbola. The current/concentration relations at other membrane potentials also showed similar saturation. This was not expected from the independence-constant field theory; the dashed line in Figure 3 plots, for example, the sum of $I(K)+ I(Na)$ predicted by the Goldman-Hodgkin-Katz equation for current at -150 mV, when K^+ at varying concentration is the only cation outside.

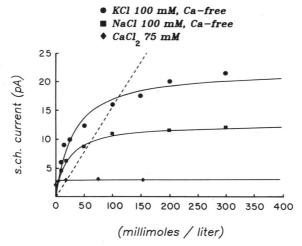

Fig. 3: Current/concentration relationship for K^+, Na^+ and Ca^{2+} at the membrane potential of -150 mV. The dashed line plots the current/concentration relation expected for K^+ by applying the Goldman-Hodgkin-Katz equation.

The asymptotic value of the current was greater for K^+ than for Na^+ at all membrane potentials and was much lower in the presence of Ca^{2+} alone. Also the half-saturating external cation concentration followed the sequence $K^+ > Na^+ > Ca^{2+}$ alone. Thus, among these three cations, K^+ is able to pass at the highest rate through the s.a. channel, while Ca^{2+} permeates the channel with the greatest difficulty.

Among the other cations, Li^+ and Ba^{++} proved to be less permeant than K^+ or Na^+ through the s.a. channel, but more permeant than Ca^{2+}. La^{+++} acted as an effective channel blocker at the concentration of 100 $\mu M/l$.

In a second group of experiments, the current/concentration relation was investigated when the most permeant cation, i.e. K^+, was present at varying concentrations in the patch pipette, in the presence of constant Ca^{2+} concentrations, at different membrane potentials. The presence of Ca^{2+} did not affect the current asymptotic value while it increased the half-saturating K^+ external concentration in a dose-dependent way. In Fig. 4, for example, currents recorded at -150 mV in the presence of $50\mu M$ Ca^{2+} outside are directly compared with those recorded in Ca-free solutions. The half-saturating K^+ concentration resulted to be 94 mM in the presence of Ca^{2+} against 30 mM when only K^+ was present. It seems reasonable therefore to assume that competitive inhibition between Ca^{2+} and K^+ (or Na^+, data not shown) takes place in the s.a. channel, probably because they bind to the same active site.

A satisfactory interpretation of the present measurements could be that shown schematically in Fig. 5, comprising a pore with two energy barriers and an energy well in between, this representing the cation binding site inside the channel. The assumption is made that the binding site must be empty before it can be occupied by any ion. The height of the barriers and the depth of the well may be different for the different cations, thus determining their migration rate across the channel, according to the free-energy variations experienced in each case.

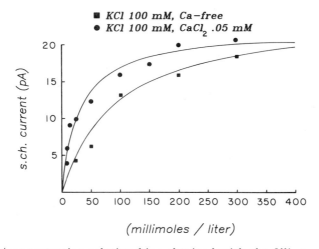

Fig. 4: Current/concentration relationships obtained with the filling solutions indicated in the inset. The patch membrane potential was -150 mV.

The current flowing through the channel can be treated, according to the Eyring theory (1963), as the rate of a reaction between the various cations present at both sides of the membrane and the binding site inside the channel. The four voltage-dependent hopping rate constants, f(in) and f(out) for binding and k(in) and k(out) for unbinding represent the probability of an ion to get over each barrier coming from a side or from the other. By optimizing the model parameters to a number of data, we obtained binding rate constants f(in) and f(out) of the same order of magnitude for K^+, Na^+ and Ca^{2+}, while the unbinding rate constants k(in), k(out) for Na^+ and K^+ resulted to be more than tenfold greater than those for Ca^{2+}.

From the rate constants, the apparent dissociation constant and the mean occupancy time of the s.a. channel can be estimated. Ca^{2+} presented much smaller dissociation constant (50 μM) and much longer occupancy time (400 nsec) than Na^+ (2 mM and 20 nsec respectively) or K^+ (3.2 mM and 12 nsec respectively). In other words, the binding site of the s.a. channel in the frog oocyte displays more "affinity" for Ca^{2+} than for Na^+ and K^+, and this fully explains the difference in conductance experienced in the presence of different cations outside.

The permeability sequence found for the monovalent cations (K>Na>Li) corresponds to the I or the II Eisenmann sequences (1983), suggesting a weak electrostatic field strength around the negative binding site inside the channel.

Current/concentration and current/voltage relations simulated by applying the model assumptions and the estimated parameters resulted to be quite similar to those obtained experimentally. In particular, in the presence of external Ca^{2+}, the channel occupancy increases and the inward flow of Na^+ or K^+ is competitively inhibited; the channel selectivity for Na^+ and K^+ is also masked. By converse, in the absence of external Ca^{2+}, the inward current carried by Na^+ or K^+ increases. Thus the model provides a satisfactory interpretation of the seeming paradox of an inward current increasing following removal from the external medium of a permeant, inwardly driven cation.

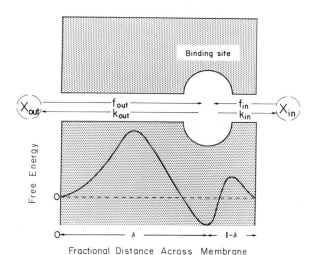

Fig. 5: Schematic representation of the s.a. channel and its possible energy profile (lower curve). X(out) and X(in) represent the outer and inner concentrations respectively of the ion X.

REFERENCES

Cooper, K. E., Tang, J. M., Ree, J. L and Eisenberg, R. S., 1986, A cation channel in frog lens epithelia responsive to pressure and Calcium, *J. Membr. Biol.*, 93:259-269.

Dumont, J. N., 1972, Oogenesis in Xenopus laevis: stages of oocyte development in laboratory maintained animals, *J. Morphol.*, 136:153-180.

Eisenman, G. and Horn, R., 1983, Ionic selectivity revisited, The role of kinetic and equilibrium processes in ion permeation through channels, *J. Membr. Biol.*, 76:197-225.

Eyring, H. and Eyring, E. M., 1963, Chapter 4. "Modern chemical kinetics," Reinhold Comp., New York.

Falke, L., Edwards, K. L., Pickard, B. G. and Misler, S., 1987, Stretch-activated anion channel in cultured tobacco cells, *Biophys. J.* 51:251a (T-Pos.18).

Guharay, F. and Sachs, F., 1984, Stretch-activated single channel currents in tissue-cultered embryonic chick skeletal muscle, *J. Physiol.*, 352:685-708.

Guharay, F. and Sachs, F., 1985, Mechanotransducer ion channels in chick skeletal muscle. The effects of extracellular pH, *J. Physiol.*, 363:119-134.

Gustin, M. C., Zhou, X. L., Martinac, B., Culbertson, M. R. and Kung, C., 1987, Stretch-activated cation channel in yeast, *Biophys. J.*, 51:251a (T-Pos.17).

Hamill, O. P., Marty, A., Neher, E., Sakmann, B. and Sigworth, F. J., 1981, Improved patch-clamp techniques for high-resolution current recording from cells and cell-free membrane patches, *Pfluegers Archiv*, 391:85-100.

Hamill, O. P., 1983, Potassium and Chloride channels in red blood cells, *In*: "Single channel recording," Sackmann, B. and Neher, E., ed., pp. 451-471, Plenum Press., New York.

Lansmann, J. B., Hallam, T. J. and Rink, T. J., 1987, Single stretch activated ion channels in vascular endothelial cells as mechanotransducer? *Nature*, 325:811-813.

Methfessel, C., Weitzmann, V., Takahashi, T., Mishina, M., Numa, S. and Sakmann, B., 1986, Patch-clamp measurements on Xenopus laevis oocyte: currents through endogenous channels and implanted acetylcholine receptor and sodium channels, *Pfluegers Archiv*, 407:577-588.

Ove, C., 1987, Mediation of cell volume regulation by Ca^{2+} influx through stretch-activated channels, *Nature*, 330:66-68.

Taglietti, V., Tanzi, F., Romero, R. and Simoncini, L., 1984, Maturation involves suppression of voltage-gated currents in the frog oocyte, *J. Cell Physiol.*, 121:576-588.

Toselli, M., Taglietti, V. and Tanzi, F., 1987, A preliminary study of the stretch-activable ionic channel in the immature frog oocyte, *Pfluegers Archiv*, 410:S14.

Yang, X. C., Guharay, F. and Sachs, F., 1986, Mechanotransducing ion channels: ionic selectivity and coupling to viscoelastic elements of the cytoskeleton, *Biophys. J.*, 49:373a.

Mechanotransduction

J. F. Ashmore
C. J. Kros

MECHANORECEPTION

J.F. Ashmore

Department of Physiology, Medical School
Bristol, U.K.

INTRODUCTION

Conversion of mechanical signals into electrical signals is one of the oldest problems faced by living forms. Many simple unicellular organisms possess a mechano-sensing mechanism, which with growing complexity has become elaborated to detect a wide variety of specialized stimuli. This chapter will describe recent work on the biophysics of the interconversion of mechanical and electrical energy in hair cells, the specialized cells of the acoustico-lateralis system of vertebrates. This system includes the cochlea, the organ of hearing; the semicircular canals, the organs concerned with the detection of angular acceleration; the saccule and utricle of the vestibular system, specialized organs concerned with the detection of linear acceleration. In some lower vertebrates, the system also includes the lateral line, the organ concerned with detection of pressure waves in an aquatic environment.

Although the morphology of each of these organs differs in a way that reflects its specific function, the underlying sensory hair cell shows remarkable similarities. Hair cells are polarized neuroepithelial cells and are found in organized arrays or rows. The apical surface of the hair cell is a membrane which is mechanosensitive. The basolateral membrane, so far not shown to be mechanosensitive, usually has a normal membrane potential established across it. This surface contains a range of membrane conductances which determine the voltage response of the cell to the current injected through the apical transduction channels, and hence determines the kinetics of neurotransmitter release from the synapse.

The chapter is not an exhaustive review of hair cell function. Instead, selected biophysical features of hair cells will be described using data from several different species. Particular emphasis will be placed on forward and reverse mechanotransduction in hair cells, and how such factors may be integrated into an understanding of mammalian cochlear function. The appendix contains a short discussion of the mathematics of simple cochlear models as a framework to integrate the cellular data.

Elementary mechanoreceptive channels

Since 1984, with the experimental reports of single channel activity sensitive to membrane stretch (Guhuray and Sachs, 1984), ionic channels which are mechano-sensitive have appeared in a wide variety of tissues. Many ionic channels which are involved in electrical signalling, such as the acetylcholine channel itself, also possess properties which are modified by the mechanical environment, but to a more limited extent. At the single channel level stretch-activated channels were described in cultured myotubes (Guhuray and Sachs, 1984). They have now appeared in tissues as varied as oocytes (Sakmann et al., 1984), snail heart cells (Sigurdson and Morris, 1986), endothelial cells (Olesen et al., 1988), dorsal root ganglion cells (Yang et al., 1986), skeletal muscle (Brehm et al., 1984), glia (Ding et al., 1988) and lens epithelial cells (Cooper et al., 1986). Most of these are channels which are cation selective, but whose selectivity is often regulated by Ca^{2+}. In some cases, such as in the lens, a volume regulation function may be envisaged. In other cases such as that of aortic endothelial cells the function is more likely to be part of a mechanosensory feedback loop.

Hair cells as specialized mechanoreceptive cells

In more complex mechanosensory organs, mechano-reception is performed by the hair cells. The apical membrane of a vertebrate hair cell is organised into a set of between 50-100 stereocilia formed into a stepped array. Individual stereocilia are between 0.3 and 0.8μm in diameter, and between 1 and 10μm long, depending on origin. The array of stereocilia may either be hexagonally ordered, or may be organized into rows, typically 3 deep, as found in the inner hair cells of the mammalian cochlea. Mechanically the individual stereocilium is stiff, but has a weak point where it thins out at its insertion into the cuticular plate. Each stereocilium contains an actin core (Flock and Cheung, 1977; Tilney et al., 1980), which because of the crosslinking between filaments confers a stiffness on the structure (Howard and Ashmore, 1986).

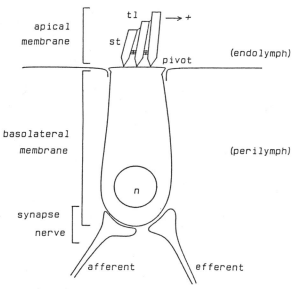

Fig. 1: Structure of the vertebrate hair cell. The cell forms part of an epithelium dividing endolymph (K approx 150mM) from perilymph, (usual extracellular fluid). st, stereocilia; tl, tip link; n, nucleus. The direction for transducer channel activation is shown (+). Some but not all hair cells have both afferent and efferent innervation.

The first cellular event in mechanoelectric transduction in hair cells is the deflection of the stereocilia around their basal insertion. This occurs when a shearing force is applied to the tip of the bundle, and although there is little direct evidence other than observation of the entire bundle motion, it appears that such force is shared quite equally between all stereocilia. Even though stereocilia are linked together (Csukas et al., 1987) the nature of the pivot will ensure that there will be movement between adjacent stereocilia. In the chick, where vestibular cells with both long and short stereocilial bundles can be isolated, the magnitude of the transducer current is proportional to the angular deflection of the stereocilium rather than its absolute deflection (Ohmori, 1987). This observation again supports the idea that it is relative stereocilial shear which gates the transducer channel.

Invertebrates also have developed specialized cells which superficially resemble the vertebrate hair cells. The specialized sensory surface is built from microtubule-based cilia which appear to have a comparable sensitivity, although studies have not been so extensive as in the vertebrate cell. In the statocyst of *Hermissenda*, experiments which manipulate the stiffness of the cilium as a means of controlling the sensory coupling have suggested that the transduction site may be located at the base of the cilium (Stommel et al., 1980).

FORWARD TRANSDUCTION

Mechanical Sensitivity of the transducer channel

At auditory threshold, the amplitude of the sound pressure wave is 0dB SPL, or 20μPa at 1kHz. Since the acoustic impedance of air is 430Ns.m^{-3}, the wave amplitude at this frequency is 7.4 pm. The simple estimate indicates that, even when the effects of the middle ear are included, the adequate stimulus has to be detected and transformed at molecular scales and below. Study of hair cell transduction is thus a question of how the sound is coupled to the hair cell and the nature of the first molecular steps in hearing.

The transduction step can be studied *in vitro*, either by studying isolated epithelia of hair cells (Hudspeth, 1983; Russell et al., 1986) or by studying the isolated cells, using patch clamp techniques (Lewis and Hudspeth, 1983; Ohmori, 1984; Ashmore and Meech, 1986; Santos-Sacchi and Dilger, 1988; Art et al., 1987). In principle, such experiments allow direct manipulation of the sensory hair bundle in a way which is not possible in the intact organ where the stimulus may be coupled in a quite complex manner.

Such direct experiments, most extensive in lower vertebrates, support the conclusion that a) pivotal deflection of the stereocilial hair bundle is the stimulus and b) that deflections of the order of nanometers are adequate to produce a change in the electrical properties of the cell. There is a required polarity to the deflection, for the deflection has to be towards the tallest stereocilium to produce inward current (Shotwell et al., 1981). The measured sensitivity of the hair cell in isolation is less than the sensitivity inferred from studies in the intact cochlea, although with improved techniques the differences appear to be decreasing (Fettiplace and Crawford, 1989).

The dynamic range of the cells, i.e the range over which the cell signals displacement of the bundle tip is of the order of 100nm (Corey and Hudspeth, 1983; Crawford and Fettiplace, 1985; Russell et al., 1986). A threshold signal, adequate to produce a 1mV change in the cells would require a deflection of approximately 1 nm in cochlear hair cells.

Rate of activation of hair cell transduction

Most of the measurements have been made on hair cells of lower vertebrates, especially frogs, turtles and chicks. The hair cells in these preparations are likely to be specialized to operate at low frequencies, below about 1kHz to match the auditory range of the animals. Even so it is clear that the latency of channel opening in these cases is short. Corey and Hudspeth (1982) studied this question using the saccular epithelium of the frog, and measured transepithelial currents during a step displacement of the stereocilia of the hair cells. They found a latency of $40\mu s$ at room temperature. The temperature dependence of the rate is also quite weak, with $Q_{10} = 2.5$ between $1-38°C$. This latency is too short to involve the chemical intermediate steps found in photoreceptors, for example. It also suggests that amplification of the signal, if important in audition is likely to involve earlier steps than at the mechano-electric transduction stage itself. Recent studies of the kinetics suggest that there is a rate limiting step with a time constant of about $500\mu s$ (Corey and Hudspeth, 1983; Fettiplace and Crawford, 1989).

The problem of rates becomes much more critical when trying to understand hearing at frequencies above 1kHz. This is the case in most forms of mammalian hearing. For example, in the case of some bats, with well documented auditory ranges approaching 100kHz, does the stereocilial bundle gate current into the cell on a cycle-by-cycle basis? Are the molecular kinetics of transduction in mammals different? Or is the linkage between the sound stimulus and the hair cell modified in such cochleas?

Permeability of the transducer channel

In vivo the apical surface of a hair cell faces a solution with a high concentration of K^+ (endolymph) in the scala media, one of the interior compartments of the cochlea. Leakage of the solution is prevented by tight junctions around the apical portions of the cell. The composition of endolymph has been measured in the guinea pig cochlea by microsampling, and found to have a composition (in mM): K, 160; Na, 14; Ca, $30\mu M$ (Bosher and Warren, 1977). In other acoustico-lateralis organs, the predominant ion facing the apical surface is also K. Thus endolymph has much the composition of intracellular fluid. The majority ion, in this case K, will not be in equilibrium across the apical membrane, but will be driven into the cell because there exists a potential across the transducer channel consisting of the difference between the intracellular potential of the hair cell and the potential in the endolymph.

In the mammalian cochlea, such a potential difference can amount to +80mV (endolymph) + 50mV(intracellular potential) = 130mV across the transducer. In other hair cells, it may not be as large since the endolymph is at a lower potential, usually between 0 and +20mV. The effect of K entry into the cell through the transducer may be to reduce the metabolic demands on the cell since it can then maintain a large K gradient across the basolateral membrane by loading itself through the transducer. In the mammalian cochlea therefore the high metabolic requirements of the Na/K pump have been moved away from the transduction mechanism to a transport epithelium (the stria vascularis), on the cochlear wall.

Ionic selectivity of the transducer channel

The majority cation which permeates the the transducer channel is thus K. To gain more information about the channel itself the selectivity sequence has been measured in chick vestibular hair cells using whole cell dialysis (Ohmori, 1985). The observed permeability sequence, measured using CsCl inside the cell and the test ion outside is: $Li > Na > K > Rb > Cs > choline > TMA >> TEA$. The permeability sequence corresponds to Eisenman sequence IX for alkali cations, which suggests that the selectivity filter for the ions is at a region of high field strength. It is interesting to note that the cation selective stretch activated channel found in pectoral muscle (Guhuray and Sachs, 1984) has a different selectivity. This channel is selective for K over Na, by a factor of 4.

Divalent cations also permeate the hair cell transducer channel. In chick vestibular hair cells, Ohmori found the following permeability ratios, based on the the shift of transducer current reversal potential, E_{rev} (relative to internal Cs): Ca,4.65; Sr,2.4; Ba,2.3; Mn,2.1. For comparison, using equivalent measurements, Edwards and others found that Ca has a lower permeability ($P_{Ca}/P_{Na} = 1.4$) in the mechanosensitive membrane of stretch receptors (Edwards et al., 1981). In the case of the vertebrate light sensitive channel, there is a comparable selectivity sequence (Hodgkin et al., 1985). In addition, divalents permeate the photoreceptor channel (Yau and Nakatani,1984) but cause a block for monovalents at physiological levels. The high permeability for Ca of such sensory channels distinguishes them from the agonist gated channels such as the acetylcholine or the glutamate channel. The external Ca concentration in the endolymph facing the transducer varies between $30\mu M$ (mammals: Bosher and Warren, 1977) and $230\mu M$ (frog saccule: Corey and Hudspeth, 1979) and may well depend upon species. The Ca ionic flux is therefore probably low, although conceivably adequate to influence intracellular levels.

In hair cells, a minimum external Ca concentration is required for hair cell mechanotransduction (Sand, 1975; Corey and Hudspeth, 1983). At present, it is not known whether this is a requirement of the channel itself or, more probably, of the linkage of the channel to the stereocilial microshear. Nevertheless, in the case of other mechanosensitive conductances, low Ca_0 enhances the selectivity of the channel for monovalents (lens: Cooper et al., 1986). Such low levels (in the micromolar range) have not been investigated in hair cells. Other features of the channel, notably the high Ca permeability, suggests that the pore may have a double binding site for Ca of the form suggested by Hess and Tsien for the L-type Ca channel (Hess and Tsien, 1984). However, more information is required about the molecular requirements of the transducer for Ca in hair cells, particularly since cellular experiments on hair cell gating kinetics have not been done at physiological Ca_0 levels. It would be particularly interesting to look for evidence of multiple ion occupancy of the channel as in the case of the "conventional" Ca channels.

The site of transduction in hair cells

There are only two direct physiological experiments on the site of transduction and they come to opposite conclusions. Hudspeth (1982), using micropositioned extracellular electrodes to investigate current source and sinks in frog saccular epithelia found that the transducer current was better described by a model in which current sinks were placed at or near the tips of the stereocilia.

A different conclusion has been reached by Ohmori (1988). In experiments which studied divalent ion permeation through the transducer, using FURA-2 loaded chick hair cells to monitor entry, it was found that manganese, which permeates the transducer channel but not voltage-gated Ca channels, quenched the visualized FURA-2 fluorescence near the base of the stereocilia when the bundle was deflected. He concluded that Mn was entering preferentially near the stereocilial base.

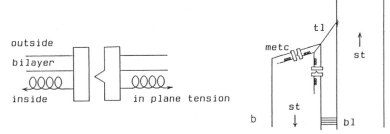

Fig. 2: Mechanosensitive channel organization. a) Channel of the type sensitive to stretch in the plane of the membrane. Possible cytoskeletal connections shown as springs. b) Schematic arrangement of mechano-electric transducer channels (metc) in the hair cell stereocilia, showing putative connections to the tip link (tl). Additional basal linkages (bl) between stereocilia may allow stresses to be transmitted basally as well. Shear (arrowed) between adjacent stereocilia will activate the metc. An alternative scheme, functionally equivalent, is given in Howard et al. (1988), where the link connects to the metc externally.

Both experiments are open to technical objections. Nevertheless, the tip-transducer model coincided with the report from high resolution scanning electron microscopy on the presence of extracellular linkages between stereocilia running between the tips of stereocilia in each row to the adjacent row (Pickles et al., 1984). Such filaments are in a position to convert shear between the stereocilia into a force on the transducer channel. Nevertheless, there are further linkages with a complex structure near the bundle pivot (Csukas et al., 1987); thus there is an ultrastructural substrate for a more proximal transducer site as well.

Taken at face value, a model with linkages between the stereocilia and the channel predicts a small non-uniform stiffness of the stereocilial bundle as the transducer channel switches between open and closed configurations. These forces have been measured (Howard and Hudspeth, 1988) as a change in the stiffness of the hair bundle as it is deflected around its pivot. Developments of these models estimate that the force required to gate a transducer channel open would be about 300fN or about 1/4 that of the actin-actin binding force.

The number of transduction sites

The mechanoelectric transduction channel is the main, if not the only, channel on the apical hair cell surface. An indirect argument is that, because of the nonequilibrium conditions across the apical membrane, any other channel would introduce a shunt conductance in the cell and depolarize it: there is no firm evidence for that possibility. There are no reports of other channels on the apical surface when that membrane is patched for whole cell recording (Roberts and Hudspeth, 1987).

The number of transduction channels can be estimated in two ways: either by comparing the single channel conductance with the maximal mechanically-activated conductance. It has not been possible to measure single transducer channels because of the difficulties of activating them, and only one report exists of unit-like events activated by bundle displacement (Ohmori, 1985). These observations depend upon the precise mechanical control of the stimulus, and a more likely explanation was that the bundle was being intermittently deflected.

The more indirect method of counting sites is to estimate the unit conductance from the permeability of the transducer to progressively larger ions. The permeability of the transducer channel to TMA but not TEA (Corey and Hudspeth, 1982) suggest that the elementary pore size is about 0.6nm. This method produces estimates for the unit conductance of an aqueous pore of about 25pS, comparable to the acetylcholine receptor. This observation, coupled with estimates for the transducer activated conductance of between 2-4nS, suggests that the total number of transducer sites per cell would be about 100, and that therefore there may only be a few ion channels per stereocilium (Ashmore & Russell, 1983; Hudspeth, 1983).

The alternative method of estimating the number, by fluctuation analysis, has been performed by Holton and Hudspeth (1986) in frog saccular hair cells. In this type of experiment, the fluctuations in the transduction current were measured as the bundle was deflected, and provide an estimate of the number of channels contributing to the current. The estimates derived from this method are between 7-238 channels per cell. This again supports the idea that there are only a few channels per stereocilium.

Asymmetry of mechanoelectric transduction

Deflection of the hair bundle produces an inward current at normal resting potentials, reversing near zero when the hair cell is uniformly bathed in extracellular fluid. The current flow is a nonlinear function of the bundle deflection however, at least for deflections in excess 10nm. In frogs (Hudspeth, 1983), turtle (Crawford and Fettiplace, 1985) and mouse (Russell et al., 1986), the resting position of the bundle is at a position in which the current is only partially activated. Only about 10-20% of the channels are open on average. For larger excursions, the current saturates in both directions. One consequence is that when stimulated by a sinusoidal displacement at high frequencies, the asymmetry of transduction will therefore produce a mean current flow into the cell. There will therefore be a depolarization of the membrane at stimulation frequencies beyond the cut off set by the membrane time constant.The most successful models at explaining this nonlinearity have been those where the channel is coupled to the stereocilial shear by a fine linkage, which buckles at forces less than that required to pivot the stereocilium. Hence pulling the link would be more effective than pushing it.

The only cell type which may prove to be an exception is the outer hair cell of the mammalian cochlea. In this case, the evidence is indirect, and obtained by recording from the intact cochlea (Russell and Sellick, 1983; Dallos, 1985; Russell et al.,1986). In the cochlea the ionic currents activated by sound are believed to generate the cochlear microphonic, one of the classical cochlear potentials. The harmonic components of the cochlear microphonic, which reflect the asymmetry in the underlying generator, although measurable, are small.

NON-TRANSDUCING CHANNELS IN HAIR CELLS

Ionic channels on the basolateral membranes

So far this chapter has concentrated on the ionic current which flows through the apical membrane of hair cells. The basolateral membrane of hair cells has also received attention, however, because the properties of this membrane determines the membrane potential, and hence transmitter release, from the whole cell. The membrane of verte-brate hair cells contain channels which are specific for potassium. The main types of channel described so far are summarized in Table I

The function of the K(Ca) channel appears to be to provide a tuned membrane resonance (see Kros, this volume), at least in the lower vertebrate hair cells. In mam-malian hair cells, although a K(Ca) conductance is present, there is little evidence for an electrical resonant behaviour.

REVERSE TRANSDUCTION

Physical devices for transducing displacement into electrical signals can often be operated in reverse. Thus piezoelectric crystals can be used as sensors and as mechanical force generators. The underlying physical principle is that the force and electric field on a charge are proportional. A similar idea has entered hair cell physiology, under the name of "bidirectionality", based more on the analogy with the physical process of auditory stimulation than the underlying physiology (Weiss, 1982). Despite this, there is evidence that some hair cells, and possibly all, can both act as sensory and motor cells. Where a hair cell differs from the equivalent physical device is that the molecular transducing mechanism itself does not reverse; indeed the transducer channel appears to be quite voltage-insensitive. Instead the sites producing forward and reverse transduction are distinct, although the cell as a whole operates bidirectionally.

Table 1: Ionic channels in hair cells basolateral membranes

$I_{K(Ca)}$	Most hair cells; Frog, saccule [1] canals [7]; turtle [2], papilla; chick [5] mammalian cochlea [3,4]
I_A	Frog, saccule [1] and canals[7]
I_{Na}	Alligator papilla [6]; chick[5]
Inward rectifier	Outer hair cells[3]
I_{Ca}	Most hair cells[1,2,4,8,6]

REFERENCES: 1) Lewis and Hudspeth, 1983; 2) Art et al., 1987; 3) Ashmore and Meech, 1986; 4) Kros and Crawford, 1988; 5) Mann and Fuchs, 1985); 6) Evans and Fuchs, 1987; 7) Housley et al., 1988; 8) Santos Sacchi & Dilger, 1988.

Reverse transduction in lower vertebrates

Direct observations on cells *in vitro* have shown that appropriately stimulated hair cells can generate forces. Turtle hair cells from the auditory papilla can be studied in isolated epithelia, and the stereocilia deflected directly (Crawford and Fettiplace, 1985). Simultaneous measurement of the stereocilial deflection showed that some of the stereocilia are spontaneously moving even in the absence of sound . The amplitude of these movements is small (typically 10nm). When the stereocilial bundle was deflected, with a constant applied force from a flexible probe, the displacement of the bundle showed an oscillatory response. This mirrored the membrane potential in the cell. Such observations indicate that turtle hair cell membrane potential and bundle forces are tightly coupled.

Observations on saccular hair cells have also been reported using high resolution optical methods to detect motion of the bundle in excess of the Brownian motion. Brownian motion would normally be expected of a free standing microscopic structure (Denk and Webb, 1989). The measurements also indicate that in frog hair cells there is a mechanism which can generate motile forces. However, the observations, being restricted to the bundle itself, do not point to where the forces are produced. Both in turtle and in frog, movements of the bundle could arise from body forces within the cell and then appear, because of the cell geometry, as a movement of the bundle. This possibility seems to be the case in the mammalian system.

Reverse transduction in the mammalian hair cells

Cochlear anatomy: The mammalian cochlea contains two morphologically distinct hair cells: 1) inner hair cells which are innervated by the afferent nerve; and 2) outer hair cells which are 3-4 times more numerous, but have only a poor afferent innervation. They have a much more pronounced efferent innervation (Spoendlin, 1978). The position of outer hair cells in the cochlear partition also suggests that they might be able to modify the partition mechanics.

Mammalian hair cell ultrastructure: Flock and Cheung pointed out that outer hair cells contain actin. The experiments used myosin S1 fragments to label the actin filaments in ultrastructure (Flock and Cheung, 1977; Tilney et al., 1980). The findings have subsequently been confirmed using specific antibodies. The main observation was that actin is in the stereocilia and in the cuticular plate region. A weak immunoreactivity is also found throughout the cell cytoplasm (Flock et al., 1986), and can be localised at the electron microscope level to the cell cortex, associated with the plasma membrane. Such findings correlated well with the then popular idea that cells of the brush border epithelia, which also show an array of villi inserted into a cuticular plate type structure, were motile. Such cells were reported to alter the position of their microvilli on Ca stimulation (Mooseker, 1976), supporting the proposition that hair cells were motile. The brush border observations have since been revised; the evidence that both actin and myosin are present in outer hair cells has however been taken up as a basis for length changes in the outer hair cells on a slow time scale.

Motility in mammalian outer hair cells

The evidence for electrical to mechanical conversion is best in mammalian hair cells. Outer hair cells, but not inner hair cells, exhibit a form of reverse transduction which has implications for cochlear function. The first observations reported that injection of current into isolated outer hair cells caused them to lengthen with hyperpolarizing and shorten with depolarizing currents (Brownell et al., 1985). The length changes are large

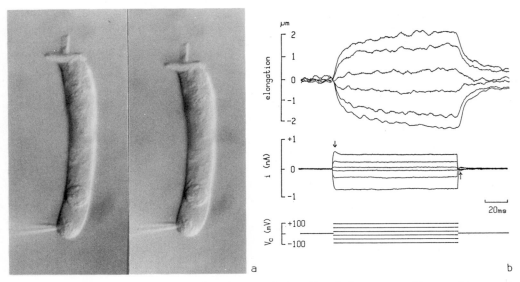

Fig. 3: Cell length changes in isolated outer hair cells of the mammalian cochlea. a) Outer hair cell stimulated via a patch pipette at a command potential of -150mV and +50mV. Cell length 70μm. b) Length changes measured by a photosensor (top) during a set of command pulses (bottom) so that elongation occurs during hyperpolarization. Middle trace, whole cell current. The length changes are graded with potential. Arrows point to a small transient current at onset and offset which corresponds to a fixed charge movement described later. Cell insolution with extracellular ionic composition. (From Ashmore (1987), reproduced from J.Physiol.)

enough to be seen and measured using a light microscope. The length changes are not phasic but are maintained for the duration of the stimulus (which allows photography of the cell; Holley and Ashmore, 1988a) and graded with it. As will be discussed below, the site of the motile response is likely to be associated with the basolateral membrane.

Good temporal control of the membrane potential and current can be obtained using the whole cell variant of the patch recording technique to record from cells which have been isolated from the organ of Corti and maintained for short periods (up to 6 hours) in tissue culture. With low access resistance pipettes, (less than 4megohms), the total changes in length can be shown to be about 4-5% of the cell length in guinea pig outer hair cells where most of the experiments have been performed (Ashmore, 1987). Since cells are changing length whilst surrounded by a viscous fluid medium, the observations indicate that the cells must be able to generate forces along their length.

The stimulus for force generation

In these experiments, the ionic distinction between the apical stereocilial surface and the basolateral surface is abolished, and the stereocilia remain undeflected. The available evidence supports the idea that it is membrane potential, rather than current, which controls cell length changes. The evidence comes from experiments in which both intracellular and extracellular media are changed.

Replacing the contents of the pipette from the normal K intracellular medium to one with with Na or Cs reduces the input conductance of the cell by a factor of nearly 3. It also shifts the zero current potential of the cell to near 0mV from -55 to -65mV found with K in the pipette (Ashmore, 1987; 1988). However the potential dependence of the length change is virtually unaffected. A comparable experiment has been performed exploiting the Ca permeability of the cell (Santos-Sacchi and Dilger, 1988). In this case Ba replaced Ca in the external medium; it was possible to reverse the direction of the current, producing an inward Ba current over a limited potential range, without altering the direction of the cell delete length change. It thus seems unlikely that a calcium current is involved in the cell length change. Indeed 1mM cadmium, sufficient to block any voltage dependent calcium current, has no affect on length changes in the cell held under voltage clamp (Ashmore and Meech, 1986).

The rate of reverse transduction

The force generating step in outer hair cells is found to be too fast to be generated by biochemical intermediate steps as found in muscle. The rate at which length changes occur is measured by a differential photodiode measuring the elongation of the cell relative to a fixed point, such as that formed by the patch site. With such measuring techniques, the length change in the long apical cells is a low pass filtered version of the voltage command pulse with a typical time constant of 1.5ms (Ashmore, 1987).

There is a possible small absolute delay of 100μs or less. Such absolute delays may well arise from the the time taken to space clamp the relatively large cell. The data is consistent with a near instantaneous activation of the cell forces. Since the length change is linearly graded with potential for small steps near -50mV, an equivalent experiment can be performed using sinusoidal potential commands to measure the frequency response. These experiments show that the cells from the apical region of the cochlea, normally involved in frequency detection at frequencies up to about 2-3kHz in the guinea pig, can respond to frequencies above 5kHz. Beyond these frequencies the bandwidth of the patch clamp becomes limiting.

Isolated experiments have been also been reported to indicate that hair cells can change length at frequencies of 30kHz, whenstimulated with large extracellular fields gradients (Zenner et al., 1987). Responses at such frequencies are consistent with a linear behaviour of the cell, although in such experiments it is not possible to be certain of the precise stimulus being applied to the cell. There seems little doubt that outer hair cells exhibit a force generating mechanism which is fast enough to be involved in most types of auditory processing because it spans the range of acoustic stimuli, but it must remain an open question whether all cochleas, including those in the bats specialized to detect ultrasonic stimuli used in echolocation, use the same mechanism uniformly.

The mechanism of rapid force generation

Induced length changes, which can be maintained over several hours, withstand a wide range of biochemical manipulations designed to disrupt many other forms of cell motility (Holley and Ashmore, 1988a). This alone argues against internal enzyme cascades. Using the patch pipette, agents, even if not membrane soluble, can be introduced into the cell whilst it is being stimulated. These experiments, essentially negative in nature, rule out: involvement of microtubules; anaerobic and aerobic ATP metabolism; actin assembly and disassembly; and a calcium requirement, since the pipette calcium can be heavily buffered (using BAPTA) to chelate free cytoplasmic Ca. In all these ca-

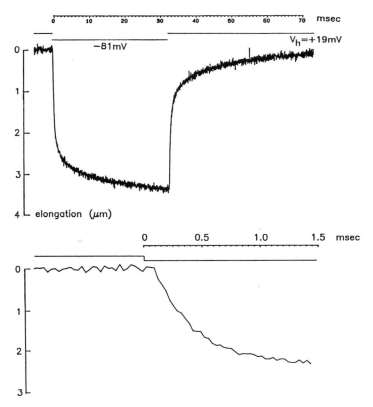

Fig. 4: Rate of elongation of outer hair cell. Bottom trace is the top trace on an expanded timescale, showing a delay within instrumentation resolution and movement initiated within $100\mu sec.$. Photosensor, recording during a command step from $+19$ to $-81mV$, had a bandwidth 0-40kHz. Signal averaged records, 250x. Whole cell recording; series resistance of pipette, 2 megohms. Pipette contained (mM): K, 140; BAPTA, 10; Mg, 2; HEPES, 5, to pH 7.3 with sucrose to 330mOsM. Cell length, $65\mu m$. Experiments performed at room temperature.

ses there is no detectable effect on the cell length change. The temperature dependence is weak with $Q_{10} = 1.33$, again arguing against elaborate biochemical cascades (Ashmore and Holley, 1988). The mechanism must therefore be different from motile mechanisms which have been described in other cells, with the possible exception of a rapid pressure change accompanying the action potential which can be measured in the squid axon (Tasaki and Iwasa, 1982).

Whatever the nature of the motor, it is distributed along the basolateral membrane. This conclusion is reached by patching the cell at the base, and observing that the movement of each part of the basolateral cortex is in proportion to the distance from the base (Holley and Ashmore, 1988a). Cells can be inflated by increased intracellular pressure applied to the patch pipette. Under these abnormal conditions, the cells round up. However close inspection shows that portions of the cell cortex dimple when the cell is polarized. The simplest explanation is therefore that the motile mechanism is a) associated uniformly with basolateral cell membrane and/or cortical structures and b) is triggered by changes in membrane potential.

Ultrastructural basis for reverse transduction

Hair cells are cylindrical cells, varying in length from about $20\mu m$ to over $90\mu m$ depending upon cochlear position and species. The passive forces which maintain the shape of the cells depend upon cytoskeletal specializations in the cell cortex. Thus when the membrane is extracted from the cells, either in the intact organ of Corti or in isolated cells, the cell ghost retains its shape (Flock et al., 1986; Holley and Ashmore, 1988b). It is against this cytoskeleton that any forces generated in the cell have to act. It seems likely that the main function of this cytoskeletal array, in which thick filaments and thin filaments are arranged in a helical coil, is to organize forces generated in the plane of the membrane into forces directed along the axis of the cell.

An early suggestion (Flock, 1983), was that the lateral array of cisternae placed along the side of the cell resembled the sarcoplasmic reticulum of skeletal muscle, and were part of a motile machinery. In guinea pigs, there are about 3-4 layers of cisternae in the outer hair cells. This has given rise to the idea that they are characteristic of this cell type. However rudimentary cisternae are also found in inner hair cells which are not motile. Although the cisternal apparatus shows similarities to the sarcoplasmic reticulum found in muscle, it does not seem to be the calcium store for contraction of the type found in muscle. Its function, conceivably related to regulating internal Ca or pH in the long term, remains presently unclear. Another feature of the outer hair cell ultrastructure is a system of pillars, about 40nm long, arranged in a 30x50nm array all around the basolateral membrane (Saito 1983; Flock et al., 1986; Bannister et al., 1988). These structures run between the cisternae and the plasma membrane. Freeze fracture shows a large number of dense particles on the p-face, with a density about the same as that of the pillars, but the same density of material is not present on the e-face (Forge, 1989). Thus pillar associated proteins do not seem to penetrate the membrane in the way that the acetylcholine receptor penetrates the postsynaptic membrane.

Charge movement in hair cell membranes

One of the handles on the coupling between membrane potential and protein confor-mational change in several systems has been the observation of charge movements. Such charge movements are observed in skeletal muscle (Adrian and Almers 1976) following a voltage step, and are believed to be the first events in excitation coupling. The Na channel in the squid axon (Armstrong et al., 1974), the calcium channel (Kostyuk et al., 1981), and rhodopsin (measured as the early receptor potential, the ERP) exhibit detectable charge movements as an early signal of protein conformational change. A similar charge movement may be detected in outer hair cell membranes.

Equal and opposite command voltage steps produce capacitance charging currents which when summed do not cancel. Complete cancellation of charging transients would be expected of a simple RC membrane. Instead there is a transient lasting for approxi-mately 3ms whose magnitude depends on the holding potential. The underlying charge movement would be equivalent to the outward movement of positive charge, or inward movement of negative charge. Measurement of the potential dependence of the charge movement indicates that it is equivalent to a single elementary charge, e, being displaced across the membrane (or equivalently a charge e/f being transferred across a fraction f of the membrane). This charge dependence is different from that found in skeletal muscle (Adrian et al., 1976), where the charge is $2e$. It is further evidence that the system is not the same as found in muscle.

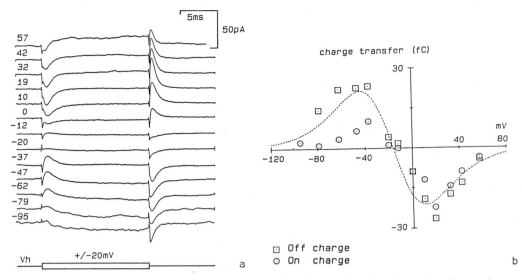

Fig. 5: Charge movements in outer hair cells. a)Currents flowing through cell membrane on averaging 100 equal and opposite command pulses (20mV). A linear membrane would yield a flat trace at all holding potentials. The transients show that the membrane capacitance was voltage dependent. Whole cell recording conditions. Pipette contained 140mM CsCl. Holding potential shown to L. of records. b) Total charge transferred at onset (circles) and at offset (squares) of command. The dotted line is the prediction of a Boltzmann distribution of charges in the membrane, with maximum transferable charge 350fC.

The charge movement is not abolished when both interior and exterior solutions are replaced by impermeant ions, such as TEA sulphate, although a substantial component of the ionic current is suppressed. The transient is therefore likely to be due to membrane-bound charge movements rather than due to movement of ionic charges through the membrane.The time course of the underlying charge movement closely matches that of the simultaneously measured length change, both at the onset and at the offset of the step. If the number of unit charges is calculated that would give rise to the observed current, it is found that in a cell $70\mu m$ long, about 1.610^{+6} elementary charges would be required. This is close to the number of pillars associated with the basolateral membrane, and it is thus tempting to associate the charge movement with a repeating structure, such as the pillars, in the outer hair cell membrane.

THE MAMMALIAN COCHLEA

Forward transduction is the basis of auditory nerve coding in the mammalian cochlea. The main arguments for reverse transduction in hair cells arise from the observed physiology, and the growing belief that there are active processes injecting energy into the cochlear mechanics. It is now thought that the outer hair cells form the subsystem which make this possible (see for example, Kim,1986), and the term "the cochlear amplifier" is used to summarize the observations.

Arguments for a "cochlear amplifier"

Sensitivity to sound in the mammalian cochlea: At auditory threshold, the movement of the basilar membrane in the mammalian cochlea is greater than originally believed. Von Bekesy's measurements of basilar membrane motion in the cochlea from cadavers suggested that the threshold displacement was about 10^{-11}m (von Bekesy, 1960). The more recent measurements suggest that this figure should be revised upwards by about a factor of 100, (40dB), to provide a threshold movement of 0.3nm (Khanna and Leonard, 1982; Sellick et al., 1982; Robles et al., 1986).

Selectivity of auditory nerve fibres: The great selectivity of the auditory outflow indicates that the simple cells. Although early attempts to explain this finding were constrained by Bekesy's apparently impregnable data, which showed that the peak motion of the basilar membrane was not precisely localised, and therefore stimulated many hair cells, the more recent data indicate that the membrane itself appears to have a vibration pattern which is locally enhanced (Kim, 1986). This enhancement requires elements which can directly affect the mechanics of the basilar membrane.

Outer hair cells in cochlear mechanics: There are in addition several pieces of evidence which point towards the outer hair cells as being critical in an understanding of the mammalian cochlea. The first is that hearing losses, often associated with noise exposure, with ototoxic drugs, or with aging processes, involve pathologies in which outer hair cells are missing or grossly disturbed (for references see Ashmore, 1987). An observation made by Mountain is that passing electrical current across the cochlear partition produces sounds which can be measured in the ear canal (Mountain, 1980). The direction of the current was such as to stimulate the outer hair cells. Mountain subsequently also found that electrical stimulation could interact with distortion products measured in the ear canal. Many of the arguments which have been advanced in favour of a reverse transduction step in the cochlea, to provide a "cochlear amplifier", have involved discussion of the inherent nonlinearities of cochlear processing. What generates the nonlinearity, even near threshold, is unknown.

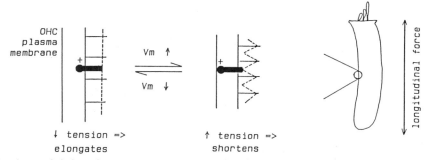

Fig. 6: A model for charge movements in the OHC basolateral membrane. Depolarization moves positive charge outward. A linkage is shown which couples charge displacement to cytoskeletal components within the plane of the plasmalemma generating tension along the cell. The restoring forces are not shown.

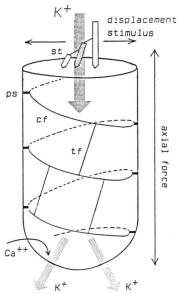

Fig. 7: Schematic organization of the mammalian outer hair cell. Shear displacements of the stereocilia generated by sound within the organ of Corti gate K entry into the cell via the metc. The main conductances of the cell are shown. The longitudinal forces are generated by charge movements which are generated by the induced potential changes in the cell membrane. Axial forces arise when the planar forces act against a system of circumferential fibres (cf) and thin filaments (tf) formed within the cell and linked to the plasma membrane by a lattice of pillar structures (ps).

With the observation of high frequency force production by outer hair cells *in vitro*, the case for the involvement for outer hair cells becomes even stronger. Of all the elements in the cochlea, the outer hair cells are the only elements to show the ability to generate forces possibly sufficient to alter the mechanics of the partition. They are placed in the cochlear partition so that the longitudinal force which they can produce is directed against the basilar membrane.

The final section and the appendix are concerned with the integration of outer hair cell mechanisms into a description of normal hearing.

COCHLEAR MODELLING

There are considerable difficulties in assembling the data obtained from cellular studies into a coherent model of the cochlea. The main requirement of any model is to explain how recent measurements of the basilar membrane mechanics arise. Knowing the mechanics, it should then be possible to predict the input to the inner hair cells, and thence provide a description of the afferent outflow from the cochlea.

One of the problems has been that relatively little is known about the mechanical properties of the other components of the partition. As shown in the Appendix, the behaviour of the basilar membrane motion can be determined by its point impedance,

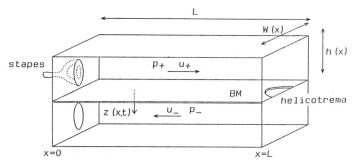

Fig. 8: Two compartment 1-D model of the cochlea. The basilar membrane, BM, is shown separating the two compartments, stimulated by sound at the stapes oval window. The helicotrema ensures that there is no net fluid pressure across the BM.

i.e. the quantity which characterizes the mechanical properties of the membrane at each point along the length. It is this quantity which is determined by the properties of the outer hair cells. Recent work has attempted to rectify this situation by measuring point impedances of the basilar membrane, using a probe placed in a position to deflect the basilar membrane (Miller, 1985) and then to work back to properties of the constituent partition components.

These experiments do not address the question of the dynamic components of the partition impedance, which will include the visco-elastic properties of the membrane. It seems most probable that reverse transduction, force production of outer hair cells, needs to be included. The original measurements made in the 1930's and 1940's (von Bekesy, 1960) used large deflections produced by extremely loud sounds under conditions in which the hair cells were certainly not functioning. The more recent experiments, using Mossbauer sources, are restricted to measurements of a very limited region of the basilar membrane near to the basal end, which is dictated by access to the recording site. For measurements of the group velocity of the travelling wave down the basilar membrane, measurements at two sites at least are required (Lighthill, 1981). In the absence of extensive experimental data, there has been a proliferation of cochlear models which attempt to reproduce the mechanical data.

The Appendix gives a derivation of the simplest model for cochlear mechanics as a framework into which data can be assembled. This is a linear model which under certain limiting conditions can be solved analytically. In general such solutions will not be possible, especially when the effects of cochlear force generators are included.

CONCLUSION

In the last 10 years, there has been great progress in what we know about mechanoreception. This is due largely to the growth of cellular recording methods. The problems which continue to beset the field pull in opposite directions. In one direction, the molecular structures of the mechano-sensitive channel and the molecular basis of reverse transduction are problems which require the appropriate molecular biology. The difficulty in pursuing the biochemistry of audition has been that the amount of tissue available is small. Although there has been recent improvement in the isolation techniques, refined for working with small samples (Shepherd et al., 1988; Holley, 1989), hair cells and their proteins do not share with photoreceptors the property of abundance.

The other direction is towards systems physiology. Here the programme is one of integrating data from recent measurements to construct models of the cochlea whose essential features can be understood without relying on extensive numerical computation. It is necessary, for example, to make a satisfactory correspondence between the microanatomy and any model parameters, a feature not shared by all cochlear models. Is there one mechanism, involving active feedback mechanisms from hair cells, or are there several mechanisms involved to cover the entire mammalian frequency range? The aim of such a programme would be to explain auditory frequency discrimination, auditory range and the sensitivity of the cochlea. This a necessary step towards understanding the processing which precedes higher auditory function.

ACKNOWLEDGEMENTS

The work was supported by the Medical Research Council. I thank Dr Matthew Holley for help with the photography and Dr Gary Housley for comments on the manuscript.

APPENDIX

BASIC COCHLEAR MECHANICS : THE 1-D MODEL

Definitions:

z(x,t): displacement of the basilar membrane, (BM), at x
x: position along cochlea, x=0,base; x=L,apex
ρ : fluid density
$p_{+/-}(x,t)$: pressure in upper/lower compartment
$u_{+/-}(x,t)$: fluid velocity in upper/lower compartment
h(x): height of compartment
w(x): width of compartment
m(x,t): mass of membrane
r(x,t): damping of membrane
s(x,t): stiffness of membrane

MACROMECHANICS

This model is one of the classical treatments of propagation of the cochlear travelling wave on the basilar membrane (Peterson and Bogert, 1950; for review see Viergever, 1986). Consider a long membrane separating upper and lower compartments containing a fluid. The equations of continuity for fluid in the compartments are then

$$\frac{\partial}{\partial x}(hwu_+) = w\frac{\partial z}{\partial t} \tag{1a}$$

$$\frac{\partial}{\partial x}(hwu_-) = -w\frac{\partial z}{\partial t} \tag{1b}$$

and the forces on the fluids are described by

42

$$\frac{\partial p_+}{\partial x} = -\rho \frac{\partial u_+}{\partial t} \qquad (2a)$$

$$\frac{\partial p_-}{\partial x} = -\rho \frac{\partial u_-}{\partial t} \qquad (2b)$$

If the pressure across the partition is $P = p_+ - p_-$ and for simplicity we assume that the width of the compartments is constant, the relation between P and z, the membrane displacement from (1a,b) and (2,b) is

$$\frac{\partial^2 P}{\partial x^2} = 2\frac{\rho}{h}\frac{\partial^2 z}{\partial t^2} \qquad (3)$$

The simplest boundary conditions are

$$\frac{\partial P}{\partial x} = 2\rho \frac{\partial^2 z_s}{\partial t^2} \quad at\ x = 0 \qquad \frac{\partial P}{\partial x} = 0 \quad at\ x = L \qquad (3a)$$

where z_s is the movement of the stapes footplate, the input to the cochlea; and at x=L a condition which corresponds to free movement of fluid through the helicotrema between the compartments, although this condition can be relaxed to include damping with no subtantial change to the underlying physics.

Eqn (3) is the basic macroscopic equation. At this point there is no assumption about the nature of the membrane itself. It is necessary to augment (3) and before we can solve for z or P.

MICROMECHANICS

Elements of the basilar membrane (BM) are individually resonant elements. The assumption takes the form of a local relation between P(x,t) and z(x,t):

$$P = m(x,t)\frac{\partial^2 z}{\partial t^2} + r(x,t)\frac{\partial z}{\partial t} + s(x,t)z \qquad (4)$$

This local relation is equivalent to specifying a point impedance for the basilar membrane.

When m,r and s are time-independent (for the full case, see Diependaal et al., 1987) Eqn (4) is readily recognised as the equation for a forced damped harmonic oscillator, with forcing term P(x,t), at each point x, with each oscillator possessing a characteristic resonant frequency f_0 given by

$$f_0 = 1/2(s/m)^{1/2} \qquad (5)$$

and resonating with a quality factor, Q_r. This term, given by $Q_r = f_0/r$, expresses the sharpness of the resonance. The smaller the damping the sharper is the resonance. Thus the cochlear model is one whose physical interpretation is that of an array of resonating elements spaced along the length of the cochlea. Each element exhibits a damped resonance (if r(x)>0). The effect of the fluid, expressed in (3), is to couple neighbouring oscillators together, and to spread the excitation from one oscillator to the next; the maximum amplitude of the basilar membrane will occur near that oscillator

43

with f_0 closest to the stimulating sound frequency. The process has also been discussed in terms of energy flow along the cochlea (Lighthill, 1981).

Making the further assumption that the mass $m = m_0$ and damping $r = r_0$ are constant but that the stiffness is an exponential function of position,

$$s(x) = s_0 exp(-a\ x) \qquad (6)$$

then eqns (3) and (4) can be transformed to give piecewise analytical solutions for P and z which provide some insight into the processes (de Boer, 1980). It is also straightforward to solve (3) in the time domain (Diependaal et al., 1987) under more general conditions. A simple algorithm for the envelope of the travelling wave, which can be implemented on a microcomputer, is given in Neely and Kim (1986). The results of such a calculation for the envelope of the travelling wave are shown in Figure 9 for three different frequencies. The dotted line shows enhancement that is required to coincide with the experimental results of Sellick et al. (1982).

The assumption (6) is the theoretical formulation of von Bekesy's findings that the stiffness of the BM decreases from base to apex. The exponential dependence is used to explain the observed physiological place map of acoustic frequency onto the cochlear spiral: the exponential in (6), when inserted into (5), ensures each doubling of the stimulus frequency will map onto a linear increment of the cochlea, with highest frequencies mapping at the base, x=0, and the lowest frequencies at the apex, x=L.

MODIFICATIONS

It was realized at an early stage that this model does not explain the sharpness of the cochlear tuning: for each point, the BM amplitude, z(x), or equivalently the velocity, dz/dt, is not sharply localized enough to explain the auditory fibre tuning. (The simplest model of the inner hair cells suggests that the fibre activity should be closely related to the amplitude z by considering the geometry of the partition). Since 1982, with the new Mossbauer measurements of the BM motion (which measure dz/dt), there has been renewed interest in the micromechanics of the partition. One suggestion has been that the dissipative term r(x)dz/dt in Eqn (4) should be adjusted to reflect an input of energy to the system: that is, any reasonable model should result in a negative value for r(x) near the peak (Neely and Kim, 1986), equivalent to feeding energy in to the motion to overcome dissipative forces. Such models, of which the Neely and Kim version is one, are linear and can reproduce the BM results quite well, but the form of r(x) has to be included on an *ad hoc* basis.

Dimensionality: The above model is 1-dimensional, for it includes only the longitudinal axis of the basilar membrane. The physical assumption that underlies the reduction of a 3-D cochlea to 1-D may be violated near the resonance peak: the wavelength of the travelling wave becomes shorter than the height of the channel (de Boer, 1980). Extensions to 2-D and 3-D have been proposed. Such models can be solved numerically in the time domain, necessary when nonlinear effects are included (Diependal et al., 1989). The qualitative differences in the numerical results have been discussed in the context of cochlear energy flow (Lighthill, 1981).

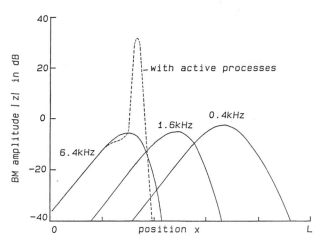

Fig. 9: Basilar membrane amplitude from Eqns (4) and (5). The used were: $m_0 =$ 0.3mg.mm^{-2}; $r_0 = 400$mg.mm^{-2} ms^{-1} and s $= 10000$ exp (-0.4x) mg mm^{-2}mg^{-2}; L$=25$mm with x in mm. Solution computed by the algorithm in Neely and Kim (1986). Figures by each figure indicate the frequency in Hz of the stimulation. The dotted line on the 6.4kHz envelope indicates the required BM amplitude to fit the experimental data.

Effect of cochlear spiralling : No significant effects arise when the 3D spiral organization is included, because the scales critical fluid wavelengths are short compared to the scale of the spiral (Viergever, 1986).

Inclusion of other resonant structures: It is assumed above that the input to the inner hair cells is proportional to z(x), or possibly dz(x)/dt. Coupling in the cochlear partition is mediated via the tectorial membrane, which itself may be a resonant structure (Zwislocki and Kletsky, 1979). Under suitable assumptions this resonance would appear as a term in the point impedance, the relationship between P and z in (4).

Nonlocal effects within the cochlea partition: Little experimental information exists about this possibility at present, although reverse transduction in outer hair cells is likely to be able to generate sufficient forces to control viscous damping terms (Ashmore, 1987). Longitudinal coupling has been investigated under various assumptions (partition damping, Jau and Geisler, 1983; fluid damping, Lutkohner and Jager, 1986; partition stiffness, Wickesberg and Geisler, 1986; decoupling of radial zones of the partition, Kolston, 1988).

Nonlinear effects: The effect of nonlinearities within the partition are poorly understood. There have been studies which include nonlinear terms in eqn (4), for example by including a damping term r(x,t) quadratic in z (Duifhuis et al., 1986). It has been argued that the effects of such a term are to limit the inherent instability of cochlear models with active feedback. At present there are too many degrees of freedom in such approaches to allow satisfactory agreement with experiment.

REFERENCES

Adrian, R. H., and Almers, W., 1976, Charge movement in the membrane of striated muscle, *J. Physiol.*, 254:339-360.

Armstrong, C. M., and Bezanilla, F., 1974, Charge movement associated with the opening and closing of the activation gates of the Na channels, *J. Gen. Physiol.*, 63:533-552.

Art, J. J, and Fettiplace, R., 1987, Variation of membrane properties in hair cells isolated from the turtle cochlea, *J. Physiol.*, 358:323-348.

Ashmore, J. F., 1987, A fast motile response in guinea pig outer hair cells: the cellular basis of the cochlear amplifier, *J. Physiol.*, 388:323-347.

Ashmore, J. F., and Meech, R. W., 1986, Ionic basis of membrane potential in guinea pig outer hair cells, *Nature*, 322:368-371.

Ashmore, J. F., 1988, Ionic mechanisms in hair cells of the mammalian cochlea, *Prog. Brain Res.*, 74:3-10.

Ashmore, J. F., and Holley, M. C., 1988, The temperature dependence of a fast moltile response in isolated hair cells of the guinea pig cochlea, *Q. J. Exp. Physiol.*, 73:143-145.

Ashmore, J. F., and Russell, I. J., 1983, The physiology of hair cells, pp 149-180, *in*: "Bioacoustics: a comparative approach", B. R, Lewis, ed., Academic, London.

Bannister, L. H., Dodson, H. C., Astbury, A. R., and Douek, E. E., 1988, The cortical lattice: a highly ordered system of subsurface filaments in the guinea pig cochlear outer hair cells, *Prog. Brain Res.*, 74:213-219.

Bosher, S. K., and Warren, R. L., 1978, Very low calcium content of cochlear endolymph, an extracellular fluid, *Nature*, 273:377-378.

Brehm, P., Kullberg, R., and Moody-Corbett, F., 1984, Properties of non-junctional acetylcholine receptor channels on innervated muscle of Xenopus Laevis, *J. Physiol.*, 350:631- 648.

Brownell, W. E., Bader, C. R., Bertrand, D., and De Ribaupierre, Y., 1985, Evoked mechanical responses of isolated cochlear hair cells, *Science*, 227:194-196.

Chandler, W. K., Rakowski, R. F., and Schneider, M. F., 1976, A non-linear voltage-dependent charge movement in frog skeletal muscle, *J. Physiol.*, 254:245-284.

Cooper, K. E., Tang, J. M., Rae, J. L., and Eisenberg, R. S., 1986, A cation channel in frog lens epithelia responsive to pressure and calcium, *J. Membr. Biol.*, 93:259-269.

Corey, D. P., and Hudspeth, A. J., 1983, Kinetics of the receptor current in bullfrog saccular hair cells, *J. Neurosci.*, 3:962-976.

Corey, D. P., and Hudspeth, A. J., 1979, Response latency of vertebrate hair cells, *Biophys. J.*, 26:499-506.

Crawford, A. C., and Fettiplace, A. C., 1985, The mechanical properties of the ciliary bundles of turtle hair cells, *J. Physiol.*, 364:359-380.

Csukas, S. R., Rosenquist, T. H., and Mulroy, M. J., 1987, Connection between stereocilia in auditory hair cells of the alligator lizard, *Hearing. Res.*, 30:147-156.

Dallos, P., 1985, Response characteristics of mammalian cochlear hair cells, *J. Neurosci.*, 5:1591-1608.

Dallos, P., Santos-Sacchi, J., and Flock, A., 1982, Intracellular recordings from cochlear outer hair cells, *Science*, 218:582-584.

De Boer, E., 1980, Reflections on reflections, *J. Acoust. Soc. Am.*, 67:882-890.

Denk, W., and Webb, W. W., 1989, Simultaneously recording fluctuations in bundle position and of intracellular voltage reveals transduction properties of saccular hair cells, *In*: "Mechanics of Hearing", J. P, Wilson, and D. T., Kemp, ed., Plenum, New York.

Diependaal, R. J., Duifhuis, H., Hoogstraaten, H. W., and Viergever, M.A., 1987, Numerical methods for solving one-dimensional cochlear models in the time domain, *J. Acoust. Soc. Am.*, 82:1655-1666.

Diependaal, R. J., 1989, Time domain solutions for 1D, 2D and 3D cochlear models, *In*: "The Mechanics of Hearing", J. P., Wilson, and D. T., Kemp, ed., Plenum, New York.

Ding, J. -P, Yang, X. -C, Bowman, C. L., and Sachs, F, 1988, A stretch activated ion channel in rat astrocytes in primary cell culture, *Soc. Neurosci. Abstr.*, 14:1056.

Duifhuis, H., Hoogstraten, H. W., van Netten, S., Diependaal, R. J., and Bialek, W., Modelling the cochlea partition with coupled Van der Pol oscillators, pp.290-297, *in*: "Peripheral auditory mechanisms", J. B., Allen et al., ed., Lecture Notes in Biomathematics, Vol 64, Springer-Verlag, Berlin.

Edwards, C., Ottoson, D., Rydqvist, B., and Swerup, C., 1981, The permeability of the transducer membrane of the crayfish stretch receptor to calcium and other ions, *Neurosci.*, 6:1455-1460.

Evans, M. G., and Fuchs, P. A., 1987, Tetrodotoxin-sensitive, voltage-dependent sodium currents in hair cells from the alligator cochlea, *Biophys. J.*, 52:649-652.

Fettiplace, R., and Crawford, A. C., 1989, Mechanoelectric transduction in turtle hair cells, *In*: "Mechanics of Hearing", J. P., Wilson, and D. T., Kemp, ed., Plenum, New York.

Flock, A, 1983, Hair cells, receptors with a motor capacity, pp.2-9, *in* "Hearing - physiological bases and psychophysics", R., Klinke, and R., Hartmann, ed., Springer-Verlag, Berlin.

Flock, A., Flock, B., and Uhlfendahl, M., 1986, Mechanisms of movement in outer hair cells and a possible structural basis, *Arch. Otolaryngol.*, 243:82-90.

Flock, A., and Cheung, H., 1977, Actin filaments in sensory hairs of inner ear receptor cells, *J. Cell. Biol.*, 75:339-343.

Forge, A., 1989, The lateral walls of inner and outer hair cells, *In*: "Mechanics of Hearing", J. P., Wilson, and D. T., Kemp, ed., Plenum, New York.

Fuchs, P. A., and Mann, A. C., 1986, Voltage oscillations and ionic currents in hair cells isolated from the apex of the chick's cochlea, *J. Physiol.*, 371:31P.

Guhuray, F., and Sachs, F., 1984, Stretch activated single ion channels in tissue-cultured embryonic chick skeletal muscle, *J. Physiol.*, 352:685-701.

Hess, P., and Tsien, R. W., 1984, Mechanism of ion permeation through calcium channels, *Nature*, 31:453-455.

Hodgkin, A. L., MacNaughton, P. A., and Nunn, B. J., 1985, The ionic selectivity and calcium dependence of the light-sensitive pathway in toad rods, *J. Physiol.*, 358:447-468.

Holley, M. C., 1989, Purification of mammalian cochlear hair cells using small volume Percoll density gradients, *J. Neurosci. Methods*, 27:219-224.

Holley, M. C., and Ashmore, J. F., 1988a, On the mechanism of a high frequency force generator in outer hair cells isolated from the guinea pig cochlea, *Proc. Roy. Soc. Lond. B.*, 232:413-429.

Holley, M. C., and Ashmore, J. F., 1988b, A cytoskeletal spring in cochlear outer hair cells, *Nature*, 335:635-637.

Holton, T., and Hudspeth, A. J., 1986, The transduction channel of hair cells from the bullfrog characterized by noise analysis, *J. Physiol.*, 375:195-227.

Housley, G. D., Norris, C. H., and Guth, P. S., 1988, Single electrode whole cell voltage and current clamp of potassium channels in hair cells isolated from the crista ampullaris of the frog, *Soc. Neurosci. Abstr.*, 14:947.

Howard, J., Roberts, W. M., and Hudspeth, A. J., 1988, Mechanoelectric transduction by hair cells, *Ann. Rev. Biophys. Chem*, 17:99-124.

Howard, J., and Ashmore, J. F., 1986, The stiffness of the frog saccular hair bundle, *Hearing. Res.*, 23:93-104.

Howard, J., and Hudspeth, A. J., 1988, Compliance of the hair bundle associated with gating of the mechanoelectrical transduction channels in the bullfrog's's saccular hair cells, *Neuron.*, 1:189-199.

Hudspeth, A. J., 1983, Mechanoelectrical transduction by hair cells of the acoustico lateralis sensory system, *Annu. Rev. Neurosci.*, 6:187-215.

Khanna, S. M., and Leonard, D., 1982, Basilar membrane tuning in the cat cochlea, *Science*, 215:305-306.

Kim, D. O., 1986, Active and nonlinear cochlear biomechanics and the role of the outer hair cell subsystem in the mammalian auditory system, *Hearing. Res.*, 22:105-114.

Kolston, P. J., Sharp mechanical tuning in a cochlear model without negative damping, *J. Acoust. Soc. Am.*, 83:1481-1487.

Kostyuk, P. G., Krishtal, O. A., and Pidoplichko, V. I., 1981, Calcium inward current and related charge movements in in the membrane of snail neurones, *J. Physiol.*, 310:402-422.

Lewis, R. S., and Hudspeth, A. J., 1983, Voltage- and ion-gated conductances in solitary vertebrate hair cells, *Nature*, 304:538-541.

Lighthill, J., 1981, Energy flow in the cochlea, *J. Fluid. Mech.*, 106:149-213.

Lutkohner, B., and Jager, D., 1986, Stability of Active Cochlear Models: "need for a second tuned filter", Hall, J. L., Hubbard, A., Neely, S. T., and Tubis, A., Springer-Verlag, Berlin.

Miller, C. E., 1985, Structural implications of basilar membrane compliance measurements, *J. Acoust. Soc. Am.*, 77:1465-1474.

Mooseker, M. S., 1975, Brush border motility, *J. Cell. Biol.*, 71:417-433.

Mountain, D. C., 1980, Changes in endolymphatic potential and crossed olivocochlear bundle stimulation alter cochlear mechanics, *Science*, 210:71-72.

Neely, S. T., and Kim, D. O., 1986, A model for active elements in cochlear biomechanics, *J. Acoust. Soc. Am.*, 79:1472-1480.

Ohmori, H., 1985, Mechano-electric transduction currents in isolated vestibular hair cells of the chick, *J. Physiol.*, 359:189-217.

Ohmori, H., 1984, Studies of ionic currents in the isolated vestibular hair cells of the chick, *J. Physiol.*, 350:561- 581.

Ohmori, H., 1987, Gating properties of the mechano-electrical transducer channel in dissociated vestibular hair cells of the chick, *J. Physiol.*, 387:589-609.

Ohmori, H., 1988, Mechanical stimulation and Fura-2 fluorescence in the hair bundle of dissociated hair cells of the chick, *J. Physiol.*, 399:115-138.

Peterson, L. C., and Bogert, B. P., 1950, A dynamical theory of the cochlea, *J. Acoust. Soc. Am.*, 22:369-381.

Roberts, W. M., and Hudspeth, 1987, Spatial distribution of ion channels in hair cells of the bullfrog's sacculus, *Biophys. J.*, 51:8a.

Robles, L., Ruggero, M. A., and Rich, N. C., 1986, Basilar membrane mechanics at the base of the chinchilla cochlea: I. Input out functions, tuning curves and response phases, *J. Acoust. Soc. Am.*, 80:1364-1374.

Russell, I. J., and Sellick, P., 1983, Low frequency characteristics of intracellularly recorded receptor potentials in guinea pig cochlear hair cells, *J. Physiol.*, 388:179-206.

Russell, I. J., Cody, A. R., and Richardson, G. P., 1986, The responses of inner and outer hair cells in the basal turn of the guinea pig cochlea and the mouse cochlea grown in vitro, *Hearing. Res.*, 22:196-216.

Sakmann, B., Methfessel, C., Mishina, M., Takahashi, T., Takai, T., Kurasaki, M., Fukuda, K., and Numa, S., 1985, Role of acetylcholine receptor subunits in gating of the channel, *Nature*, 318:538-543.

Saito, K., 1983, Fine structure of the sensory epithelium of guinea-pig organ of Corti: subsurface cisternae and lamella bodies in the outer hair cells, *Cell. Tiss. Res.*, 229:467-481.

Sand, O., 1975, Effects of different ionic environments on the mechanosensitivity of lateral line organs in the mudpuppy, *J. Comp. Physiol.*, 102:27-42.

Santos-Sacchi, J., and Dilger, J. P., 1988, Whole cell currents and mechanical responses of isolated outer hair cells, *Hearing. Res.*, 35:143-150.

Sellick, P., Patuzzi, R., and Johnstone, B. M., 1982, Measurement of the basilar membrane motion in the guinea pig using the Mossbauer technique, *J. Acoust. Soc. Am.*, 72:131-141.

Shepherd, G. M. G., Barres, B. A., and Corey, D. P., 1988, Bundle blot purification of hair cell stereocilia: electro-phoretic protein mapping, *Soc. Neurosci. Abstr.*, 14:799.

Shotwell, S. L., Jacobs, R., and Hudspeth, A. J., 1981, Directional sensitivity of individual vertebrate hair cells to controlled deflection of their hair bundles, *Ann. N.Y. Acad. Sci.*, 374:1-10.

Sigurdson, W. J., and Morris, C. E., 1986, Stretch activation of a K channel in snail heart cells, *Biophys. J.*, 49:163a.

Spoendlin, H., 1978, The afferent innervation of the cochlean, *In*: "Evoked Electrical Activity in the Auditory Nervous System", R. F., Naunton, and C., Fernandez, ed., pp. 21-39, Academic Press, New York.

Stommel, E. W., Stephens, R. E., and Alkon, D. L., 1980, Motile statocyst cilia transmit rather than directly transduce mechanical stimuli, *J. Cell. Biol.*, 87:657-662.

Tasaki, I., and Iwasa, K., 1982, Rapid pressure changes and surface displacements in the squid axon associated with the production of action potentials, *Jap. J. Physiol.*, 32:69-81.

Tilney, L. G., DeRosier, D. J., and Mulroy, 1980, The organization of actin filaments in the stereocilia of cochlear hair cells, *J. Cell. Biol.*, 86:244-259.

Viergever, M. A., Cochlear macromechanics: A review *In*: "Peripheral Auditory Mechanisms", J. B., Allen, B., Hall, A., Hubbard, S. T., Neely, and A., Tubis, ed., Springer-Verlag, Berlin.

Von Bekesy, G., 1960, "Experiments in Hearing", McGraw-Hill: New York.

Weiss, T. F., 1982, Bidirectional transduction in vertebrate hair cells: a mechanism for coupling mechanical and electrical processes, *Hearing Res.*, 7:353-360.

Wickesberg, R. E., and Geisler, C. D., 1986, Longitudinal stiffness coupling in a 1-dimensional model of the peripheral ear *In*: "Peripheral Auditory Mechanisms", J. B., Allen, J. L., Hall, A., Hubbard, S. T., Neely, and A., Tubis, ed., Springer-Verlag, Berlin.

Yang, X. C., Guhuray, F., and Sachs, F., 1986, Mechano-transducing ion channels: ionic selectivity and coupling to viscoelastic components of the cytoskeleton, *Biophys. J.*, 49:373a.

Yau, K.-W, and Nakatani, K., 1984, Cation selectivity of light sensitive conductance in retinal rods, *Nature*, Lond., 309:352-354.

Zenner, H. P., Zimmermann, U., and Gitter, A. H., 1987, Electrically induced fast motility of isolated mammalian auditory sensory cells, *Biochem. Biophys. Res. Comm.*, 149:304-308.

Zwislocki, J., and Kletsky, E. J., 1979, Tectorial membrane, a positive effect on frequency analysis in the cochlea, *Science*, 204:639-641.

ELECTRICAL PROPERTIES OF THE BASOLATERAL MEMBRANE

OF HAIR CELLS - A REVIEW

C.J. Kros

The Physiological Laboratory
Cambridge, United Kingdom

INTRODUCTION

Hair cells, as found in the lateral line, vestibular and auditory systems of all classes of vertebrates, share a number of obvious similarities. All of them are situated in a modified epithelium separating two very different extracellular fluids: perilymph, surrounding the basolateral part of the cells and endolymph, in contact with the apical part of the cells from which the stereociliary bundle, the mechanoreceptive element of the hair cell, projects. Sodium is the major cation in the perilymph, as in most extracellular fluids elsewhere in the body; potassium is the major cation in the endolymph. The fluids are separated by tight junctions between the cells of this modified epithelium.

When the hair bundle is in its neutral, unstimulated position, the apical part of the cell is permeable to small cations, resulting *in vivo* in a small inward current carried mainly by potassium ions. Movement of the stereociliary bundle towards the kinocilium (or its remnant the basal body in cells in which the kinocilium degenerates in development) increases this transducer current. Movement of the bundle away from the kinocilium causes a decrease in the transducer current.

This modulation of the transducer current results in the generation of a receptor potential, which will in turn modulate the release of neurotransmitter from the hair cell and thus cause the activation of the afferent nerve fibres which make synaptic contact with the hair cell. The receptor potential is not directly proportional to the transducer current. It will be shaped by the passive electrical properties of the cell (membrane capacitance and input resistance) and also by time- and voltage- dependent conductances if these are activated in the membrane potential range of the receptor potential. These mechanisms shaping the receptor potential are believed to depend largely on the properties of the cell's basolateral membrane. Receptor cells in general extract only certain features from the stimulus offered to them. The receptors respond only to a limited range of intensities and frequencies of stimulation. One of the ways in which different hair cells vary in their response to a stimulus is by having different ionic conductances in the basolateral membrane. In this review I will discuss what is known about the ionic conductances in the basolateral membrane of vertebrate hair cells.

The first evidence that voltage-sensitive conductances might be important in modifying the response of hair cells from the cochlea of a lower vertebrate to injection of transducer current was provided by Crawford and Fettiplace (1981a). They used the isolated half-head of the red eared turtle. In this preparation the cochlea could be stimulated acoustically and intracellular voltage responses could be recorded from individual hair cells; it was also possible to inject currents into the hair cells. They found that current injection into a hair cell produced a potential step with an oscillation superimposed on it: an electrical resonance. The frequency of this resonance, for small current steps, was close to the characteristic frequency (c.f.: the frequency of sound stimulation to which the cell is most sensitive, i.e. for which it responds to the smallest intensity of stimulation) of the cell as assessed by sound stimulation. Cells had characteristic frequencies from 86 to 425 Hz in this study, which was close to the range spanned by the c.f.'s of auditory nerve fibres (30 to 700 Hz) and also of hair cells (70 to 670 Hz), as found in an earlier study (Crawford and Fettiplace, 1980). One explanation that they offered for the basis of this electrical resonance, was the presence of time- and voltage-dependent conductances in the cell's membrane, since similar phenomena had been observed in squid axons for small current injections (Hodgkin and Huxley, 1952). In the squid, these oscillations had been shown to arise largely from the properties of the delayed rectifier potassium conductance. At this stage the nature of the underlying conductances in the hair cells was not known, but they did notice a pronounced outward rectification in the steady-state current-voltage curves derived from current injection experiments (Crawford and Fettiplace, 1981a), and also that the addition of tetraethylammonium ions (TEA) or 4-aminopyridine (4-AP), both blockers of potassium channels, abolished the resonance (Crawford and Fettiplace, 1981b).

Similar electrical resonances have subsequently been demonstrated in hair cells in a number of other vertebrates: the frog's sacculus (Lewis and Hudspeth, 1983; Ashmore, 1983; Hudspeth and Lewis, 1988a,b) and amphibian papilla (Ashmore and Pitchford, 1985; Roberts, Robles and Hudspeth, 1986; Pitchford and Ashmore, 1987), the tall hair cells (more or less comparable to mammalian inner hair cells) of the chick's cochlea (Fuchs and Mann, 1986; Fuchs, Nagai and Evans, 1988) and the tall hair cells from the alligator's cochlea (Fuchs and Evans, 1988). These observations suggest that electrical resonance may be a fairly universal mechanism by which many lower vertebrates enhance their frequency selectivity.

The development of the patch-clamp technique (Hamill, Marty, Neher, Sakmann and Sigworth, 1981) has made it possible to study the ionic conductances in isolated hair cells. Lewis and Hudspeth (1983) and Hudspeth and Lewis (1988a,b) demonstrated, in the bullfrog's sacculus, the presence of a calcium-activated potassium conductance, a non-inactivating calcium conductance and an A-type potassium conductance. They proposed that the first two conductances were the most likely candidates for generating electrical resonance, because they were the only conductances that were activated in the voltage range in which the oscillations occur. The A-current was largely inactivated at and above the resting potential (-65 to -58 mV), and was therefore not involved in the generation of the electrical resonance. It is probably also present in a non-resonant subpopulation of hair cells in the alligator's cochlea (Fuchs and Evans, 1988). Lewis and Hudspeth also found that reducing the extracellular calcium concentration, or addition of TEA to the extracellular solution, affected both the calcium-activated potassium conductance and the electrical resonance, thus showing that this conductance was necessary for normal electrical resonance.

Art and Fettiplace (1987) correlated the properties of the calcium-dependent potassium conductance, as measured under voltage clamp, with the resonance frequency and sharpness of tuning of hair cells isolated from the turtle's cochlea. Their model of this conductance predicted the oscillatory response of the cells to current injection quite well. A more elaborate model, explicitly taking the calcium conductance into account as well, is provided by Hudspeth and Lewis (1988b). This establishes that the interplay of the calcium current and the calcium-activated potassium current is not only necessary but also sufficient for a sharply tuned electrical resonance in these cells.

Non-inactivating calcium currents and calcium-dependent potassium currents have also been demonstrated in tall hair cells isolated from the cochlea of the chick (Fuchs, Nagai and Evans, 1988) and alligator (Fuchs and Evans, 1988) and in hair cells isolated from the chick's vestibulum (Ohmori, 1984). No electrical resonance has been demonstrated in the latter preparation.

The mean conductance of the ionic channels underlying the calcium-dependent potassium currents was reported by Hudspeth and Lewis (1988a) to be 154 pS, in inside-out patches. Art and Fettiplace (1987) found a mean value of 102 pS, in cell-attached patches. In both cases the outside of the cell's membrane faced a potassium-rich extracellular solution. These unitary conductances are comparable to those of the large-conductance calcium-activated potassium channels described in numerous preparations (Marty, 1983; Blatz and Magleby, 1987).

Other ionic conductances (which presumably reside in the basolateral membrane) found in lower vertebrate hair cells include an inwardly rectifying potassium conductance, normally activated only at potentials more hyperpolarized than -80 mV to -100 mV, in chick vestibular hair cells (Ohmori, 1984). It seems less consistently present in other hair cells, but was occasionally found in turtle cochlear hair cells (Art and Fettiplace, 1987) and hair cells isolated from the "basal" region of the chick's cochlea (Fuchs, Nagai and Evans, 1988) and may well be present in hair cells in the bullfrog's sacculus (Corey and Hudspeth, 1979). This inward rectifier was prominent in tall hair cells isolated from the apical tip of the alligator's (Fuchs and Evans, 1988) and the chick's cochlea (Fuchs, Nagai and Evans, 1988). In these chick cells about half of the outward current that activated in the membrane potential range of the receptor potential seemed not to be calcium-dependent, suggesting the presence of yet another type of potassium conductance contributing to the shape of the receptor potential. These cells did not normally oscillate, but showed large action potentials in response to current injection. In low-frequency, spiking tall hair cells isolated from the apical region of the alligator's cochlea, sodium currents were found which activated in the membrane potential range of the receptor potential (Evans and Fuchs, 1987). They probably contribute to the upshoot of the action potentials found in these cells. Finally, many hair cells receive an efferent innervation. In the turtle cochlea the main effect of efferent nerve stimulation on the hair cells is an increase of a potassium conductance, probably mediated by release of acetylcholine from the efferent nerve fibres (Art, Fettiplace and Fuchs, 1984). The conductance has not been characterized in detail, since voltage clamp data are lacking.

Membrane properties of mammalian hair cells inferred from in vivo recordings

The extreme vulnerability of mammalian hair cells has precluded recordings from isolated cells until very recently, I will therefore first discuss what inferences have been made about membrane properties of these cells from *in vivo* recordings.

Two features of the mammalian cochlea which set it apart from the hearing organs of lower vertebrates are its sensitivity to much higher frequencies (in the guinea-pig extending psychophysically to well over 40 kHz (Heffner, Heffner and Masterton, 1971) and the presence of two distinct classes of hair cells, the inner hair cells and the outer hair cells. Birds and Crocodilia have interesting intermediate properties in both respects (Dooling, 1980; Turner, 1980). Anatomical studies (Spoendlin, 1972; Morrison, Schindler and Wersäll, 1975) show that 90-95% of the afferent fibres of the auditory nerve make contact only with inner hair cells; no responses to sound have ever been recorded from auditory nerve afferents known to innervate outer hair cells (Liberman, 1982; Robertson, 1984). Various lines of indirect evidence suggest that the outer hair cells function to sharpen the tuning of the inner hair cells, probably by an effect on cochlear micromechanics (Brown and Nuttall, 1984; Liberman and Dodds, 1984; Patuzzi and Sellick, 1984; Patuzzi, Sellick and Johnstone, 1984; Patuzzi, Yates and Johnstone, 1989). The extended frequency range of the mammalian cochlea compared to hearing organs of lower vertebrates prompts consideration of the passive membrane properties of inner hair cells, i.e. input resistance and capacitance, because they possibly place limits on the mechanisms that could determine frequency selectivity in the mammalian cochlea. These passive properties will be considered first, after which evidence for the presence of ionic conductances in mammalian hair cells will be discussed.

Russell and Sellick (1977, 1978) were the first to publish results from intracellular recordings in inner hair cells. Their surgical approach allowed *in vivo* recordings from hair cells from the first turn of the cochlea of the guinea-pig. The cells showed sharp tuning to acoustic stimuli (as good as that found in auditory nerve fibres by Evans (1972)) with c.f.'s of about 14 to 20 kHz. Q_{10dB}'s were between 6 and 11. The Q_{10dB} is the quality factor, defined as the characteristic frequency divided by the response bandwidth measured 10 dB above the threshold intensity for a just detectable response of the cell. The resting potentials were -25 to -45 mV. Large receptor potentials could be recorded, showing a depolarizing DC component of up to 27 mV and an AC component which was strongly frequency dependent and was attenuated, as the frequency increased, by 6 to 9 dB per octave with respect to the DC component. They attributed this attenuation to the effect of the membrane capacitance and resistance, acting as a low-pass filter (6 dB/octave being expected for a first order filter). The cell's time constant was estimated from the ratio of the in-phase and 90^0 out-of-phase membrane potential changes in response to sinusoidal current injections (Pinto and Pak, 1974), and ranged from 0.31 to 0.76 ms in seven cells. In three of these cells the input resistance was measured (46 to 61 MΩ) and the cell's capacitance (7.8 to 15.8 pF) was calculated from the ratio of the time constant and the input resistance. It was reported that the time constant becomes faster in more depolarized cells.

In a subsequent paper (Russell and Sellick, 1983) they concentrated on responses to low-frequency tones in more first turn hair cells with high c.f.'s. AC responses of up to 30 mV peak-to-peak were observed in response to tones of 100 to 300 Hz, largely in the depolarizing direction. Estimates for time constants and corner frequencies (-3 dB point) were again made using the method of Pinto and Pak (1974). Corner frequencies varied from 178 to 837 Hz in 18 cells. They noted that in auditory nerve fibres the decrease in "phase-locking" (firing of the auditory nerve is most likely to occur in a particular phase of the stimulating waveform, this is called phase-locking) is 6 dB per octave above 1 kHz, implicitly suggesting that the inner hair cell filter may be responsible for this.

A comparison of the frequency-dependence of phase-locking in the guinea-pig auditory nerve and the filtering properties of the inner hair cell's membrane was made by Palmer and Russell (1986) and Russell and Palmer (1986). No direct measurements of the time constant of the cells were reported in these papers. Instead, the inner hair cell data were here presented as the ratio of the AC and the DC components of the receptor potential. The frequency-dependence of this ratio can give an indirect measure of the cell's time constant on the assumption that the kinetics of the transducer current have a much higher corner frequency than that determined by the cell's membrane. They concluded that phase-locking in the guinea-pig is consistent with the suggestion that the limiting factor is the time constant of the inner hair cell. The authors did, however, also note a clear discrepancy between the corner frequencies of the neural data and those of the inner hair cells. The receptor potentials had a systematically lower corner frequency (on average about 500 Hz from their figures of the pooled AC/DC ratios) than the corner frequencies of phase-locking (about 1000 Hz). They suggested that this discrepancy may have arisen from the limited bandwidth of the intracellular recording system used to make the hair cell measurements. This is a problem inherent to all recordings with high-resistance intracellular microelectrodes which can be circumvented by using low-resistance patch-pipettes.

Cody and Russell (1987) extended these earlier results to include a substantial number of measurements on outer hair cells. The corner frequency of the outer hair cell's membrane filter was estimated to be 1200 Hz, based on voltage responses of the cells to small current pulses (0.1 nA). For the inner hair cells the same method gave a corner frequency of at least 700 Hz.

Dallos, Santos-Sacchi and Flock (1982) and Dallos (1985a,b) presented results from *in vivo* recordings from inner and outer hair cells in the third and fourth turns of the guinea-pig's cochlea. These cells were found to have c.f.'s of 800 to 1000 Hz, as opposed to 15 to 25 kHz for the first turn cells studied by Russell and his colleagues. They were less sharply tuned than first turn inner hair cells. This does not necessarily imply poorer recording conditions because the same trend is found in auditory nerve fibres (Evans, 1972). The Q_{10dB} can be estimated from Fig. 16. in Dallos (1985a) as 1.7 for one cell. This is within the range of 1 to 4 reported by Evans (1972) for auditory nerve fibres with c.f.'s below 2 kHz. Dallos gives corner frequencies for the hair cell's membrane of 470 Hz for inner hair cells and 1250 Hz for outer hair cells. These results are based on the ratio of the frequency response curve of a cell's receptor potential measured intracellularly to the frequency response of the extracellular potential measured close to the cell (Dallos, 1984). These corner frequencies agree rather well with those of Russell and his colleagues discussed above.

In summary, it appears that the passive membrane properties of the inner hair cell are probably responsible for the upper limit of phase-locking in the auditory nerve. Palmer and Russell (1986) and Russell and Palmer (1986) mention that phase-locking is no longer evident above 3.5 kHz. This has the consequence that if guinea-pig inner hair cells were to employ the type of electrical resonance seen in hair cells of lower vertebrates, this mechanism would not obtain above this frequency. Above 3.5 kHz alternative mechanisms for achieving frequency selectivity are obligatory, and probably involve mechanical feedback from outer hair cells.

Now I will discuss what is known, from *in vivo* recordings, about the presence of ionic conductances in the basolateral membrane of mammalian hair cells. Russell (1983) used current injections of up to 2 nA in the recording electrode, in basal turn inner hair cells with resting potentials from -30 to -55 mV. It was concluded that the current-voltage relations were relatively linear over a range of absolute potentials from -150 mV to +40 mV, the membrane resistance varied from 76 to 99 MΩ in various cells. These results implied that there were no time- and voltage-dependent conductances operating in inner hair cells, and that the only change in resistance during sound stimulation would be due to the transducer. This observation supported Davis' (1965) resistance microphone theory of the cochlea. Brown and Nuttall (1984) obtained similar results for injected currents ranging from -1.1 to +1.1 nA. They also studied the effects of stimulation of the crossed olivocochlear bundle (COCB) on the inner hair cell's receptor potential and found that this decreased the Q_{10dB} of the frequency tuning curve without changing the cell's input resistance, supporting the notion that COCB-stimulation influences the tuning of inner hair cells via an effect on cochlear micromechanics.

Later the same groups reported evidence for voltage-dependent conductances in inner hair cells. Nuttall (1985) noted non-linear current-voltage curves in response to current injection in some, but not all inner hair cells. The non-linearity in these cells was outwardly rectifying. Russell, Cody and Richardson (1986) came to the same qualitative conclusion but also provided more quantitative information. They reported inner hair cell conductances of 13 to 40 nS at the resting membrane potential (-25 to -45 mV), decreasing to nearly half this value in some inner hair cells when the membrane potential was hyperpolarized beyond -60 mV. They noted no further increase in conductance for depolarization from the resting potential. Outer hair cells (resting potentials -75 to -100 mV) showed a pronounced outward rectification and an additional inward rectification when hyperpolarized beyond -90 mV.

Dallos (1986) described the effects of current injection on the responses of third and fourth turn inner and outer hair cells to tones. Hyperpolarizing current gave an increase in the AC receptor potential (to about 220% for a current of -2 nA) and depolarizing current a decrease (to about 50% for a current of +2 nA). In outer hair cells depolarizing current was considerably more effective in reducing the receptor potential (to about 30% for a current of +1 nA) than hyperpolarizing current was in increasing it (to about 120% for a current of -1.5 nA).

All these current injection experiments are potentially affected by the non-ohmic behaviour of high-resistance intracellular microelectrodes, behaving differently inside and outside cells. This problem and the potentially large leakage conductance around intracellular microelectrodes, which would shunt any rectification due to time- and voltage-dependent conductances, thus making the rectification seem less pronounced than it would otherwise be, warrant some caution in interpreting these results.

Dallos (1984, 1985a) suggested that third and fourth turn inner hair cells may possess a low-quality electrical resonance, with a frequency of about 1200 Hz, somewhat higher than the c.f. of the cells in this region. His argument is based on modelling the ratio of the frequency response curves measured intracellularly in inner hair cells and extracellularly just outside the cells. It incorporates a number of unverified assumptions about the way in which extracellular potentials are related to the potentials measured intracellularly. Therefore more direct evidence is necessary to see whether this suggestion can be substantiated.

Response properties of low-frequency cochlear nerve fibres suggest a similar possibility. Tuning curves of fibres with c.f.'s below about 1000 Hz almost invariably show a secondary mimimum about half an octave to an octave above the main c.f. (Liberman and Kiang, 1978; Evans, 1983, 1989). The impulse responses of fibres with low c.f.'s, derived from reverse correlation functions (De Boer, 1973), show a ringing with two frequency components, of which the lower-frequency component persists considerably longer, suggesting it is more sharply tuned than the higher-frequency component (Evans, 1989). These data were obtained in cats; the detection of the secondary minima in the tuning curves depends on a sufficiently fine frequency resolution in the method used to measure the tuning curves. Comparable data are not available for the guinea-pig.

Membrane properties of mammalian hair cells derived from in vitro recordings

Recordings from isolated mammalian hair cells have only become possible very recently, with the realization that tissue-culture media prolong the survival of the cells compared to more conventional saline solutions (Zenner, 1984). The study of isolated cells can provide useful information additional to that derived from recordings *in vivo*, which show how hair cells interact in their natural environment. The principle advantage of using isolated cells is that they can be voltage clamped, using the whole-cell recording mode of the patch-clamp technique. This allows the study of time- and voltage-dependent conductances in much more detail than would ever be possible with *in vivo* recordings. Similarly, the underlying ionic channels in patches of basolateral cell membrane can be studied using one of the other modes of the patch-clamp technique. Caution is still necessary with respect to the quality of the isolated hair cells, which is quite variable. The isolation procedure (involving opening of the bone of the cochlea and mechanical and enzymatic means of dissociating the cells) is potentially traumatic. Moreover, the apical surface of the isolated cells is bathed in the wrong, perilymph-like extracellular solution. Damage may manifest itself morphologically (Lim and Flock, 1985) and by low resting potentials and large leakage conductances.

Flock and Strelioff (1984) recorded from coils isolated from the organ of Corti, where the hair cells were still in their normal anatomical relationship with respect to one another, but the tectorial membrane had been removed. In recordings made with high-resistance intracellular microelectrodes at room temperature they found a mean membrane potential in 17 inner hair cells of -23 mV (range -10 to -44 mV). For one cell an input resistance was reported of 100 MΩ. Membrane potentials in outer hair cells were smaller than -15 mV. Brownell (1984), also using conventional high-resistance micropipettes, reported resting potentials in isolated inner hair cells of -30 to -50 mV, and of -6 to -12 mV in isolated outer hair cells.

Russell and his colleagues developed a cochlear explant preparation from neonatal mice (Russell, Cody and Richardson, 1986; Russell, Richardson and Cody, 1986; Russell and Richardson, 1987). Isolated coils were cultured for up to 5 days before recordings were made. In these preparations the tectorial membrane has usually not yet covered the hair cells (apart from a very thin membrane, the minor tectorial membrane, which disappears in the cultures), making them especially suitable for studies requiring intact hair bundles. The recordings were made with high-resistance intracellular microelectrodes. Stable resting potentials of -30 to -70 mV were found for inner hair cells and of -40 to -70 mV for outer hair cells (Russell, Richardson and Cody, 1986). This shows that inner hair cells could be more hyperpolarized under these conditions than was found previously *in vivo*. On average, however, the inner hair cell resting potential was -38 mV, close to that found *in vivo*. The mean resting potential of the outer hair cells was -57 mV, rather more depolarized than that found in the guinea-pig *in vivo* (Russell and Richardson, 1987). Current injection experiments showed some outward rectification in inner hair cells, with a mean slope conductance of 17 nS for membrane potentials more hyperpolarized than -40 mV and 31 nS for membrane potentials more depolarized than -30 mV. In outer hair cells a more pronounced outward rectification was found, in line with the *in vivo* findings quoted in the previous section, which could be abolished by adding 20 mM TEA to the extracellular medium, suggesting that the outward rectification is caused by a potassium conductance. There was no evidence for inward rectification at potentials more hyperpolarized than -90 mV unlike the finding in guinea-pig outer hair cells *in vivo* by Russell, Cody and Richardson (1986).

Patch-clamp recordings of guinea-pig outer hair cells have revealed some properties of the underlying conductances. All these experiments were done at room temperature. Ashmore and Meech (1986a) and Ashmore (1986, 1987), using outer hair cells isolated from the third and the fourth turn, found that the resting potential and input conductances were strongly dependent on the intracellular calcium buffer. EGTA gave low resting potentials between -15 and -40 mV and a high input conductance (25-40 nS). Use of 10 mM BAPTA resulted in slowly developing more hyperpolarized resting potentials (mean -60 mV) and a lower input conductance (about 3 nS). This conductance did not reduce on hyperpolarizing the cells further and can be interpreted as a linear leak conductance. Voltage clamping the cells in the presence of BAPTA from a holding potential close to the resting potential showed a slowly developing (over about 20-30 ms) outward current, activated at membrane potentials more depolarized than 0 mV. It was attributed to a calcium-dependent potassium conductance. The current was reduced by addition of 1 mM cadmium ions to the extracellular solution; it was, however, not reported whether there was any effect of lowering the calcium concentration in the extracellular solution. Santos-Sacchi and Dilger (1986, 1988), also in guinea-pig outer hair cells, found evidence for a time- and voltage-dependent outward current, activating for depolarizations above -40 mV, a much more hyperpolarized level than found by Ashmore and Meech (1986a). The currents rose sigmoidally to a peak in 10-20 ms and then partially inactivated more slowly. The currents were reduced by extracellular TEA, barium ions and cadmium ions and by replacing intracellular potassium with caesium ions. Again the effect of lowering the extracellular calcium concentration on this current is not discussed. The underlying conductance is therefore a time- and voltage-dependent potassium conductance and possibly a calcium-dependent one. The mean resting potential was -45 mV, the input conductance (approximately equal to the linear leak conductance) was 9.2 nS, with EGTA as the calcium chelator. They also demonstrated an inward current on reducing the potassium current, apparent after subtraction of the leak conductance, activated at potentials more depolarized than -30 mV. Replacing extracellular calcium with 50 mM barium increased the inward current, resulting in a net inward current between +10 and +40 mV. These barium currents did not inactivate. This strongly suggests the presence of a time- and voltage-dependent calcium current in outer hair cells.

Gitter, Zenner and Frömter (1984, 1986) and Gitter and Zenner (1988) also recorded from isolated outer hair cells, mostly from the apical coil. Using patch-clamp electrodes they found stable resting potentials for at least 5 minutes with a mean of -45 mV and a most hyperpolarized value of -69 mV. High-resistance microelectrodes give less stable recordings with a rapid decay of the initial resting potential, however in some cases they measured a mean resting potential of -42 mV. No whole-cell voltage clamp experiments were described. The membrane capacitance of outer hair cells was measured directly by Ashmore an Meech (1986a) and Ashmore (1987) and had a range of 24-32 pF (mean 27 pF).

Recordings of single ionic channels of isolated outer hair cells have produced evidence for the presence of calcium-dependent potassium channels in the basolateral cell membrane (Ashmore and Meech, 1986a,b). They found two channel types with mean unitary conductances in symmetrical KCl solutions of 233 and 45 pS. The probability of opening of the large channel increased with depolarization, though not very steeply (e-fold per 30 mV). In addition they provided some evidence for a third potassium channel with a unitary conductance of about 30 pS which became evident following hyperpolarization of the patch. Gitter, Zenner and Frömter (1986) and Gitter, Frömter and Zenner (1986), reported evidence for a potassium channel in the basolateral membrane of the outer hair cell with a unitary conductance of 149 pS in asymmetrical solutions, the opening probability of which increased with depolarization. This is probably the same channel as the large calcium-dependent potassium channel found by Ashmore, although these authors did not test the calcium-dependence. In addition they found some evidence in the basolateral cell membrane for two other types of potassium channels with unitary conductances of 20 and 200 pS and two types of chloride channels with unitary conductances of 40 and 22 pS, the latter channel showing an increasing opening probability with depolarization. They also reported the presence of ionic channels on the apical surface of the cell, but they did not characterize them. Santos-Sacchi and Dilger (1986) found bursting channels which were not voltage-dependent with a unit conductance of 80 pS on the basolateral cell membrane of outer hair cells.

Reports of recordings from inner hair cells using the patch-clamp technique have only appeared very recently. Kros and Crawford (1988, 1989) published evidence for two types of time- and voltage-dependent potassium currents and a comparatively small linear leak conductance (1.0-2.5 nS), in inner hair cells isolated from the apex of the guinea-pig cochlea. The potassium conductances activated steeply over the potential range of about -60 to -20 mV; in the range of the receptor potential of inner hair cells. The largest of these potassium conductances was also very fast: at 35-38°C the current was virtually completely turned on in less than a millisecond at all potentials. This is faster than currents in any other hair cell preparations (see e.g. Lewis and Hudspeth, 1983; Art and Fettiplace, 1987). This suggests an important physiological role for these fast potassium currents in inner hair cells, because they could modify the receptor potential on a time scale as fast as a few kHz. Ashmore (1988) provided evidence for a potassium current activating at membrane potentials more depolarized than about -60 mV. It reached full activation at -50 to -40 mV, with a mean maximum slope conductance of 23 nS. This is about ten times smaller than the maximum slope conductance reported by Kros and Crawford (1988). Values for the resting potential were not given by Ashmore, although he does suggest that the intracellular potassium concentration may be below 30 mM, suggesting very depolarized resting potentials, contrasting with the mean resting potential of -68 mV reported by Kros and Crawford (1988). He did find evidence in cell-attached patches for potassium-selective ionic channels in the basolateral membrane with a unitary conductance of about 95 pS. Gitter and Zenner (1988) found evidence for three types of ionic channels in the basolateral cell membrane of inner hair cells, in cell attached patches with a rather peculiar solution in the pipette (95 mM NaCl, 50 mM KCl), making comparison of the unitary conductances with single-channel results from others difficult. The mean unitary conductances were 10, 21 and 50 pS. Kros and Crawford (1989) found, with a high-potassium extracellular solution in the pipette, evidence for at least two types of potassium channel in cell-attached patches, with unitary conductances of 120 and 10-20 pS. The membrane capacitance of isolated inner hair cells was reported to be about 9 pF (Kros and Crawford, 1988, 1989); the range was 6 to 15 pF.

These results of *in vitro* recordings from mammalian hair cells show that very different findings still appear from different laboratories. Leakage conductances are often considerably higher than in isolated hair cells from lower vertebrates and cannot, at least in the case of outer hair cells, be attributed to the transducer conductance (Ashmore, 1988; Santos-Sacchi and Dilger, 1988). This might indicate a less than optimal physiological state of the cells. Nevertheless, the results obtained in the last few years show that mammalian hair cells do possess various time- and voltage-dependent ionic conductances. In outer hair cells the potassium currents are rather slow and small, and they activate outside the range of outer hair cell receptor potentials as measured *in vivo* (Cody and Russell, 1987; Dallos, 1985a), so that at the moment their physiological significance is uncertain. In inner hair cells the potassium currents are large and fast, and do activate in the range of the receptor potential. Could these potassium currents in mammalian inner hair cells influence the receptor potential as dramatically as do the calcium- activated potassium currents in lower vertebrate hair cells? This is just one of the new questions that have to be answered in this exciting area of sensory physiology.

I would like to thank Dr. A.C. Crawford and Dr. M.G. Evans for commenting on an early version of this manuscript.

REFERENCES

Art, J.J. and Fettiplace, R., 1987, Variation of membrane properties in hair cells isolated from the turtle cochlea, *J. Physiol.*, 385:207-242.

Art, J.J., Fettiplace, R. and Fuchs, P.A., 1984, Synaptic hyperpolarization and inhibition of turtle cochlear hair cells, *J. Physiol.*, 356:525-550.

Ashmore, J.F., 1983, Frequency tuning in a frog vestibular organ, *Nature*, 304:536-538.

Ashmore, J.F., 1986, The cellular physiology of isolated outer hair cells: implications for cochlear frequency selectivity, *in*: "Auditory Frequency Selectivity," B.C.J. Moore and R.D. Patterson, ed., pp. 103-108, New York, Plenum Press.

Ashmore, J.F., 1987, A fast motile response in guinea-pig outer hair cells: the cellular basis of the cochlear amplifier, *J. Physiol.*, 388:323-347.

Ashmore, J.F., 1988, Ionic mechanisms in hair cells of the mammalian cochlea, *in*: "Progress in brain research", Vol.74, W. Hamann and Iggo A., ed., pp. 1-7.

Ashmore, J.F. and Meech, R.W., 1986a, Ionic basis of membrane potential in outer hair cells of guinea pig cochlea, *Nature*, 322:361-371.

Ashmore, J.F. and Meech, R.W., 1986b, Three distinct potassium channels in outer hair cells isolated from the guinea-pig cochlea, *J. Physiol.*, 371:29P.

Ashmore, J.F. and Pitchford, S., 1985, Evidence for electrical resonant tuning in the hair cells of the frog amphibian papilla, *J. Physiol.*, 364:39P.

Blatz, A.L. and Magleby, K.L., 1987, Calcium-activated potassium channels, *Trends in Neurosciences*, 10:463-467.

Boer, E., 1973, On the principle of specific coding, *J. Dyn. Syst. Meas. Control*, Trans., ASME, 95-G:265-273.

Brown, M.C. and Nuttall, A.L., 1984, Efferent control of cochlear inner hair cell responses in the guinea-pig, *J. Physiol.*, 354:625-646.

Brownell, W.E., 1984, Microscopic observation of cochlear hair cell motility, *Scanning Electron Microscopy*, 3:1401-1406.

Cody, A.R. and Russell, I.J., 1987, The responses of hair cells in the basal turn of the guinea-pig cochlea to tones, *J. Physiol.*, 383:551-569.

Corey, D.P. and Hudspeth, A.J., 1979, Ionic basis of the receptor potential in a vertebrate hair cell, *Nature*, 281:675-677.

Crawford, A.C. and Fettiplace, R., 1980, The frequency selectivity of auditory nerve fibres and hair cells in the cochlea of the turtle, *J. Physiol.*, 306:79-125.

Crawford, A.C. and Fettiplace, R., 1981a, An electrical tuning mechanism in turtle cochlear hair cells, *J. Physiol.*, 312:377-412.

Crawford, A.C. and Fettiplace, R., 1981b, Non-linearities in the responses of turtle hair cells, *J. Physiol.*, 315:317-338.

Dallos, P., 1984, Some electrical circuit properties of the organ of Corti. II, Analysis including reactive elements, *Hearing Res.*, 14:281-291.

Dallos, P., 1985a, Response characteristics of mammalian cochlear hair cells, *J. Neuroscience*, 5:1591-1608.

Dallos, P., 1985b, Membrane potential and response changes in mammalian cochlear hair cells during intracellular recording, *J. Neuroscience*, 5:1609-1615.

Dallos, P., 1986, Neurobiology of cochlear inner and outer hair cells: intracellular recordings, *Hearing Res.*, 22:185-198.

Dallos, P., Santos-Sacchi, J. and Flock, 1982, Intracellular recordings from cochlear outer hair cells, *Science*, 218:582-584.

Davis, H., 1965, A model for transducer action in the cochlea, *Cold Spring Harbor Symp. Quant. Biol.*, 30:181-189.

Dooling, R.J., 1980, Behavior and psychophysics of hearing in birds, *in:* "Comparative studies of hearing in vertebrates," A.N. Popper and R.R. Fay, ed., pp. 261-288, New York, Springer Verlag.

Evans, E.F., 1972, The frequency response and other properties of single fibres in the guinea-pig cochlear nerve, *J. Physiol.*, 226:263-287.

Evans, E.F., 1983, How to provide speech through an implant device, Dimensions of the problem: an overview, *in:* "International conference on cochlear implants # 25," M.M. Merzenich and R.A. Schindler, ed.

Evans, E.F., 1989, Cochlear filtering: a view seen through the temporal discharge patterns of single cochlear nerve fibres, *in:* "Cochlear mechanisms Structure, function and models," J.P. Wilson and D.T. Kemp, ed., New York, Plenum Press, to be published.

Evans, M.G. and Fuchs, P.A., 1987, Tetrodotoxin-sensitive, voltage-dependent sodium currents in hair cells from the alligator cochlea, *Biophys. J.*, 52:649-652.

Flock, Å. and Strelioff, D., 1984, Studies on hair cells in isolated coils from the guinea pig cochlea, *Hearing Res.*, 15:11-18.

Fuchs, P.A. and Evans, M.G., 1988, Voltage oscillations and ionic conductances in hair cells isolated from the alligator cochlea, *J. Comp. Physiol.*, A 164:151-163.

Fuchs, P.A. and Mann, A.C., 1986, Voltage oscillations and ionic currents in hair cells isolated from the apex of the chick's cochlea, *J. Physiol.*, 371:31P.

Fuchs, P.A., Nagai, T. and Evans, M.G., 1988, Electrical tuning in hair cells isolated from the chick cochlea, J. Neuroscience, 8:2460-2467.

Gitter, A.H., Frömter, E. and Zenner, H.P., 1986, Membrane potential and ion channels in isolated mammalian outer hair cells, *Hearing Res.*, 22:29.

Gitter, A.H. and Zenner, H.P., 1988, Auditory transduction steps in single inner and outer hair cells, *in:* "Basic Issues in Hearing," H. Duifhuis, J.W. Horst and H.P. Wit, ed., pp. 32-39, London, Academic Press.

Gitter, A.H., Zenner, H.P. and Frömter, E., 1984, "Patch clamp studies on mammalian inner ear hair cells", Workshop on noise analysis and related techniques, Leuven, Belgium.

Gitter, A.H., Zenner, H.P. and Frömter, E., 1986, Membrane potential and ion channels in isolated outer hair cells of guinea pig cochlea, *ORL J. Otorhinolaryngol. Relat.*

Spec., 48:68-75.

Hamill, O.P., Marty, A., Neher, E., Sakmann, B. and Sigworth, F.J., 1981, Improved patch-clamp techniques for high-resolution current recording from cells and cell-free membrane patches, *Pfl ügers Arch.*, 391:85-100.

Heffner, R., Heffner, H. and Masterton, B., 1971, Behavioural measurements of absolute and frequency difference thresholds in guinea-pig, *J. Acoust. Soc. Am.*, 49:1888-1895.

Hodgkin, A.L. and Huxley, A.F., 1952, A quantitative description of membrane current and its application to conduction and excitation in nerve, *J. Physiol.*, 117:500-544.

Hudspeth, A.J. and Lewis, R.S., 1988a, Kinetic analysis of voltage- and ion-dependent conductances in saccular hair cells of the bull-frog, Rana catesbeiana, *J. Physiol.*, 400:237-274.

Hudspeth, A.J. and Lewis, R.S., 1988b, A model for electrical resonance and frequency tuning in saccular hair cells of the bull-frog, Rana catesbeiana, *J. Physiol.*, 400:275-297.

Kros, C.J. and Crawford, A.C., 1988, Non-linear electrical properties of guinea-pig inner hair cells: a patch-clamp study, *in*: "Basic Issues in Hearing," H. Duifhuis, J.W. Horst and H.P. Wit, ed., pp. 27-31, London, Academic Press.

Kros, C.J. and Crawford, A.C., 1989, Components of the membrane current in guinea-pig inner hair cells, *in*: "Cochlear mechanisms. Structure, function and models," J.P. Wilson and D.T., Kemp, ed., pp. 189-196, New York, Plenum Press.

Lewis, R.S. and Hudspeth, A.J., 1983 Voltage- and ion-dependent conductances in solitary vertebrate hair cells, *Nature*, 304:538-541.

Liberman, M.C., 1982, Single-neuron labeling in the cat auditory nerve, *Science*, 216:1239-1241.

Liberman, M.C. and Dodds, L.W., 1984, Single-neuron labeling and chronic cochlear pathology. III. Stereocilia damage and alterations of threshold tuning curves, *Hearing Res.*, 16:55-74.

Liberman, M.C. and Kiang, N.Y.S., 1978, Acoustic trauma in cats, *Acta Otolar Suppl.*, 358:1-63.

Lim, D.J. and Flock, Å., 1985, Ultrastructural morphology of enzyme-dissociated cochlear sensory cells, *Acta Otolar.*, 99:478-492.

Marty, A., 1983, Ca^{2+}-dependent K^+ channels with large unitary conductance, *Trends in Neurosciences*, 6:262-265.

Morrison, D., Schindler, R.A. and Wersäll, J., 1975, A quantitative analysis of the afferent innervation of the organ of Corti in guinea pig, *Acta Otolar.*, 79:11-23.

Nuttall, A.L., 1985, Influence of direct current on dc receptor potentials from cochlear inner hair cells in the guinea pig, *J. Acoust. Soc. Am.*, 77:165-175.

Ohmori, H., 1984, Studies of ionic currents in the isolated vestibular hair cell of the chick, *J. Physiol.*, 350:561-581.

Palmer, A.R. and Russell, I.J., 1986, Phase-locking in the cochlear nerve of the guinea-pig and its relation to the receptor potential of inner hair cells, *Hearing Res.*, 24:1-15.

Patuzzi, R. and Sellick, P.M., 1984, The modulation of the sensitivity of the mammalian cochlea by low frequency tones, II, Inner hair cell receptor potentials, *Hearing Res.*, 13:9-18.

Patuzzi, R., Sellick and Johnstone, B.M., 1984, The modulation of the sensitivity of the mammalian cochlea by low frequency tones, III, Basilar membrane motion, *Hearing Res.*, 13:19-27.

Patuzzi, R., Yates, G.K. and Johnstone, B.M., 1989, Outer hair cell receptor current and its effect on cochlear mechanics, *in*: "Cochlear mechanisms, Structure, function and models," J.P. Wilson and D.T. Kemp, ed., New York, Plenum Press.

Pinto, L.H. and Pak, W.L., 1974, Light-induced changes in photoreceptor membrane resistance and potential in Gecko retinas. I. Preparations treated to reduce lateral interactions, *J. Gen. Physiol.*, 64:26-48.

Pitchford, S. and Ashmore, J.F., 1987, An electrical resonance in hair cells of the amphibian papilla of the frog Rana temporaria, *Hearing Res.*, 27:75-83.

Roberts, W.M., Robles, L. and Hudspeth, A.J., 1986, Correlation between the kinetic properties of ionic channels and the frequency of membrane-potential resonance in hair cells of the bullfrog, *in*: "Auditory Frequency Selectivity," B.C.J. Moore and R.D. Patterson, ed., pp. 89-95, New York, Plenum Press.

Robertson, D., 1984, Horseradish peroxidase injection of physiologically characterised afferent and efferent neurones in the guinea pig spiral ganglion, *Hearing Res.*, 15:113-121.

Russell, I.J., 1983, Origin of the receptor potential in inner hair cells of the mammalian cochlea - evidence for Davis' theory, *Nature*, 301:334-336.

Russell, I.J., Cody, A.R. and Richardson, G.P., 1986, The responses of inner and outer hair cells in the basal turn of the guinea-pig cochlea and in the mouse cochlea grown in vitro, *Hearing Res.*, 22:199-216.

Russell, I.J. and Palmer, A., 1986, Filtering due to the inner hair-cell membrane properties and its relation to the phase-locking limit in cochlear nerve fibres, *in*: "Auditory Frequency Selectivity," B.C.J. Moore and R.D. Patterson, ed., New York, Plenum Press.

Russell, I.J. and Richardson, G.P., 1987, The morphology and physiology of hair cells in organotypic cultures of the mouse cochlea, *Hearing Res.*, 31:9-24.

Russell, I.J., Richardson, G.P. and Cody, A.R., 1986, Mechanosensitivity of mammalian auditory hair cells in vitro, *Nature*, 321:517-519.

Russell, I.J. and Sellick, P.M., 1977, Tuning properties of cochlear hair cells, *Nature*, 267:858-860.

Russell, I.J. and Sellick, P.M., 1978, Intracellular studies of hair cells in the mammalian cochlea., *J. Physiol.*, 284:261-290.

Russell, I.J. and Sellick, P.M., 1983, Low-frequency characteristics of intracellularly recorded receptor potentials in guinea-pig cochlear hair cells, *J. Physiol.*, 338:179-206.

Santos-Sacchi, J. and Dilger, J.P., 1986, Patch clamp studies of isolated outer hair cells, *in*: "Advances in Auditory Neuroscience: IUPS satellite symposium on hearing," p. 23.

Santos-Sacchi, J. and Dilger, J.P., 1988, Whole cell currents and mechanical responses of isolated outer hair cells, *Hearing Res.*, 35:143-150.

Spoendlin, H., 1972, Innervation densities of the cochlea, Acta Otolar., 73:235-248.

Turner, R.G., 1980, Physiology and bioacoustics in reptiles, *in*: "Comparative studies of hearing in vertebrates," A.N. Popper and R.R. Fay, ed., pp. 205-237, New York, Springer Verlag.

Zenner, H.P., 1984, Short-time culture of isolated outer hair cells from guinea pig cochlea, *in*: "ARO seventh midwinter research meeting - abstracts," pp. 12-13.

Chemotransduction

E. Bignetti, A. Cavaggioni and R. Tirindelli

K. E. Kaissling

E. Bignetti, S. Grolli and R. Ramoni

THE CHEMISTRY OF CHEMORECEPTION

IN VERTEBRATE SENSORY SYSTEMS

E. Bignetti, A. Cavaggioni, R. Tirindelli

Istituto di Biologia Molecolare and Istituto di Fisiologia
Universita' di Parma, Italy

In a first section we will try to understand the organization of chemoreception from the chemical-structure/sensation relationship. In a second section we will briefly touch on the biochemical evidence for olfactory and taste receptors. In the last section we will study a family of proteins that are models for olfactory receptors.

CHEMORECEPTION

Chemoreception is a collective term for the phenomena in which there is an interaction between a chemical and an organism resulting in an action which is meaningful to the organism. Thus, in chemoreception there is a relationship between the chemical parameters of the chemical, which is considered as a stimulus, and the meaningful action, which is the response. The system of the organism involved is referred to as the receptor system. Broadly speaking, there are two kinds of chemoreception : an external chemoreception in which the stimulus belongs to the world outside the organism, and an internal chemoreception in which the stimulus is produced within the organism. The two types of stimuli differ substantially in variety because there is an enormous diversity in the external chemicals as compared to the small number of chemical species that have been selected within the organism by evolution. This difference is reflected in the specificity of the interaction, very high for internal ligands - an example is the coenzyme/apoenzyme interaction leading to the holoenzyme - whereas the great variety of external ligands requires either a restricted number of receptor systems with coarse specificity or a great number of highly specific receptor systems. We will concern ourselves only with external chemoreception in vertebrate animals, excluding bacterial chemotaxis and the immune system. We are thus left with olfaction and taste, namely, the chemical senses. Having defined our task we now turn to consider the relationship between the structural parameters of the stimuli and that particular type of meaningful action which is the sensation.

Specialist versus generalist receptor systems

The organization of sensory systems is not unique. We can envisage specialist as well as generalist types of organization. This distinction is not well defined in theoretical terms but has a practical value. It is best understood by referring the organization of a classical specialist system: the visual system. The visual system of vertebrates is based on the interaction of a narrow band of the electromagnetic radiation spectrum with a small number of wavelength selective pigments, one for night-vision, three for colour-vision. Each pigment is found only in one cell type. The monochromacy of scotopic vision and the trichromacy of colour vision are the direct consequence of the limited number of pigments. The absence of one or more pigments in the retina, leads to degeneracy of vision which then is di- or mono-chromatic. In a specialist organization stimulus representations at any level may be expressed as the sum of suitably chosen coordinates in the same way as a vector in a Euclidean space can be described by the components x, y and z. In a generalist type of organization no coordinate system can be found to express the stimulus representations in a Euclidean space and correlations are the only useful parameters. An example will make the difference clear. Considering the colour, a blue colour sensation given by a light source is independent on the intensity and will remain blue no matter how dim or bright the light source is as long as the emission spectrum is not changed. Consider now the odour: a pleasant cucumber odour sensation given by 2-nonenal is appreciated only at low concentration. At high 2-nonenal concentration a very unpleasant odour is perceived which is completely different in quality.

The great majority of external chemicals interact with the chemical senses. Considering olfaction, all volatile molecules are rather uniform in their chemical parameters: they do not have or have one and rarely more than one polar group, they do not bear charge, the molecular weight is between 15 and 300, the shape is rarely defined as different spatial configurations are possible. Molecules of this type do not offer the handles necessary for highly specific interactions with the olfactory system of vertebrates. Insects, however, have developed very specific olfactory modalities for pheromones, see the lecture of Professor Kaissling in this book. A great number of odorants with high threshold for detection such as alcohols, aldehydes, ketones, or small molecules with a strong polar character as NH_3 or H_2S interact with any type of cell giving changes, e.g., of the membrane potential and it is unlikely that the chemical senses have developed higly specialized receptors for them. At the cellular level electrophysiological studies have shown that olfactory neurons of vertebrates are not excitable by one odorant only, but respond to a variety of chemically non-related odorants. At the subjective level odours can rarely or never be completely resolved into a small number of elementary components.

By and large we reach the conclusion that the organization of the olfactory system is based on the collective interaction of a population of receptor systems with a variety of odorants, and this is a generalist type of organization. But the case for specific modalities in vertebrate olfaction cannot be dismissed without consideration of specific anosmias. This original approach goes back to the idea of "odeurs fondamentales" (primary odours) of Guillot based on cases of specific anosmias for single odorants. The first case he reported was a subject who could not perceive the violet-component of the ionone-odour. He made also the observation of a specific but temporary anosmia and of an anosmia which started during puberty. This is interesting and again underlines the difference with the visual system in which a specific defect in color vision is an inherited condition which lasts for life. We now have a list of 8 well characterized specific anosmias in man: for 5alpha-androstan-3-one (Ia), the urinous olfactory modality in 40% of subjects; for isobutyric acid (IIa), the sweaty modality in 9% of Caucasian subjects; for pentadecanolide (IIIa), the musky modality in 7% of Caucasian subjects; for l-pyrroline (IVa), the spermous modality in 20% of subjects; for trimethylamine (Va), the fishy modality in 7% of subjects; for isovaleric aldehyde (VIa), the malty modality in 36% of subjects; for carvone (VIIa), the minty modality; for cineole (IIXa), the camphor modality.

Ia   IIa   IIIa $(CH_2)_{14}$ C=O   IVa

Va   VIa   VIIa   IIXa

The anosmia for pentadecanolide is hereditary with a single recessive autosomal character. It is likely that there are more than 8 primary odour modalities and a thorough study suggests that there may be at least 32 primary odours for man. In the majority of the cases there is a relative defect, rather than absolute, as the detection threshold is higher than normal. In animals, anosmias are more rare than in man but a strain of mice anosmic to isolaveric acid has been isolated. Although the demonstration of primary odours does not necessarily imply, it is likely that the primary odours are coded by somewhat specific receptors more concentrated in some and less in other receptor cells. Similar defects have been described also in taste. There is a relative defect in the bitter modality evoked by phenylthiourea, but not by quinine in some subjects. Inbred strains of mice differ greatly in their ability to taste certain bitter substances; a single autosomal gene is responsible for the ability to taste the bitterness of sucrose octoacetate and strychnine, but the ability to taste quinine is strongly influenced by another gene, and that of raffinose acetate by still another gene.

As a conclusion, in olfaction and taste we may assume a continuum of sensory modalities and of receptor molecules ranging from the generalist type at one end, extending to the specialist type, for which specific defects have been demonstrated, at the other end.

The olfactory stimuli

Chemical molecules are described by specifying atoms, groups, bonds, their spatial relationship - that is what we try to represent when writing down a chemical formula - as well as by the macroscopic properties, air/water partition coefficient, boiling point and so on. The olfactory system of vertebrates is not interested in this wealth of chemical detail and the ordinary chemical properties are irrelevant to the sense of smell. Mainly the morphology, namely, the size and shape of the chemical stimulus is important. Polar atoms are not important *per se* but because by making hydrogen bonds or dipole-dipole interactions with the receptor they limit the freedom of movement of the remaining part of the molecule, and in doing so, tend to define a spatial profile of the molecule. Accordingly, a broad distinction can be made between molecules that are not polar, and those that have one or two polar groups. For a non-polar molecule the fit with the receptor site is defined by the size, as the odorant can freely rotate, translate and assume different configurations. For instance, molecules with spherical or quasi-spherical shape and with a diameter of about 7 A give rise to a camphoraceous odour:

I   II   III

In III the polar group is a secondary feature as compared to the overall molecular shape. Replacement of H by F and of CH_2 by Br does not affect the camphoraceus odour because the shape is the same. For molecules with one polar group, the fit is defined by complementarity of the receptor site and the oriented profile of the odorant. For instance IV is a strong odorant but not V; observe the difference in profile as the orientation of one bond is changed:

IV   V

Another example is menthol, a substituted hexane with 3 asymmetric carbon atoms; only the stereoisomer in which all the substituent groups are co-planar with the hexane ring has a strong minty odour.

The best studied family of odourants with one polar group is that of VI and VI bis:

VI   VI bis

X can be C=O, O, S, NHorNCH$_3$ X-Y is O-CO

With n between 7 and 10 the polar interaction with the receptor is expected to be more important than the hydrophobic interaction and the odour is minty; with n between 9 and 13 the polar interaction is less important and the hydrophobic moiety can assume an almost spherical configuration of about 7 A, and the odour is camphoaceus; with n between 14 and 17 the molecule can assume a larger elongated shape, and the odour is now a strong musk. The oriented profile of the molecule may undergo conformational adaptation to the binding site. Therefore molecules similar in shape but different in rigidity may have different odour intensities. For example VII is a strong musk, but VIII is only a weak musk:

VII   VIII

Conversely flexible molecules can adopt the shape of more rigid molecules within the binding site as indicated by the musk odour of both IX and X:

IX   X

In order to interact with the receptor group, polar atoms or groups should not be screened by hydrophobic groups. So the isochroman musk XI is very strong but XII, in which the position of a methyl group has been changed, is almost devoid of any quality

70

XI   XII

In molecules with two polar groups the distance between the groups may influence the quality of the odour. Surprisingly, at first sight, molecules with the polar groups very close each other or very far apart are more similar in odour than molecules in which the polar groups are at an intermediate distance. On close inspection, however, this agrees with the hypothesis that the only relevant feature is the oriented profile. Consider for instance the musk odorants lactones XIII and oxalactones XIV and XV. XIII and XV are very strong musks, but XIV only a weak musk:

$$\begin{bmatrix} (CH_2)_{15} \\ O \quad\quad C=O \end{bmatrix} \quad \begin{bmatrix} (CH_2)_4-O-(CH_2)_{10} \\ O \quad\quad\quad\quad\quad C=O \end{bmatrix} \quad \begin{bmatrix} (CH_2)_7-O-(CH_2)_7 \\ O \quad\quad\quad\quad\quad C=O \end{bmatrix}$$

XIII   XIV   XV

Clearly the $-(CH_2)_{15}-$ group of XIII and the $-(CH_2)_7 - O - (CH_2)_7-$ group of XV assume the same profile whereas in XIV the oxygen, which is only four carbon atoms distant from the lactone group, cooperates to the polar interaction as a complex polar group $-O - (CH_2)_4 -O- CO_2$, and the $-(CH_2)_{10}$ has now a shorter profile. Chirality effects in vertebrate olfaction are rare. An example of difference between enantiomers are the + and - carvones VIIa and XVI, the - form having spearmint odour and the + form odour of caraway oil:

VII a   XVI

The taste stimuli

In molecules giving taste sensations, the polar groups are the most important feature. The molecular size and shape are less important as we know sweet tasting molecules ranging in molecular weight from 75 (glycine) to 21,000 (thaumatin). Taste molecules are non-volatile and water soluble and have usually more than one polar group. We may assume that some polar groups of the taste molecule make polar bonds with the receptor molecules. Clearly, the interactions in taste are likely to be much more selective than in olfaction and to depend on the exact structure of the taste molecule. The subjective taste modalities are only four, sweet, bitter, salty and acid. Sweet and bitter modalities are related and most bitter/sweet stimuli have at least 2 polar groups in the molecule, although many natural organic molecules taste bitter and only a few are sweet. Sweet stimulants exhibit the combination of two adjacent features: an acidic proton and a free electron pair; we write symbolically AH-B. Also the spatial orientation of the AH-B groups is important as the distance between the H and B orbitals is about 3A. Adjacent OH groups of sugars in gauche conformation satisfy these criteria but not in the anti-conformation - too far apart - nor in the eclipsed-conformation - because of internal hydrogen bonding.

71

gauche   anti   eclipsed

Possibly two hydrogen bonds with the receptor sites are formed

$$\text{ligand}\begin{bmatrix}\text{AH......B}\\\text{B......AH}\end{bmatrix}\text{receptor}$$

In aldopyranoses, the C3-C4 OH groups both in equatorial configuration are responsible for the intense sweetness of glucopyranose (XVII).

The equatorial-axial configuration (still a gauche configuration) is more disposed to form an internal H-bond with the ring oxygen, and (XVIII) galactopyranose is less sweet than glucopyranose. beta-d-fructose (IXX) is the sweetest monosaccharide known; the AH-B pair is probably the combination of the equatorial C1OH and of the axial C2OH. The reason for the intense sweetness is that C5OH is disposed to form an hydrogen bond with the oxygen atom of the ring and this leaves the C2OH free to make an external, rather than an internal, hydrogen bond. Beta-d-fructose, however, is unstable and, when heated, is transformed by mutarotation into the beta-furanose form (XX) which has no OH group in the gauche form. Notably, fructose is less sweet when heated in solution. The importance of external polar bonding is in agreement with the notion that hydrophobic substitutions decrease sweetness and confer bitterness to sugars, e.g. sucrose octoacetate is strongly bitter. Chirality effects are expected to be strong in sweet/bitter stimulants with 3-polar non-coplanar groups. Examples are amino-acids: the l-form of hydrophobic amino acids is bitter whereas the d-form is flat or sweet. The bitter modality is more complex. A class of bitter tasting molecules is characterized by an AH-B pair at an orbital distance of 1.5 A at which a strong hydrogen bond connects the two groups. beta-maltopyranose (XXI) is a bitter monosaccharide probably because there is the possibility of a strong hydrogen bond between the O atom of the ring and the OH of C1. Obviously the receptor site should have a strongly acidic proton to break the internal bond. In substituted thioureas the AH-B pair is probably represented by the $S-NH_2$ group, e.g., XXII with S as the substituent atom is very bitter.

XXII X=S
XXIII X=O

But XXIII, dulcine, in which the substituent is an O atom, is 250 times sweeter than sucrose. This shows how closely related the sweet and bitter modalities are, and how important the nature of the polar groups is in taste. Sugars have detection thresholds in the mM range, much higher than most odorants, but some sweet tasting proteins have a low detection threshold. Thaumatin is a protein with molecular weight 21,000 which is 10^5 times sweeter than sucrose on molar base. It is extracted as a complex of thaumatins from the tropical fruit Thaumatococcus danielli. Its sweetness is abolished by denaturation, and increased by Al^{3+}. Monellin is another sweet protein with molecular weight 11,500 isolated from the tropical fruit Discoreophyllum cumminsii. It is a heterodimer and the isolated monomers are not sweet; recombination of the dissociated monomers is accompanied by a partial reappearance of sweetness. The amino- acid sequence of the two proteins shows no homology. The structures resolved by X-rays show no tertiary similarity either.

Passing now to the sour modality, all molecules with one or more -COOH groups or -SO$_3$ groups taste sour. It is not clear whether the active species is H^+ or the dissociated on the non-dissociated acid group but dicarboxylic organic acids, e.g., cytric acid, taste more sour than monocarboxylic acids. The rest of the molecule is not relevant to the sour modality. A generalist type of receptor is most likely to be involved. The salty modality is more complex and is never pure. Some chemical salts are salty and others are sweet, such as lead acetate, and even salty salts have a sweet component at low concentration, e.g., NaCl and to a lesser extent KCl. Many salts of divalent cations are bitter, e.g., CaCl$_2$. Amiloride, a diuretic drug which blocks Na reabsorbtion in the kidney tubules, reduces the response to NaCl applied on the tongue. It is thus likely that, at least for NaCl, a mechanism of Na exchange common to other cells, is part of the sensory mechanism. Finally we consider some hot "tasting" substances found in nature although this is not a taste modality. Beta-polygodial (XXIV)

XXIV

is a dialdhyde found in the leaves of some african shrubs, very long lasting hot tasting for man and antifeedant for insects. The alpha-isomer in which the distance between the two aldehyde groups is increased, is tasteless and displays no antifeedant activity. Interestingly the beta-isomer, but not the alpha-isomer can react with primary amino-groups.

Olfactory versus taste chemoreception

The difference between olfactory stimuli and taste stimuli has been briefly outlined in the above sections. We should realize that also the sensory modalities and the meaningful purposes for which these chemosystems evolved, are different. The complexity of the olfactory sensations is very great as an expert profumier can distinguish hundreds of aromas, among which a very few if any have the character of primary odours and the majority are composite aromas. The taste system is a much poorer system as the quality of the information is concerned, namely, the four modalities we mentioned, easily distinguished one from the other. This difference is also reflected in the different appreciation of sensations arising from the two systems. Taste is related to the choice of food and to start reflexes that initiate digestion; the dominant character of olfactory sensations is hedonic, ranging from pleasant to disgusting with a powerful drive on emotions and behaviour. The taste nerves terminate in the lower brain stem where neurovegetative reflexes are integrated, whereas the olfactory nerves are directly connected to the rhinencephalon, and indirectly to the limbic system and hypothalamus where emotions and hormonal responses are shaped. A third chemosensor organ, the vomero-nasal organ is present in many vertebrate species with the exception of higher primates and man. This organ seems to detect water soluble substances but its anatomical and functional organization are more similar to the olfactory than to the taste system. This organ is used by reptiles, e.g., to track the prey as the animal samples the soil with the tongue and in lower mammals it is involved in functions relevant to the sexual sphere. Some of these functions are hormonal. The most dramatic and subtle effect is the interruption of pregnancy in the mouse which can be evoked in particular circumstances by stimulation of the vomero-nasal organ with presentation in the cage of the urine of a male mouse different from the partner mouse. The ability to discriminate odours by the olfactory system is further highlighted by the ability that mice have of discriminating exclusively on the base of urine, the major histocompatibility group of conspecifics. The chemical stimuli involved in these extraordinary performances are not known but it is possible that urinary proteins are involved.

THE PROBLEM OF OLFACTORY AND TASTE RECEPTORS

A full comprehension of the chemistry of chemoreception would require that in addition to stimulant molecules also receptor molecules are chemically defined. A receptor molecule in chemoreception is a molecule which i) is present in receptor cells, ii) interacts chemically with odorant or taste molecules and iii) is implicated in sensory transduction. The fact that chemical information from the outside world is purposefully channeled in different sensory modalities indicates that there are receptor molecules. Direct demonstration of receptor molecules, however, is difficult because of the paucity of sensory cells, about $30 \cdot 10^6$ in the olfactory mucosa and $2 \cdot 10^3$ in taste buds of man, and of the small number of receptor molecules probably present in the cells, about 10^6 per cell. In addition, olfactory cilia are difficult to obtain pure and even more so taste cells. Despite these limitations, the problem has been tackled by several laboratories. A 95 KD protein, present in olfactory cilia, olfactin, has been demonstrated in the frog, but its function is still hypothetical. Binding studies with radioactive odorants on the whole tisse have led to promising results but its localization in the receptor cells has not been demonstrated as yet. A similar caution applies also to studies with recombinant DNA.

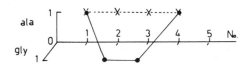

Fig. 1: Homology in proteins. See text.

The fact that many receptor are expected to be of the generalist type, renders useless the common tools of the experimenter who likes to work with the sharp edge of chemical specificity. In addition different types of receptors may coexist in the same sensory modality. For instance, the bitterness of sugars may possibly be due to a specialist type of interaction with membrane receptor molecules, whereas the bitterness of quinine may be due to a pharmacological block of currents through Ca-activated K-channels, and that of xanthines to interference with the metabolism of cyclic nucleotides. It is a common opinion that the study of chemoreceptor molecules has not yet passed the test of demon-

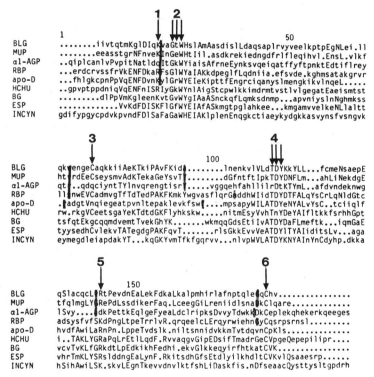

Fig. 2: Significant homology defines a supra-family of proteins. BLG: beta-lactoglobulin; MUP: mouse major urinary protein; alpha 1-AGP: alpha 1-acid glycoprotein of serum; RBP: rat retinol-binding protein; apo-D: apolipoprotein D of serum; HCHU: human alpha 1-microglobulin; ESP: epididimal sperm-binding protein; BG: frog olfactory-gland protein; INCYN: tobacco hornworm insecti-cyanin. Amino-acids are in progressive order starting from the amino-terminal residue aligned for best homology. The arrows point to highly conserved residues. Matches of 3 or more residues are shown in upper case. Abbreviations for the amino acids are: A, Ala; C, Cys; D, Asp; E, Glu; F, Phe; G, Gly; H, His; I, Ile; K, Leu; M, Met; N, Asn; P, Pro; Q, Gln; R, Arg; S, Ser; T, Thr; V, Val; W, Trp; and Y, Tyr. The position of exon-intron junctions is indicated by shaded bars. From Ali and Clark, J. Mol. Biol. (1988), 199.

strating their existence unambiguously, although there are promising prospects. When scientific progress is slow or blocked it is wise to take a step back. Looking for models of rather than trying to isolate chemoreceptors may help to resolve these difficulties.

RECEPTOR MODELS

A supra-family of proteins is defined, in general, by homology in the amino-acid sequence. Homology is a topological term which has to do with paths in space. Two paths are homologous if together they bind a region of space, e.g., making a loop. This is in equivalence relationship which divides the paths into mutually exclusive classes, the homology classes. A protein can be represented as a path in a space of n + 1 coordinates where n is the number of different amino acids found in nature, about 20, and the (n + 1)th dimension is the progressive residue number in the amino-acid sequence. If two proteins have at least two identical amino-acids in the same position of the sequence, they are homologous. Significant homology occurs whenever the homology is statistically significant, i.e., can not be due to chance. Homology does not provide a measure of the similarity/dissimilarity. For this, a metric, the divergence, can be introduced by considering the number of non-identical residues in the same position of the sequence. Two proteins differing in 20% of the residues have a 20% divergence relative to identity. Fig. 1 shows, as an example, the paths described by tetrapheptydes ala-ala-ala-ala (x) and ala-gly-gly-ala (.). They together bind a region and are homologous. They have a 50% divergence relative to identity.

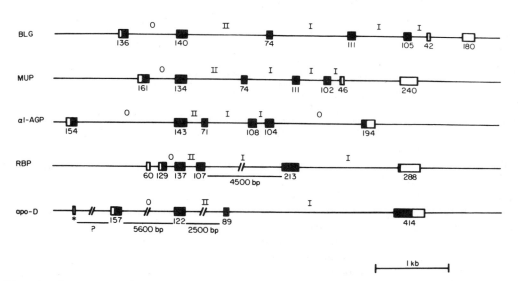

Fig. 3: Comparison of gene structures and exon-intron arrangement in the genes of proteins of the supra-family. Exons are shown as open boxes, and their coding regions are shaded. Protein abbreviations as in Fig. 2. From Ali and Clark, J. Mol. Biol. (1988), 199.

A supra-family of proteins is known many of which bind hydrophobic molecules.
Members of this supra-family are the retinal binding protein (RBP) which carries and
delivers retinol to some cells, the apolipoprotein-D (apo-D) of HDL which binds to
lecithin: cholesterol acyltransferase and which is possibly involved in the transport of
cholesterol, the tobacco hornworm insecticyanin (INCYN) which binds biliverdins and
is possibly involved in camouflage, as well as the sperm binding protein (ESP) of the
epididymal fluid, the HCHU protein of human serum possibly involved in neutrophil
chemotaxis, beta-lectoglobulins (BLG) of the cow milk which can bind retinol; among the
proteins less characterized functionally are an acidic glycoprotein of serum (alpha1 AGP),
a protein found in olfactory glands (BP) and the urinary proteins of rodents (MUP). The
significant homology in the amino-acid sequence is shown in Fig. 2. These proteins are
coded by 5-6 exons in the genome, Fig. 3, and are probably all derived from a common
ancestor. Unfortunately there was no evidence that these proteins bind volatile odorants,
until Paolo Pelosi in 1981 found a new protein, the pyrazine-binding protein (PBP) in
the nasal mucosa of the cow, which binds 2-isobutyl-3-methoxypyrazine (XXIV), a very
low threshold odorant for man. The analysis of the amino-acid sequence of this protein
showed homology to the above mentioned supra-family (Fig. 4). This observation esta-

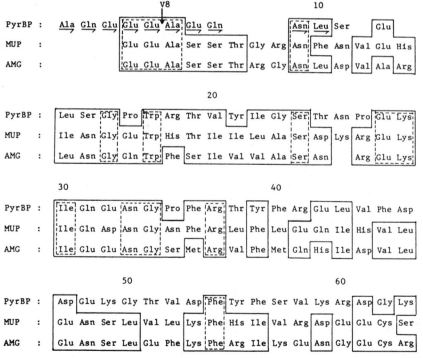

Fig. 4: Highly significant homology in a sub-set of the supra-family. PyrBP: pyrazine-
binding protein (PBP in the text); MUP: mouse major urinary proteins; AMG:
rat urinary proteins. Boxes include similar residues and dotted boxes identical
residues. Only the first 63 residues are represented for every protein. The arrow
indicates a cleavage point utilized in sequencing. From Cavaggioni et al., 1987,
FEBS Letters 212.

blished a link between an odorant binding protein and the supra-family. Close inspection of the divergence between the PBP and the other numbers of the supra-family showed that the divergence was least with the urinary proteins of rodents. These proteins deserve a few introductory words. Proteinuria is not a physiological phenomenon in animals except in mice and rats. This is an interesting exception because a proteinuria is a negative term in the nitrogen balance, and very negative indeed for these animals that, in the wild state, live on the brink of starvation. Even more puzzling is the fact that this proteinuria is present only in adult male rodents, not in females, nor in young animals, nor in castrates.

These proteins are synthesised in the liver under androgen steroid stimulation and are filtered freely through the kidneys because of their small dimension (m.w. 18,000). These proteins are considerably different in the mouse and in the rat. But either in mouse or rat, several forms of the proteins are present, differing only in a few residues. A certain degree of polymorphism in the different forms has been reported among animals of different strains, and their differential expression seems to be influenced by hormones. Proteins, close cognate of the urinary ones, are expressed constitutionally, i.e., both in male and female rodents, also in submaxillary, lacrymal and preputial glands. The urinary proteins are coded by a multi-gene family of about 35 genes localized in chromosome 4, about half of which are pseudogenes in the mouse. There is evidence for phylogenetic pressure for divergence between rat and mouse genes. There is also evidence that, within the mouse, the genes of the multi-gene family are under a conservative pressure, Fig. 5. This is demonstrated because the evolutionary unit of the gene is a 45 kb unit formed of a true gene and of a pseudogene in tandem with head to head arrangement; while random mutations accumulated in time in the pseudogene, only a few and non-randomly distributed mutations were retained in the true gene.

About 2% of chromosome 4 is made up of these units. Taken together these observations suggest an important function for urinary proteins, but which? Behaviourist know that the urine of male rodents has several behavioural as well as hormonal effects or conspecifics that are abolished by removal of the olfactory system, e.g., pregnancy interruption and puberty acceleration. It is also possible that non-volatile urinary components are involved in these effects, possibly through stimulation of the vomero-nasal organ. Now, the observation that the urinary proteins are homologous to an odorant binding protein, PBP, suggests that proteins of this subset of the supra-family bind also odorants. Experimental evidence shows the urinary proteins as well as the PBP bind some odorants with the lowest detection threshold for man, namely, of the order of one

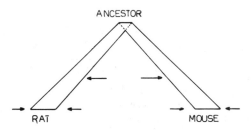

Fig. 5: Phylogenetic relationship between urinary proteins of rodents. Arrows indicate phylogenetic pressure, see text.

odorant molecule per 10^{13} water molecules. Most interestingly, the spectrum of the high affinity odorants is not restricted to a chemical class. The highest affinity odorant found is geosmin (XXV), with an OH as polar group and the decalin system as the hydrophobic moiety, but other high affinity odorants are thymol (XXVI), with an aromatic ring as the hydrophobic majority, 2-isobutyl-3-methoxypyrazine (XXIV) in which the polar group is an aromatic heterocycle, 2-nonenal (XXVII) with a flexible hydrocarbon chain, and beta-ionone (XXVIII) with a rigid polyene chain.

The common feature is a polar atom or group and a hydrophobic moiety of 6-12 carbon atoms. Clearly these proteins are models for a generalist type of receptor. On close inspection of the binding pattern of urinary proteins of mouse as well as of rat and PBP, it is evident that the patterns are not identical, but, e.g., beta-ionone binds to rat but not to mouse urinary proteins. Not every odorant, moreover, is bound, e.g., neither the urinous odorant 5 alpha-androstan-3-one (Ia), nor the sweaty odorant isovaleric acid (VIa). Clearly these proteins are models for one class of receptor molecules only. Since several of these proteins can be obtained pure in gram quantities, the amino-acid sequences are known, and X-ray diffraction from crystals can be attempted, they lend themselves to the study of structure/activity relatioships at the molecular level.

We can speculate that the odour binding activity of the urinary proteins explains the divergence between mouse and rat proteins, as at the beginning of species differentiation early male mice wanted their urinary love messages to be sniffed and understood only by female mice - with female rats the affair was not so interesting; it explains also the conservation of proteins within the species as an olfactory species-specific alphabet evolved. These proteins probably help to concentrate male pheromones in the urine and either release the pheromones in air or reach the vomero-nasal organ by licking with the odorant bound. Proteins homologous to the cow PBP have been found also in rat and frog olfactory glands (BP of Fig. 2). The functional significance of these nasal proteins is obscure. Gaupp as early as in 1904 described characteristic eosinophylic granuli in the cells of olfactory glands that are always present in olfactory mucosa; because odorants have to traverse a layer of secretion he supposed a specific combination of this secretion with odorants. It is worth mentioning that both bacterial chemotaxis as well as in the immune system ligands are carried to the cell membrane by secretory molecules. A carrier role also for these proteins cannot be ruled out at present.

REFERENCES

Beets, M.G.J., 1978, Structure-activity relatioship in human chemoreception, *Applied Science Publ.*, Barking.

Bignetti, E., Cavaggioni, A., and Pelosi, P., 1987, An odorant binding protein *in*: "Discussion in Neurosciences," Hudsperth, A.J., MacLeish, P.R., Margolis, F.L., and Wiesel, T.N., ed., FESN, Geneva.

Cowley, J.J., 1978, *in*: "Biological determinants of Sexual Behaviour," Hutchinson, J.B., ed., John Whiley, Chichester.

Getchell, T.V., Margolis, F.L., and Getchell, M., Periceptor and receptor events in vertebrate olfaction, 1984, *Progress in Neurobiology*, 23:317-345.

Ghazal, P., Clark, A.J., and Bishop, J.O., 1985, Evolutionary amplification of a pseudogene, *Proc. Natl. Acad. Sci.*, USA, 82:4182-4185.

Halpern, H., The organization and function of the vomeronasal system, 1987, *Ann. Rev. Neurosci.*, 10:325-362.

Keverne, E.B., Pheromonal influences on the endocrine regulation of reproduction, 1983, *Trends in Neurosci.*, 6:381-384.

Lancet, D., Vertebrate olfactory reception, 1986, *Ann. Rev. Neurosci.*, 9:329-355.

Pelosi, P., 1985, Classificazione e misura degli odori, *in*: "Caratteristiche olfattive e gustative degli odorii," ed., Chirotti, Pinerolo.

Polak, E., Is odour similarity quantifiable? 1983, *Chemistry and Industry*, 3 January, 30-36.

ANTENNAE AND NOSES: THEIR SENSITIVITIES AS MOLECULE DETECTORS

K.-E Kaissling

Max-Planck-Institut für Verhaltensphysiologie
Seewiesen, Fed. Rep. of Germany

SUMMARY

The sensitivity of noses and antennae depends on behavioral, geometrical, physical, biochemical and physiological factors. They determine the input cross section of the organ, the relative velocity of air current and organ, the rate of adsorption of molecules at the sensory area, the fraction of absorbed molecules exciting sensitive cells, and the minimum number of nerve impulses that must be elicited by a detectable stimulus. The sensitivity to a given odorant is proportional to the capture coefficient, which is the molecule capture rate of a cell related to the stimulus concentration in air. The sensitivity is also proportional to the square root of the number of sensitive receptor cells times the critical stimulus duration divided by the rate of spontaneous impulse firing of a cell. The pheromone receptor cells in the silk moth antenna have a much higher capture coefficient for bombykol compared with cells in the human nose for mercaptane and with cells in the dog nose for α-ionone. The higher capture coefficient compensates for the much smaller number of receptor cells of the moth. For potent odorants stimulus concentrations at the receptor cells as low as 10^{-14}M build up during brief (below 1 s) threshold stimuli in the vertebrate nose.

INTRODUCTION

Olfactory organs of animals serve to detect and to distinguish biologically meaningful chemicals diluted in the surrounding medium. Whereas arthropods developed antennae, vertebrates have noses, and both organs are used by species living in water and in air. These two types of olfactory organs are of fundamentally different construction, rather like everted and inverted eyes, but reach a similar sophistication of function in terms of sensitivity to and discrimination of odors. Thus the olfactory threshold of a male silkworm moth is similar to that of a dog. A honey bee discriminates a variety of odors and subtle differences in mixtures of odor compounds to a degree that allows comparison with the human sense of smell. Examples concerning insect olfaction are summarized in Kaissling (1986, 1987). A collection of profound review articles on various aspects of vertebrate olfaction can be found in Margolis and Getchell (1988). The present study compares the function of antennae and noses on a quantitative basis for a few air-living species of insects and mammals.

Sensory Transduction
Edited by A. Borsellino *et al.*
Plenum Press, New York, 1990

Antennae and noses are complex morphological structures which collect odor molecules and lead them to the olfactory receptor cells. Finally, they sequester the stimulus molecules in order to terminate stimulation. We shall consider these steps in sequence: the collecting of stimulus molecules, the transduction of the stimulus into nervous excitation by the receptor cells, and the biochemical processes responsible for the inactivation of odor molecules. All of these processes contribute to the absolute olfactory sensitivity, which can be defined as the reciprocal magnitude of the olfactory threshold stimulus.

This quantitative comparison will be restricted to the absolute sensitivities of organisms with noses and antennae to single stimulus compounds. The ability to discriminate between odorants and odorant mixtures is more difficult to describe in quantitative terms because of the vast number of possible odorants. The power of discrimination will be mainly determined by the chemical specificities of the receptor cells, by their number and, in particular, by the number of types of these cells. In addition, it will largely depend on the ability of the CNS to distinguish patterns of excitation among the receptor cells. Very little is known about most of these factors, especially for the vertebrate nose. Therefore, a quantitative understanding of the ability to discriminate odor qualities does not yet seem possible beyond the almost trivial statement that discrimination may improve with a larger number of receptor cells. For functional comparison of antennae and noses we restrict our consideration to morphological and physiological factors determining the sensitivity for a given odor compound.

PARAMETERS DETERMINING THE SENSITIVITY OF OLFACTORY ORGANS

Usually, the sensitivity of an olfactory organ is characterized by the reciprocal olfactory threshold of a compound expressed as reciprocal threshold concentration C_t (number of odor molecules per cm^3 of air)$^{-1}$. First of all, the sensitivity depends on factors controlling the capture of odor molecules by the olfactory organ. Animals with noses sample the odor molecules within a volume V of air inhaled during a sniff. The number of molecules N entering the nose in one sniff at concentration C is

$$N = C * V \text{ (molecules) (Dimensions as used in Tables 1-3).} \tag{1}$$

In contrast to the situation in noses, the sensory areas of antennae are directly exposed to the surrounding medium. Therefore, relative movements of antenna and medium influence the rate of collection of odor molecules at a given odor concentration. Under this condition the external stimulus for an antenna may be characterized by the odor intensity, which is the product of the odorant concentration C and the relative velocity u between antenna and medium

$$J = C * u \text{ (molecules/cm}^{2*}\text{s).} \tag{2}$$

The number of molecules N available for detection is

$$N = J * \sigma * T = C * u * \sigma * T \text{ (molecules).} \tag{3}$$

In this equation σ is the "input area" of the antenna, the cross section of the air stream corresponding to the antennal outline area. The outline area is the projection area of the antenna on a plane at right angles to the direction of relative movement. T is the stimulus duration. Our notations C, N, V and T follow Mozell et al. (1984).

Equation 3 can be applied to animals with noses if we consider the product $u*\sigma*T$

as the "input volume" of the antenna and as equivalent to the volume V inhaled by the nose.

$$V = u * \sigma * T \ (cm^3).\hspace{3cm}(4)$$

For the nose σ could be the cross sectional area of the nostril and u represents the velocity of the inhaled air flow at the nostril. The duration of a sniff can be considered as stimulus duration T, usually a fraction of a second per sniff. The parameters u and σ can be controlled by lung expansion and by widening, or narrowing the nostrils.

With antennae these three parameters can depend on active movements of the animal or the antennae as well as on external conditions such as wind velocity and wind direction with respect to the antenna. The stimulus duration depends also on the distribution of molecules in the air stream. The antennae may more easily monitor extremely brief stimuli produced by quick changes of stimulus concentration within an odor plume (Rumbo and Kaissling, 1989). Because the average reaction time of a male silkworm moth responding with wing fluttering to stimulation with the female pheromone bombykol is one second (Kaissling and Priesner, 1970), in the subsequent calculations for the antenna stimulus duration T was 1 s; an air stream velocity of u = 60cm/s was used for the odor stimulation of moths.

Having established the stimulus parameters we now consider (i) the conduction of stimulus molecules to the receptor cells and (ii) the transduction process, and discuss their contribution to the efficiency of an olfactory organ. Several efficiency quotients have been introduced to represent the reduction in number of stimulus molecules N to a number N_{eff} that finally arrives at the olfactory receptor cells and produces excitation.

For the nose two quotients have been considered (Moulton, 1977):

1) The quotient Q_v gives the fraction V_{olf} of the inhaled volume V which passes the olfactory slit with the olfactory mucosa. This quotient depends both on the morphology of the nose and on the velocity of inhalation u (Stuiver, 1958; de Vries and Stuiver, 1960; Mozell et al., 1984, 1987).

$$Q_v = V_{olf}/V.\hspace{3cm}(5)$$

2) The quotient Q_c indicates the fraction C_{olf} of stimulus concentration C that is taken from V_{olf} due to adsorption at the olfactory epithelium, i.e. at the sensory area A_{sen}. The quotient may depend on the odor substance and its adsorptive properties.

$$Q_c = C_{olf}/C.\hspace{3cm}(6)$$

For moth antennae two different quotients have been measured:

1) The antennal adsorption quotient Q_{ant} (formerly called Q_{ads}, Kaissling 1971, 1987) which is the ratio between the the molecules adsorbed at the antenna N_{ant} and the input number N (see equ.3).

$$Q_{ant} = N_{ant}/N.\hspace{3cm}(7)$$

Q_{ant} might depend on the odorant and on the relative airstream velocity u. A concentration dependence is expected if the adsorption saturates but has not been found even with stimulus intensities above the physiological range (unpubl. observation).

2) The quotient Q_{sen} relates the fraction of molecules N_{sen} adsorbed at the sensory area, the surface of the olfactory hairs, to the number of molecules adsorbed on the

antenna N_{ant}. This quotient was formerly called Q_{eff} (Kaissling, 1971, 1987). Only those molecules hitting the sensory area have a chance of reaching the receptor cells.

$$Q_{sen} = N_{sen}/N_{ant}. \tag{8}$$

The product $Q_{ant} * Q_{sen}$ is analogous to the product $Q_v * Q_c$ used for the nose. Both products account for the loss of stimulus molecules between input of the organ and the adsorption of the molecules at the sensory area A_{sen}, the surface of the mucus of the cuticular hair wall.

A third quotient needs to be considered for any type of olfactory organ. Of the molecules N_{sen} adsorbed at the sensory area only a fraction N_{eff} will reach the plasma membrane of the receptor cells sensitive to the respective odorant and produce nervous excitation. We call this ratio of molecules Q_{eff}:

$$Q_{eff} = N_{eff}/N_{sen}. \tag{9}$$

This quotient depends on losses of molecules within the olfactory mucosa or the olfactory hairs. It must be a function of the number of receptor cells sensitive to the respective odorant N_s relative to the total number of receptor cells N_r. The value of Q_{eff} might also depend on mechanisms which sequester and inactivate odor molecules as discussed below.

Combining all of the quotients we find the fraction of effective molecules N_{eff} as related to the stimulus molecules N offered to the organ. This is for the nose

$$N_{eff}/N = Q_v * Q_c * Q_{eff} \tag{10}$$

and for the antenna

$$N_{eff}/N = Q_{ant} * Q_{hairs} * Q_{eff}. \tag{11}$$

Multiplication of the ratio N_{eff}/N by the input volume V gives an expression for the capture efficiency of the entire olfactory organ which can be called the effective stimulus volume V_{eff}.

$$V_{eff} = V * N_{eff}/N \ (cm^3). \tag{12}$$

The effective volume V_{eff} multiplied by the concentration C of the odorant in air gives the number of effective molecules N_{eff} of the respective odorant reaching receptor cells sensitive to this odorant and producing excitation within a given stimulus duration T.

$$N_{eff} = C * V_{eff} \ (molecules). \tag{13}$$

Defining the olfactory sensitivity S as reciprocal threshold concentration $1/C_t$ we obtain

$$S = V_{eff}/N_{efft} \ (cm^3/molecules). \tag{14}$$

Here and in following equations the suffix t denotes the threshold situation. This relationship shows clearly that the sensitivity does not depend only on the capture efficiency of the olfactory organ as expressed by V_{eff}. It is also dependent on the number of effective odor molecules N_{efft} required for excitation of receptor cells at threshold. The fewer effective molecules are necessary to produce a sensation or a behavioral response, the higher is the sensitivity of the organism.

The value of N_{efft} is mainly determined by the capability of the CNS to detect a small excitatory signal, for instance a sudden increase in number of nerve impulses coming from the receptor cells. One very important factor determining the sensitivity is, therefore, the rate of spontaneously (i.e. in the absence of stimulus molecules) fired nerve impulses of the receptor cells. The overall spontaneous activity represents a noise which determines the size of a minimum detectable signal.

The following considerations relate to a case where the nerve impulses fired spontaneously by a receptor cell at an average rate R_{isp} are Poisson (randomly) distributed in time. A Poisson process in time is also assumed for the nerve impulses elicited by threshold stimuli whose total number in all cells sensitive to the respective stimulus compound is N_{istim} during time interval T. Poisson distributions of spontaneous and stimulus-induced nerve impulses have been observed in the silkmoth (Kaissling and Priesner, 1970, Kaissling, 1971) and can be assumed for the vertebrate nose (van Drongelen et al., 1978a).

We assume here that at low stimulus concentrations the nerve impulse events from N_s receptor cells sensitive to a given odorant are superimposed by way of convergence upon a higher level cell. The impulse events at this higher level are assumed to be Poisson distributed with an average rate $R_{isp} * N_s$. The average number of impulse events occurring in a sample of duration T is $R_{isp} * N_s * T$ with a standard deviation $(R_{isp} * N_s * T)^{1/2}$ (see, e.g. van Drongelen et al., 1978a). If we take a multiple a of the standard deviation as an estimate of the minimum detectable increment $N_{istim\ min}$ of the impulse number caused by the stimulus, then

$$N_{istim\ min} = a * (R_{isp} * N_s * T)^{1/2} \text{ (nerve impulses)}. \tag{15}$$

The constant a depends on the specified probability of signal detection. In the following and in Tables 1-3 we use the threefold value of the square root as a threshold criterion ($a = 3$). Given that at threshold each stimulus-induced nerve impulse is elicited by one effective odor molecule we can use $N_{istim\ min}$ as the detectable minimum number of effective molecules and substitute $N_{istim\ min}$ for N_{efft} in equ.14. In this case the maximum sensitivity S_{max} for a given compound is described by

$$S_{max} = V_{eff}/N_{istim\ min} = V_{eff} / (a * (R_{isp} * N_s * T)^{1/2}) \text{ (cm}^3/\text{molecules)}. \tag{16}$$

This expression gives the counter-intuitive result that the sensitivity increases with decreasing number of receptor cells N_s. This apparent contradiction can be understood if we realize that a decrease of N_s at fixed V_{eff} increases the average molecule capture of a receptor cell. We introduce a "capture coefficient" F which is the capture rate N_{eff}/T per number of sensitive receptor cells N_s and per stimulus concentration C.

$$F = N_{eff}/(N_s * T * C) \text{ (cm}^3/\text{receptor cell} * s) \tag{17}$$

Because of equ. 13, F can also be regarded as the effective volume of one receptor cell cleaned from odorant molecules per unit time.

$$F = V_{eff}/(N_s * T) \text{ (cm}^3/\text{receptor cell} * s). \tag{18}$$

Replacing V_{eff} in equation 18 we find a capture coefficient for an antenna

$$F = u * \sigma * Q_{ant} * Q_{sen} * Q_{eff}/N_s \text{ (cm}^3/\text{receptor cell} * \text{s).} \qquad (19)$$

and, correspondingly, for the nose

$$F = u * \sigma * Q_v * Q_c * Q_{eff}/N_s \text{ (cm}^3/\text{receptor cell} * \text{s).} \qquad (20)$$

Clearly, at fixed capture coefficient F the sensitivity to an odorant increases with the square root of the number of sensitive receptor cells and of the stimulus duration. This becomes evident if we combine equations 16 and 18 and find

$$S_{max} = F * (N_s * T/R_{isp})^{1/2}/a \text{ (cm}^3/\text{molecules).} \qquad (21)$$

The surprising relationship of S_{max} and N_s (equ. 16) can now be understood. It results from the fact that changing of N_s, when leaving all other parameters (V_{eff}, R_{isp}, T) constant, influences both the capture coefficient F and the summed spontaneous activity $R_{isp} * N_s * T$. Thus an exclusive increase of N_s by a factor of 10 would decrease F by 10 times. Since N_s in the numerator of equ. 21 appears under the square root the sensitivity would decrease with increasing N_s.

The reader may imagine two antennae with identical V_{eff}, i.e. at the same air stream velocity (u), stimulus integration time (T), with the same geometry, number and arrangement of olfactory hairs, identical adsorption properties ($Q_{ant} * Q_{sen}$), the same fraction of effective molecules (Q_{eff}) and spontaneous impulse rate (R_{isp}). Of the two, the antenna with a larger number of sensitive receptor cells N_s would be less sensitive. This may be the reason why there is usually only one receptor cell of each type in each olfactory hairs of the extremely sensitive moth antennae (Kaissling et al., 1978; Meng et al., 1989).

An extended form of equation 21 is obtained by using equ. 19 and 20. It shows the basic parameters determining the sensitivity of an antenna

$$S_{max} = u * \sigma * Q_{ant} * Q_{sen} * Q_{eff} * (T/R_{isp} * N_s)^{1/2}/a \qquad (22)$$

$$S_{max} = u * \sigma * Q_v * Q_c * Q_{eff} * (T/R_{isp} * N_s)^{1/2}/a. \qquad (23)$$

These equations show that the sensitivity - given all other parameters are constant - is inversely proportional to the square root of the number of sensitive receptor cells N_s and of the average rate of spontaneous nerve impulses. At the same time it is proportional to the square root of the stimulus duration T over which the animal integrates nerve impulses fired by the receptor cells.

Clearly, there are physical and physiological constraints precluding an olfactory organ with just one receptor cell within each class of chemical specificity. First of all, there are physical upper limits for the capture coefficient F of a receptor cell. In addition, a large number of receptor cells of the same type may inrease the working range of stimulus intensities of the whole organ and may provide variability of dynamic and other cell characteristics.

QUANTITATIVE COMPARISON OF NOSES AND ANTENNAE

In the following we discuss the magnitudes of the various efficiency factors for the noses of humans and dogs and for the antenna of the silkworm moth *Bombyx mori*. This discussion will admittedly suffer from the very limited knowledge about many of these factors, especially in the vertebrate nose. Nevertheless, it may be worthwhile to direct attention to various problems to be solved and it may also lead to considerations of general principles in the function of olfactory organs.

Measured or estimated values of the parameters determining V_{eff} for potent odorants in humans, dog and silkworm moth *Bombyx mori* are listed in Table 1. For the nose V can be fairly well measured. Q_v and Q_c were determined using a model of the human nose and diffusion equations for the conditions in the olfactory slit with the regio olfactoria (Stuiver, 1958). The values for the dog were roughly estimated by Moulton (1977). One problem is that these quotients may differ depending on the velocity of inhalation and on the odorant (Stuiver, 1958).

For the antennae of the silkworm moth Q_{ant} and Q_{sen} were determined using radioactively labelled bombykol. The wind velocity u was experimentally adjusted to about 60 cm/s, the outline area of one antenna was $\sigma = 6$ mm^2 and T was 1 s according to the average reaction time of the fluttering response of male moths at low stimulus concentration. This amounts to an input volume V of the two antennae of 7.2 cm^3.

The value of $Q_{eff} = N_{eff}/N_{sen}$ was calculated for the silkmoth from the number of bombykol molecules adsorbed on the hairs $N_{sen} = N_{ant} * Q_{sen}$ and from the number of nerve impulses elicited N_{istim}. The latter number indicates the number of molecules N_{eff} actually reaching the cell membrane and producing excitation of the bombykol receptor cells at low stimulus concentrations. This conclusion is based on the experimental results of Kaissling and Priesner (1970), which we summarize briefly in the following paragraphs.

Earlier calculations used as threshold stimulus for *Bombyx mori* the smallest amount of bombykol on the odor source ($3 * 10^{-6}\mu g$) eliciting wing vibration in a significant percentage of male moths (22%) (Kaissling and Priesner, 1970, Kaissling, 1971, 1987). In this paper we relate all calculations to the stimulus load at which 50% of the animals responded ($1.3 * 10^{-5}\mu g$ of bombykol). Since the release of bombykol from the odor source was assumed to be proportional to the load, the number of stimulus molecules is increased by the factor 4.3 compared with values given for the 22% threshold. Proportionality of release was found using ^{3}H-labelled bombykol with loads of odor sources between $10^{-2}\mu g$ and $10^2\mu g$ and was assumed for the lower source loads.

The number of nerve impulses elicited by a one-s stimulus was significantly raised above the spontaneous firing level at $10^{-5}\mu g$ of bombykol per source and increased in proportion to the stimulus load up to $10^{-3}\mu g$ (Kaissling and Priesner, 1970). Therefore, the numbers of nerve impulses elicited at the 50% threshold are also higher by the factor 4.3 than those calculated for the 22%-threshold.

For the 50% threshold we derive from Kaissling and Priesner (1970) a value of $N_{sen} = C_t * V * Q_{ant} * Q_{sen} = 4.2 * 10^3$ molecules adsorbed on the hairs during one s of stimulation and a value of $N_{istim\ t} = 4.66 * 10^2$ nerve impulses elicited per 34,000 receptor cells (of both antennae). From these numbers and considering the observed distribution of nerve impulses obeying a Poisson distribution in time it was concluded that one molecule of bombykol was sufficient to produce a nerve impulse and did not produce more than one nerve impulse. Therefore, at stimulus concentrations near threshold the number of effective stimulus molecules N_{eff} equals the number of stimulus-induced nerve impulses N_{istim}.

Table 1

			man		dog		silkmoth	
			sec. butyl mercaptane		α-ionone		bombykol	
input volume (sniff)	$V=u*\sigma*T$	(cm³)	75	a	60	b	7.2	z
input velocity	u	(cm/s)	250	c	150	c	60	i
input cross section of both sides	σ	(cm²)	2	c	4	c	0.12	k
stimulus duration	T	(s)	0.15	a	0.1	b	1.0	l
efficiency quotients for: nose	$Q_v=V_{olf}/V$		0.07	a	0.1	b	-	
nose	$Q_c=C_{olf}/C$		0.3	a	0.1	b	-	
antenna	$Q_{ant}=N_{ant}/N$		-		-		0.27	i
antenna	$Q_{sen}=N_{sen}/N_{ant}$		-		-		0.8	m
nose and antenna	$Q_{eff}=N_{eff}/N_{sen}$		(0.1)	d	(0.1)	d	0.11	n
effective stimulus volume	$V_{eff}=V*Q_v*Q_c*Q_{eff}$	(cm³)	(0.15)	d	(0.06)	d	-	z
" " "	$V_{eff}=V*Q_{ant}*Q_{sen}*Q_{eff}$	(cm³)	-		-		0.17	z
no. of receptor cells, total	N_r	(cells)	$6*10^6$	e,f	$1*10^9$	g	$6.8*10^4$	o
no. of rec. cells, sensitive to stim. cpd. N_s		(cells)	$(6*10^5)$	h	$(1*10^8)$	h	$3.4*10^4$	p

a) Stuiver 1958. V was calculated from the inhaled volume per s for normal breathing $= 500$ cm³/s and the stimulus duration of 0.15 s which corresponds to the "critical time" of stimulation (see text). The fraction of air passing the olfactory slit depends on the flow rate and was estimated from photographed spurs of aluminium particles in a perspex model of the nose. It was between 5 and 10% for normal breathing (we set $Q_v = 0.07$). The adsorbed fraction Q_c was calculated from the dimensions of the olfactory slit and from a diffusion coefficient of 0.08 cm²/s. With normal inspiration 2% of the inhaled molecules were thought to adsorb on the olfactory epithelium ($Q_v * Q_c = 0.02$). b) Values estimated by Moulton 1977. c) Values of σ were estimated in order to obtain an estimate of u from $V/\sigma * T$. d) Values are based on the assumption that $Q_{eff} = 0.1$, i.e. that 10% of the molecules adsorbed on the olfactory mucosa reach receptor cells sensitive to the respective stimulus compound and produce excitation. e) Menco 1983. $3 * 10^6$ sensory endings/cm² of olfactory epithelium. f) Moran et al. 1982. $3 * 10^6$ sensory endings/cm². Total area of olfactory epithelium $= 2$ cm². g) This value is based on the total area of the olfactory epithelium $= 170$ cm² reported for the German shepherd dog by Lauruschkus (1942) quoted by Moulton (1977) and the density of sensory endings of $6.1 * 10^4$/mm² given by Menco 1980. h) Assumed that 10% of the receptor cells are sensitive to the respective stimulus compound. i) Based on measurements of adsorption of ³H-labeled bombykol on the antenna (Kaissling and Priesner, 1970). k) Kaissling 1971. l) Rounded average reaction time of wing vibration of the male moth at low stimulus intensities (0.86s, Kaissling and Priesner, 1970). m) Based on measurements of adsorption of ³H-labeled bombykol on the hairs (Steinbrecht and Kasang, 1972). A similar value was obtained for the Saturniid moth *Antheraea polyphemus* (Kanaujia and Kaissling, 1985). n) Calculated from the number of nerve impulses elicited by 1 s exposure to threshold stimuli of $1.3 * 10^{-5} \mu g$ of bombykol loaded on the odor source (50% behavioral threshold). The impulse number was linearly extrapolated from nerve impulse responses measured at stimulus loads of 10^{-5} (n = 866 measurements) and $10^{-4} \mu g$ (n = 895). The number of spontaneous nerve impulses was subtracted (n = 1070). The data were taken from Kaissling and Priesner (1970). o) Steinbrecht

1970. *p)* Kaissling et al. 1978. *q)* Stuiver 1958. Threshold concentrations C_t were measured for several mercaptanes in 5 persons. The value given was the one of the most sensitive individual. The average threshold concentration for all individuals was $1.3 * 10^8$ molecules/cm^3 for allyl mercaptane and $3.6 * 10^8$ molecules/cm^3 for sec. butyl mercaptane. *r)* Moulton and Marshall 1976; Moulton 1977. Threshold concentration for α-ionone was $4 * 10^5$ molecules/cm^3 for one dog and between this value and $1.3 * 10^7$ molecules/cm^3 for three other dogs tested. Neuhaus 1957 reported a value of $3.2 * 10^4$ molecules/cm^3 for α-ionone. *s)* Assumed that near threshold one nerve impulse is elicited by one stimulus molecules as found for the silkmoth, $N_{eff} = N_{ist}$. *t)* $F = u * \sigma * Q_{ant} * Q_{sen} * Q_{eff}/N_s$ for the antenna and $F = u * \sigma * Q_v * Q_c * Q_{eff}/N_s$ for the nose. *u)* Assumed value of R_{isp}, based on measurements in frog (van Drongelen, 1978b). *v)* The calculated concentrations in the mucus and in the olfactory hairs relate to molecules accumulated during stimuli of duration T. *z)* Calculated from measured values given in Tables 1 and 2.

Thus we obtain $Q_{eff} = N_{eff}/N_{sen} = 0.11$ for the response during the first s after stimulus onset. This means that about one out of 10 bombykol molecules adsorbed on olfactory hairs reaches a bombykol receptor cell and produces a nerve impulse within T, the first s after stimulus onset. In fact, Q_{eff} was 0.22 when all nerve impulses elicited by a one-s stimulus, scattered over 2 s after stimulus onset, were counted. The relationship between stimulus strength and number of nerve impulses N_{istim} deviated from linearity above a rate of 3 nerve impulses per cell and s. Correspondingly, Q_{eff} started decreasing.

The value of Q_{eff} can be estimated only roughly for the vertebrate nose. Assuming the best case - that every molecule adsorbed on the olfactory mucosa reaches a receptor cell - the quotient Q_{eff} may be estimated from the fraction of receptor cells sensitive to the stimulus compound. Assuming that 10% of the whole cell population was sensitive to the test odor and that odorant molecules hitting a non-sensitive cell were lost, we used $Q_{eff} = 0.10$. This estimate is extremely unreliable and is probably an overestimate because the fraction N_s/N_r is expected to be much smaller that in Bombyx (0.5). It is useful, however, for a first-approximation comparison of V_{eff} of noses and antennae.

The effective volume V_{eff} of the silkmoth antennae for bombykol is about 0.17 cm^3. The estimated values of V_{eff} for man (mercaptane) and dog (α-ionone) are within the same order of magnitude (Tab. 1). The latter values are of very little significance because of the uncertainties about the efficiency quotients, especially Q_{eff} for the nose. However, even if allowing for a large error in our determination it is striking that the calculated values of V_{eff} are relatively similar whereas the threshold concentrations of very potent odorants chosen for each species differ over many orders of magnitude (Tab. 2). There are also large differences in the total numbers of receptor cells in the three species compared (Tab. 1). These results suggest that besides the effective volume V_{eff} additional factors might determine the absolute sensitivity to an odorant. According to equation 16 these factors are the number of receptor cells sensitive to the odorant N_s, the spontaneous rate of nerve impulses R_{isp}, and the integration time T needed by the CNS to detect a stimulus or an increase of the rate of nerve impulses.

These factors limit the behavioral threshold of the silkmoth to bombykol. Since the spontaneously fired and the stimulus-induced nerve impulses of the bombykol receptor cells near threshold were shown to be almost randomly distributed (see also Kaissling, 1971) we can apply equation 15. The average rate of spontaneously fired nerve impulses was $R_{isp} = 0.086$ impulses per cell per s or $N_{istim\ min} = R_{isp} * N_s * T = 2924$ impulses for all bombykol receptor cells. The odor-induced signal at the 50 %-threshold of $N_{istim\ t} = 466$ impulses (extrapolated from measurements) is about three times larger than theoretical minimum signal $N_{istim\ min}$ which is $3 * 2924^{1/2} = 162$. This means that the 50 % threshold concentration of bombykol is about three times above the theoretical minimum (Table 2).

A similar calculation can be made for the nose. Although there is no information about the spontaneous impulse activity of olfactory sense cells in dog or human, we assume a spontaneous rate of $R_{isp} = 0.1$ impulses/s as found in the frog (Rana sp., van Drongelen, 1978b). Using this number and the estimated number of sensitive cells N_s we obtain a theoretical maximum of sensitivity S_{max} or a minimum threshold concentration C_{tmin} from equ. 16 (Tab 2). Since the estimated average number of effective stimulus molecules per sensitive cell N_{efft}/N_s is far below 1 it seems obvious that at least in the dog a single odor molecule is sufficient to elicit a nerve impulse (Neuhaus, 1953; Moulton, 1977).

For the dog we obtain a minimum threshold concentration in air of $C_{tmin} = 5 * 10^4$ molecules of α-ionone per cm^3 (Moulton and Marshall, 1976) only 8 times below the measured threshold value. Threshold concentrations as much as five times lower were reported for butyric acid in the dog ($9 * 10^3$ molecules/cm^3, Neuhaus, 1953; not reached by the dogs of Moulton et al., 1960). This would mean that either the spontaneous activity was (about 25-fold) smaller than assumed or the capture coefficient F was (about 5 times) higher for butyric acid than for α-ionone. Possibly, several of the quantities in equ. 23 have to be adjusted correspondingly. Apparently the dog, like the silkworm moth operates near the theoretical minimum of detection. In spite of many uncertainties it seems that the parameters determining sensitivity are resonably well estimated for the dog.

One uncertainty concerns the unknown actual number of sniffs the dog was allowed to take in the behavioral test. Of course, evaluation of stored nervous messages obtained from a larger number of sniffs could improve the signal-to-noise ratio and increase the sensitivity according to equ. 21 or 23 because of increased T.

Table 2: Values were obtained as explained in the legend of Table 1

			man		dog		silkmoth	
			sec. butyl mercaptane		α-ionone		bombykol	
threshold concentration in air,	$C_t=N_t/V$	(molecules/cm^3)	$1.3*10^7$	q	$4*10^5$	r	$2.7*10^3$	i,k
no. of effective molecules at threshold	$N_{efft}=C_t*V_{eff}$	(molecules)	$(2*10^6)$	d,s	$(2.4*10^4)$	d,s	$4.7*10^2$	n
av. no. of eff. molec. per sensitive cell	N_{efft}/N_s	(molec./cell)	(3.3)	d,h	$(2.4*10^{-4})$	d,h	$1.4*10^{-2}$	z
capture coefficient	$F=V_{eff}/N_s*T$	(cm^3/cell*s)	$(1.7*10^{-6})$	d,h,t	$(6*10^{-9})$	d,h,t	$5*10^{-6}$	t
rate of spontaneous nerve impulses	R_{isp}	(imp./s*cell)	(0.1)	u	(0.1)	u	0.086	i
min. thresh. conc.	$C_{tmin}=3*(R_{isp}/N_s*T)^{1/2}/F$	(molec./cm^3)	$(1.9*10^3)$	d,h,u,s	$(5*10^4)$	d,h,u,s	$9.5*10^2$	z
observed thresh. conc. / min. thresh. conc.	C_t/C_{tmin}		$(7*10^3)$	d,h,u,s	(8)	d,h,u,s	2.9	z

values in brackets are based on assumptions (d, h, u, s)

A very different result from that in the dog is obtained for the human response to mercaptane with a threshold concentration of $1.3 * 10^7$ molecules/cm^3 (Stuiver, 1958, see Table 2). With the assumed spontaneous activity of $R_{isp} = 0.1$ nerve impulse per cell per s and with a sniff duration of $T = 0.15$ s we obtain a theoretical minimum threshold value C_{tmin} of only $1.9 * 10^3$ molecules/cm^3 which is about $7*10^3$-fold below the observed threshold concentration (Table 2). In order to explain this discrepancy, it would be unreasonable to assume that the spontaneous activity of the cells is much higher than the assumed figure. Even a 100 times higher R_{isp} would decrease the sensitivity by only a factor of 10. Therefore, the low sensitivity as compared with the theoretical minimum should be due to a smaller V_{eff}. This could mean a smaller Q_c than assumed if mercaptane is less well adsorbed by the mucosa, or a smaller Q_{eff} if fewer of the adsorbed molecules excite mercaptane-sensitive cells.

The calculations show that the human nose could, in principle, perceive odor compounds at more than 1000-fold lower concentrations than those for mercaptane. This is expected from equ. 16 if we assume equal Q_{eff} ($= 0.1$), equal R_{isp} ($= 0.1$ impulses/s * cell) and an equal percentage of sensitive cells N_s/N_r as in the dog. In fact, the human thresholds to several groups of compounds (musks, phenols, vanillin) or for 1-p-menthene-8-thiol (Demole et al., 1982) are claimed to be below the one for mercaptane. For compilations of human olfactory thresholds see Patte et al. (1975) and Devos et al. (1989). The threshold concentration of the human nose for α-ionone is $4 * 10^8$ molecules/cm^3 (Zwaardemaker, 1914).

Strikingly, the theoretical minimum threshold concentration for bombykol in the moth is even lower than the minimum threshold for the dog, in full accordance with equ. 16. The comparison between dog and moth shows clearly the inverse relationship between maximum sensitivity and number of receptor cells.

Table 3: Values were obtained as explained in the legend of Table 1

			man		dog		silkmoth	
			sec. butyl mercaptane		α-ionone		bombykol	
sensory area, nose (=regio olfactoria)	A_{sen}	(cm^2)	2	f	170	g	-	
" " antenna (=olfactory hairs)		(hairs)	-		-		$3.4*10^4$	o
volume of 1. μm of mucus layer	$V_{muc}=A_{sen}*1\mu m$	(l)	$2*10^{-7}$	z	$1.7*10^{-5}$	z	-	
" of one hair	V_{hair}	(l)	-		-		$1*10^{-12}$	q
threshold concentration in air	C_t	(moles/l)	$2.2*10^{-14}$	q	$6.6*10^{-16}$	r	$4.5*10^{-18}$	i,k
no. of molecules adsorbed on A_{sen}	N_{sent}	(molecules)	$2*10^7$	z	$2.4*10^5$	z	-	
" " " " on 34,000 hairs	N_{sent}	(molecules)	-		-		$4.2*10^3$	z
" " " " on one hair (average)		(molecules)	-		-		0.12	z
stim. conc. in 1. μm of mucus layer	$C_{muc}=N_{sent}/V_{muc}$	(moles/l)	$1.7*10^{-10}$	z,v	$2.3*10^{-14}$	z,v	-	
" " in 34,000 hairs	$C_{hairs}=N_{sent}/V_{hairs}$	(moles/l)	-		-		$(2.1*10^{-14})$	z,v
" " in one hair at 1 molecule/hair		(moles/l)	-		-		$(1.7*10^{-12})$	z
ratio of stimulus conc. mucus/air	$C_{muc}/C_t=V_{sen}/V_{muc}$		$7.5*10^3$	z,v	35	z,v	-	
" " " " hair/air	$C_{hairs}/C_t=V_{sen}/V_{hairs}$		-		-		$4.7*10^3$	z,v

The silkmoth has only $3.4 * 10^4$ receptor cells sensitive to bombykol (on both antennae) whereas the total number of receptor cells sensitive to α-ionone in the dog, assuming $N_s/N_r = 0.1$, is more than 10,000 fold higher. This would lead to 100-fold higher sensitivity of the dog according to equ. 21 if F were the same in the dog and the moth. However, the capture coefficient F of a receptor cell in the silkmoth antenna is about 1000-fold higher than that estimated for the dogs nose, which more than compensates for the smaller number of receptor cells of the antenna (Table 3). Clearly there are advantages in having more receptor cells. The moth perceives only two stimulus compounds, the pheromone components bombykol and bombykal, with extreme sensitivity. The dog, with its much larger number of receptor cells, probably has many more types of receptors and perceives many compounds with high sensitivity.

Comparing the noses of dog and man we calculate relatively similar values of V_{eff} but much smaller values N_s in man if we assume same N_s/N_r for both species (Tab.1). Therefore, according to equ. 16, the theoretical minimum threshold concentration C_{tmin} could be lower in humans than in the dog. The fact that such low threshold concentrations have not been found for humans might mean that our nose sacrificed low thresholds in favor of better quality discrimination. It is conceivable that N_s/N_r for a given odor compound is lower in humans than in the dog. In other words, humans might have more types of receptor cells sensitive to different odorants.

STIMULUS TRANSPORT AND TRANSDUCTION

The detector function of an olfactory organ can be divided into three major steps: the adsorption of molecules on the sensory area, the transport of the adsorbed molecules towards the receptor cells and the actual transduction processes. The efficiency of the latter two steps is expressed by Q_{eff} and deserves a more detailed discussion.

It is assumed that the adsorbed molecules reach the receptor cell by passive transport (diffusion) within the mucus layer of the nose (Getchell et al., 1984) and along the cuticular wall and pore tubules of the olfactory hair of the antenna (Fig. 1). It is also assumed that the odor molecules interact with receptor molecules at the cell membrane. Therefore, we first try to estimate the odorant concentration in the vicinity of the hypothetical receptor molecules at the cell membrane.

For the nose we assume that the odor molecules adsorbed at the sensory area A_{sen}, the regio olfactoria, distribute themselves within the mucus layer, which is about 10 μm thick in humans and dogs (Menco, 1983; Moran et al., 1982). The mucus layer contains a dense packing of ciliary processes of receptor cells. We assume that the adsorbed molecules N_{sen} reach the receptor cell membrane predominantly within the outer 1 μm of the mucus layer and obtain $V_{muc} = A_{sen} * 1 \mu m$ (Table 3). Consequently, the odorant concentration C_{muc} accumulated during T in the mucus is N_{sen}/V_{muc}, which is of the order of 10^{-10}M for mercaptane in man and of 10^{-14}M for α-ionone in the dog.

The calculation of $C_{hair} = N_{sen}/V_{hair}$ for an olfactory hair of the antenna encounters two difficulties. First, we expect that the odorant molecules reach the receptor cell membrane directly via pore tubules contacting the dendrite. According to one current model the odorant molecules move into the sensillum lymph surrounding the receptor cell only after their interaction with the receptor molecules (Kaissling, 1986, 1987).

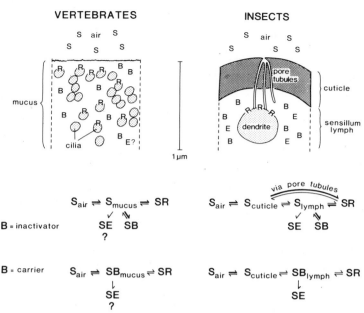

| | VERTEBRATES | INSECTS |

$$S_{air} \rightleftharpoons S_{mucus} \rightleftharpoons SR \qquad S_{air} \rightleftharpoons S_{cuticle} \rightleftharpoons S_{lymph} \rightleftharpoons SR$$

B = inactivator

$$SE \quad SB \qquad\qquad SE \quad SB$$
$$? $$

B = carrier

$$S_{air} \rightleftharpoons SB_{mucus} \rightleftharpoons SR \qquad S_{air} \rightleftharpoons S_{cuticle} \rightleftharpoons SB_{lymph} \rightleftharpoons SR$$
$$SE \qquad\qquad\qquad SE$$
$$?$$

Fig. 1: Schematic view of the first μm of sensory odorant adsorbing structures adjacent to the air space. Left: Outer layer of olfactory mucosa of the vertebrate nose with cross-sections of terminal portions of olfactory cilia (after Reese, 1965; see also Andres, 1969). Right: Portion of a cross-section of an olfactory hair of the moth antenna (after Keil, 1982), Cuticular hair wall (dark) with pore tubules in contact with cross-sectioned dendrite of olfactory receptor cell. S-stimulus molecule, R-receptor molecule at the olfactory cell membrane, B-odorant (or pheromone) binding protein, E-odorant (or pheromone) degrading enzyme. Below: Proposed reaction schemata for alternative functions of the odorant binding proteins as inactivator or carrier. Inactivator model: In vertebrates some of the stimulus molecules dissolved in the mucus will become bound to the protein before they interact with the receptors and a certain loss of stimulus molecules has to be tolerated. In the insect olfactory hair the odor molecule may reach the dendrite membrane with the receptor molecules directly via the pore tubules. In this way the stimulus molecules might be protected from binding to the protein before they interact with the receptors (Vogt and Riddiford, 1981; Kaissling, 1986, 1987). This pathway would be important only at low stimulus strength. At high stimulus strengths only a very small fraction of molecules is required for excitation. Carrier model: The odor molecules are transported towards the receptor cell membrane while bound to the binding protein (Vogt et al., 1985). For the insects this model is not consistent with experimental data (discussed in Kaissling, 1986, 1987).

So far it is impossible to estimate the volume near the receptor sites in which the odorant molecules are dissolved before they interact with the receptors. This volume may be much smaller than the volume of the hair V_{hair} which we use as a maximum estimate of the volume of the space near the receptor molecules. The second difficulty is that the concept of concentration is meaningless if just one odorant molecule is adsorbed on a hair as is the case at threshold (see above). The calculation of C_{hair} for the olfactory hair is, therefore, significant only for the purpose of rough comparison of efficiencies which can be expressed independent of concentration.

If we relate the concentration of adsorbed odorant molecules to the stimulus concentration in air we find

$$C_{muc}/C = V_{sen}/V_{muc} \qquad (24)$$

for the nose and

$$C_{hair}/C = V_{sen}/V_{hair} \qquad (25)$$

for the antenna. These ratios are similar for man and silkmoth. The value for the silkmoth is, however, a minimum value and might be considerably higher since the stimulatory molecules are distributed in a volume which is probably much smaller than V_{hair}. The ratio is relatively small in the dog compared with human because of the larger sensory area in the dog.

These considerations show that adsorption can lead to a considerable increase of stimulus concentration between air and the immediate surrounding of receptor cells. This increase can be more than 1000-fold within stimulus duration T (Table 3). The actual stimulus concentrations at the receptor sites can be as low as 10^{-14}M and are of interest in relation to the dissociation constant of the complexes between odorant and hypothetical receptor molecules, binding proteins and odorant degrading enzymes.

In discussing the efficiency of stimulus transport one has to consider possible losses of odorant molecules on their way to the sensitive receptor cells. An important factor might be binding of the odorant molecule to extracellular soluble proteins. Odorant-binding proteins have been found in the sensillum lymph of olfactory hairs of moths (Vogt and Riddiford, 1981; Vogt et al., 1985; Kaissling et al., 1985; de Kramer and Hemberger, 1987; Györgyi et al., 1988) and also in the olfactory mucus of the vertebrate nose (for reviews see Anholt, 1987; Snyder et al., 1989). Binding to these proteins might serve a carrier function but is more likely to inactivate odorants, i.e. to reduce their active concentration at the receptor sites immediately after stimulation, as proposed in a quantitative model for moths (Kaissling, 1986, 1987). In this case the stimulus molecule must be protected against binding before it interacts with the receptor, at least at low stimulus intensities. In the olfactory hair the molecules might reach the cell membrane via the pore tubules, which may also have a protective function. No protecting structures are known in the mucus layer of the vertebrate nose; here a certain loss of odorant molecules may have to be tolerated. The different situations in antenna and nose are schematized in Fig. 1.

Another possible loss causing a decrease of Q_{eff} may occur when an odorant molecule hits an insensitive receptor cell. Due to lipophilic properties of odorants they might be irreversibly taken up by any receptor cells, including the wrong cell. It would, of course, be very favorable if stimulus molecules were "reflected" by the wrong cells. Further loss of odorant molecules can be obtained by encymatic degradation (Kasang, 1971; Kasang et al., 1988; Gower and Hancock, 1982; Hancock and Gower, 1986). This process, however, occurs within minutes and, therefore, does not produce noticeable losses of odorant molecules during stimulation times of seconds or below.

The experimentally determined value of $Q_{eff} = 0.11$ for the moth antenna was obtained by counting for one-s the nerve impulses produced by a one s stimulus. However, about half of the nerve impulses occur after the first s of response (at low stimulus intensities). Therefore, Q_{eff} would double if one were to count all nerve impulses elicited. A Q_{eff} of 0.22 means that the molecules in the olfactory hair are at least partially, lost by other reasons than uptake by the second, insensitive receptor cell in the olfactory hair. The loss of molecules is still astonishingly low.

Finally, it is to be expected that Q_{eff} decreases drastically with increasing stimulus concentration. Nerve impulse firing of olfactory cells hardly exceeds impulse rates of 100/s whereas the odor intensity may increase over many decades. Such decrease of Q_{eff} may indicate a higher percentage of molecules caught by binding proteins before reaching the receptor sites. In addition the "gain" of the receptor cell itself may change by processes causing desensitization.

REFERENCES

Andres, K. H., 1969, Der olfaktorische Saum der Katze, *Z. Zellforsch.*, 96:250-274.

Anholt, R. H., 1987, Primary events in olfactory reception, *TIBS*, 12:58-62.

Demole, E., Enggist, P., Ohloff, G., 1982, 176. 1-p-Menthene-8-thiol: A powerful flavor impact constituent of grapefruit juice (*Citrus paradisi* MacFayden), *Helvet. Chim. Acta*, 65:1785-1794.

Devos, M., van Gemert, L. J., Patte F., Rouault, J. and Laffort, P, Standardized human olfactory thresholds in air, submitted to IRL-Press for publication (1988).

De Kramer, J. J. and Hemberger, J, 1987, Neurobiology of pheromone reception, *in*: "Pheromone Biochemistry," G.D. Prestwich, G.J. Blomquist, eds., Clarendon Press, Oxford, 433-472.

De Vries Hl. and Stuiver, M., 1960, The Absolute Sensitivity of the Human Sense of Smell, *in*: "Sensory Communication," W.A. Rosenblith, ed., Wiley & Sons, New York, 159-167.

Getchell, T. V., Margolis, F. L. and Getchell, M. L., 1984, Perireceptor and receptor events in vertebrate olfaction, *Progr. in Neurobiol.*, 23:317-345.

Gower, D. B. and Hancock, M. R., 1982, Investigations into the metabolism and binding of the odorous steroid, 5α-androst-16-en-3-one, in porcine nasal tissues, *in*: "Olfaction and Endocrine Regulation", W. Breipohl, ed., IRL Press Ltd., London, 267-277.

Györgyi, T. K., Roby-Shemkovitz, A. J. and Lerner, M. R., 1988, Characterization and cDNA cloning of the pheromone-binding protein from the tobacco hornworm, Manduca sexta: A tissues-specific developmentally regulated protein, *Proc. Natl. Acad. Sci. USA*, 85:9851-9855.

Hancock, M. R. and Gower, D. B., 1986, Properties of 3 α-and 3b-hydroxysteroid dehydrogenases of porcine nasal epithelium: some effects of steroid hormones, *Biochem. Soc. Transact.*, 14:1033-1034.

Kaissling, K.-E., 1971, Insect olfaction, *in*: "Handbook of Sensory Physiology", Vol.IV, 1, L.M. Beidler, ed., Springer Verlag, Berlin, 351-431.

Kaissling, K.-E., 1987, R. H., "Wright Lectures on Insect Olfaction", K. Colbow , ed., Simon Fraser University, Burnaby B.C., Canada.

Kaissling, K.-E., Kasang, G., Bestmann, H. J., Stransky, W. and Vostrowsky, O., 1978, A new pheromone of the silkworm moth Bombyx mori, Sensory pathway and behavioral effect, *Naturwissenschaften*, 65:382-384.

Kaissling, K.-E., Klein, U., de Kramer, J. J., Keil, T. A., Kanaujia, S. and Hemberger, J., 1985, Insect olfactory cells: Electrophysiological and biochemical studies, *in*: "Molecular Basis of Nerve Activity," Proc. Int. Symp. in Memory of D. Nachmansohn; J.-P. Changeux et al., eds., de Gruyter, Berlin/New York, 173-183.

Kaissling, K.-E. and Priesner, E., 1970, Die Riechschwelle des Seidenspinners, *Naturwissenschaften*, 57:23-28.

Kanaujia, S. and Kaissling, K.-E., 1985, Interactions of pheromone with moth antennae: adsorption, desorption and transport, *J. Insect Physiol.*, 31:71-81.

Kasang, G., 1971, Bombykol reception and metabolism on the antennae of the silkmoth Bombyx mori, in: "Gustation and Olfaction," G. Ohloff and A. F. Thomas, eds., Academic Press, London/New York, 245-250.

Kasang, G., von Proff, L. and Nicholls, M., 1988, Enzymatic conversion and degradation of sex pheromones in antennae of the male silkworm moth Antheraea polyphemus, Z. Naturforsch., 43c:275-284.

Keil, T., 1982, Contacts of pore tubules and sensory dendrites in antennal chemosensilla of a silkmoth: demonstration of a possible pathway for olfactory molecules, Tissue & Cell, 14:451-462.

Margolis, F. L. and Getchell, T. V., 1988, "Molecular Neurobiology of the Olfactory System", Plenum Press, New York and London.

Menco, B. P., 1980, Qualitative and quantitative freeze-fracture studies on olfactory and nasal respiratory structures of frog, ox, rat, and dog, Cell Tissue Res., 207:183-209.

Menco, B. P., 1983, The ultrastructure of olfactory and nasal respiratory epithelium surfaces, in: "Nasal Tumors in Animals and Man", Vol.I. Anatomy, Physiology, and Epidemiology, G. Reznik, S. F. Stinson, eds., CRC Press, Inc., Boca Raton, Fl., 45-102.

Meng, L. Z., Wu, C. H., Wicklein, M., Kaissling, K.-E. and Bestmann, H. J., 1989, Number and sensitivity of three types of pheromone receptor cells in Antheraea pernyi and A. polyphemus, J. comp. Physiol., 165:139-146.

Moran, D. T., Carter Rowley, III., J., Jafek, B. W. and Lovell, M. A., 1982, The fine structure of the olfactory mucosa in man, J. Neurocytol., 11:721-746.

Moulton, D. G., 1977, Minimum odorant concentrations detectable by the dog and their implications for olfactory receptor sensitivity, in: "Chemical Signals in Vertebrates," D. Müller-Schwarze, M. M. Mozell, eds., Plenum Press, New York, 455-464.

Moulton, D. G., Ashton, E. H. and Eayrs, J. T., 1960, Studies in olfactory acuity, 4. Relative detactability of n-aliphatic acids by the dog, Anim. Behav., 8:117-128.

Moulton, D. G. and Marshall, D. A., 1976, the performance of dogs in detecting α-ionone in the vapor phase, J. comp. Physiol., 110:287-306.

Mozell, M. M., Sheehe, P. R., Hornung, D. E., Kent, P. F., Youngentob, S. L. and Murphy, S. J., 1987, "Imposed" and "inherent" mucosal activity patterns, J. Gen. Physiol., 90:625-650.

Mozell, M. M., Sheehe, P. R., Swieck, S. W. jun., Kurtz, D. B. and Hornung, D. E., 1984, A parametric study of the stimulation variables affecting the magnitude of the olfactory nerve response, J. Gen. Physiol., 83:233-267.

Neuhaus, W., 1953, Über die Riechschärfe des Hundes für Fettsäuren, Z. vergl. Physiol., 35:527-552.

Neuhaus, W., 1957, Über das Verhältnis der Riechschärfe zur Zahl der Riechrezeptoren, Verh. Dtsch. Zool. Ges., 285-292.

Neuhaus, W., 1957, Wahrnehmungsschwelle und Erkennungsschwelle beim Riechen des Hundes im Vergleich zu den Riechwahrnehmungen des Menschen, W. vergl. Physiol., 39:624-633.

Patte, F., Etcheto, M. and Laffort, P., 1975, Selected and standardized values of suprathreshold odor intensities for 110 substances, in: " Chemical Senses and Flavor", E.P. Köster, and H. R. Moskozitw, eds., D. Reidel Publ. Co., Dordrecht-Holland/Boston-USA, 283-305.

Reese, T. S., 1965, Olfactory cilia in the frog, J. Cell. Biol., 25:209-230.

Rumbo, E. R. and Kaissling, K.-E., 1989, Temporal resolution of odor pulses by three types of pheromone receptor cells in Antheraea polyphemus, J. Comp. Physiol., 165:281-291.

Snyder, S. H., Sklar, P. B. Hwang, P. M. and Pevsner, J., 1989, Molecular mechanisms of olfaction, *TINS*, 12:35-38.

Steinbrecht, R. A., 1970, Zur Morphometrie der Antenne des Seidenspinners, Bombyx mori L.: Zahl und Verteilung der Riechsensillen (Insecta, Lepidoptera), *Z. Morph. Tiere*, 68:93-126.

Steinbrecht, R. A. and Kasang, G., 1972, Capture and conveyance of odour molecules in an insect olfactory receptor, *in*: "Olfaction and Taste IV," D. Schneider, ed., Wiss. Verlagsges., Stuttgart, 193-199.

Stuiver, M., 1958, "Biophysics of the Sense of Smell", Thesis, Rijksuniversiteit, Groningen.

Van Drongelen, W., Holley, A., Doving, K. B., 1978a, Convergence in the olfactory system: Quantitative aspects of odour sensitivity, *J. Theor. Biol.*, 71:39-48.

Van Drongelen, W., 1978b, Unitary recordings of near threshold responses of receptor cells in the olfactory mucosa of the frog, *J. Physiol.*, 277:423-435.

Vogt, R. G., Prestwich, G. D. and Riddiford, L. M., 1988, Sex pheromone receptor proteins, *J. Biological Chem.*, 263:3952-3959.

Vogt, R. G. and Riddiford, L. M., 1981, Pheromone binding and inactivation by moth antennae, *Nature*, (Lond.), 293:161-163.

Vogt, R. G., Riddiford, L. M. and Prestwich, G. D., 1985, Kinetic properties of a pheromone degrading enzyme: the sensillar esterase of Antheraea polyphemus, *Proc. Nat. Acad. Sci. USA*, 82:8827-8831.

Zwaardemaker, H., 1914, Geruch und Geschmack, *in*: "Handbuch der physiologischen Methodik III, 1", R. Tigerstedt, ed., S. Hirzel, Leipzig, 46-108.

MOLECULAR AND CELLULAR RECEPTORS IN OLFACTION:

A NOVEL MODEL UNDER INVESTIGATION

E. Bignetti, S. Grolli and R. Ramoni

Istituto di Biologia Molecolare
Universita' di Parma, Italy

ABSTRACT

Vertebrate olfactory neurons, the receptors cells in the nose that sense odorants, are continously dying and replaced from differentiating staminal cells located at the basal lamina of the epithelium. Eventhough the neuronal turnover in the nose represents an exceptional property in the nervous system, it is not yet understood. In analogy with the immune system behaviour, olfaction shows high plasticity and sensitivity in odorant reception and coding. On the basis of this analogy we postulate that the turnover of olfactory neurons and their spectrum of sensitivity is dependent on environmental odorants. To test this hypothesis, we follow staminal cells proliferation and differentiation in controlled conditions, i.e. *in vitro*, by means of a monoclonal antibody against an antigen of the staminal cells.

INTRODUCTION

Why is purifying an odorant receptor protein so elusive?

In considering the chemical sense of olfaction at the molecular level, we should like to refer to several stimulating reviews on this and related subjects (Koshland, 1983; Lancet et al., 1987). It appears that even in the most accessible system, i.e. the olfactory system in vertebrates, the questions about the existence of specific receptors molecules and about their chemical nature are still unsolved. On the basis of speculation, one can postulate the existence of specific receptor protein on the olfactory cilia. All such speculations, however, have been criticized to some extent. There is evidence, for example, that some central and peripheral neuronal structures, other than olfactory, can be actively stimulated by direct application of different smelling compounds. Odorants (mostly lipophilic) have also been postulated to interact directly with the natural lipids of the olfactory membranes, altering their fluidity and temporarily producing small pores for ion channelling or affecting the conductance state of ion-channels. Furthermore, not only volatile smelling compounds but also soluble molecules, natural or artificial, can elicit an olfactory response whenever directly applied to the olfactory sheet.

Sensory Transduction
Edited by A. Borsellino *et al.*
Plenum Press, New York, 1990

The lack of any direct and inequivocable experimental evidence on this subjects has induced research workers who believe in the existence of receptors proteins, to speculate about the reasons why such receptors are so elusive. Known odorants and odorants-like molecules are approximately 10^3-10^4 subdivided in several classes such as fruity, floreal, minty, etc. Therefore, two boundary hipotheses have been made (Lancet, 1986): either there are as many molecular receptors as odors or there are few receptors as classes of odors. In the first case receptors should exhibit narrow specificity and higher affinity, while, in the second case, receptors should exhibit a broad specificity and lower affinity. A loose odorant binding would results in a difficult identification and purification of the binary odorant-receptor complex.

The second hypothesis is compatible with Amoore's theory, i.e., there must be a stereochemical common denominator whitin each class of odorants through which olfactory recognition operates (Amoore, 1971).

Cilia on olfactory neurons show high densities of membrane- bound proteins. Since some of these proteins are tissue-specific they have been considered putative receptors (Lancet, 1986). A trasductory pathway in cilia has also been studied. An odorant-sensitive adenylate cyclase has been found and characterized (Pace et al., 1987; Sklar et al., 1986; Nakamura et al., 1987).

The immunological model

The olfactory system possesses extraordinary sensytivity and plasticity and has been intuitively compared to the immune system by many authors (see Reviews quoted above). We would like to point out some old and new features.

Odorants consist of so many different foreign molecules that they might behave, in the olfactory system, as antigenic structures do in the immune system. As a consequence, an olfactory receptor might work as a bifunctional protein composed of two parts: a variable regions, clonally selected for binding a wide spectrum of odorants, and a constant region, anchored to membrane and capable of propagating the chemical signal to the internal transduction machinery (Lancet, 1986).

In the past few years, research on antibody specificity has elucidated a relevant biological effect, namely the "Multispecificity Paradox". Every antibody, in fact, may carry a polifunctional binding site where antigens, other than the priming antigen, can bind in a competitive fashion. This phenomenon is amplified in a apparently omogeneous population of antibodies, so that a broad specificity for many different antigens has been observed. An even more wide spectrum of activities could appear in the presence of a mixed population of different antibodies. These effects suggest a model for understanding "specificity" and "diversity" of olfactory receptors. This model, furthermore, has the advantage that the number of individual receptors might be greatly reduced to a few classes. Moreover, groups of receptor classes might interact with a common transductory machinery, thus favouring a funnel-like mechanism in signal processing.

In addition, the olfactory system has been referred to as a "tabula rasa" (Engen, 1986) which can be shaped by odorants, so that animals' socio-sexual choices are heavily dependent on them (see Doty, 1986 for a review). Furthermore the system, even in insects, is characterized by high plasticity. Hildebrand (Hildebrand, 1987) shows that by grafting male-antennae in female moth fly, male-like behaviour can be induced.

The immunological model offers a plausible explanation for such flexibility and adaptability, at least concerning peripheral signal processing. Just as "clonal exclusion" in lymphocyte differentiation and specification operates to select antibodies adequate to external aggression, the sensitivity spectrum of olfactory system might be modulated by odorant-pressure. In other words, with time, novel odorants might stimulate the synthesis of more adequate molecular receptors thus specifying neurons. A sort of tuning mechanism of neurons towards environmental odorants implies that a mechanism of receptors replacement and up-dating must be available throughout the life-span of nose. According to this hypothesis, the turnover of olfactory neurons (Graziadei et al., 1979) might have a teleological meaning if the kinetics of staminal cells growth and differentiation into novel olfactory neurons were dependent on the "usage" of the nose. On this bases, one could explain why the regeneration of olfactory neurons is unique in the nervous system. In fact in others sensory organs where the molecular receptors have a fixed activity spectrum throughout life, the replacement of the entire sensitive cells might be energetically unfavorable.

Now, the questions obviously arising are where, how and when such tuning mechanism of receptor molecules could be inserted in the development of olfactory neurons from staminal cells. Again from the immune system, we might have some indications. We might assume that odorants (small molecules of m.w.<300) behave like haptens. As a consequence, the presence of an odorant-carrier might be required in order to present odorants to staminal cells. The interaction might induce their proliferation and differentiation into mature olfactory neurons with a more specific range of odorant sensitivity. In our opinion, the unique experimental approach to the model proposed is through cell culturing in-vitro.

RESULTS

A soluble odorant-binding protein, OBP, which is produced by sero-mucous glands of the vertebrate nose and is possibly secreted into mucus, has been identified and characterized (see Bignetti et al., 1988 for a Review and Bignetti et al., 1989, this volume). OBP is still the unique example of a well-purified and characterized nasal protein with a measurable binding competence for many odorants of various primary classes.

Monoclonal antibodies against a preparation of bovine OBP have been raised in mice in the laboratory of Professor P. Graziadei, Tallahassee, Fla. Some of the collected ascite fluids have been selected for further purification on the basis of their ability to label different components of the bovine nasal mucosa. One of them (Mab C8.235) labels the staminal cells which lie on the basal lamina of the nasal ephitelium (manuscript in preparation). This finding suggests that the Odorant-Binding Protein, or an immunologically related protein, is present in neurons before differentiation into mature receptor cells. However, the specificity of this clone is still under investigation.

The monoclonal antibody against staminal cells has been used as a tool to follow the isolation, growth and differentiation of staminal cells in culture. Therefore we have developed a physico- chemical method for staminal cells isolation from the bovine tissue (the details are described in Fig. 1). We found that most of the cells, once isolated by our method, were specifically labelled by the monoclonal C8.235 and were viable (see Fig. 2 and legend for methodological details).

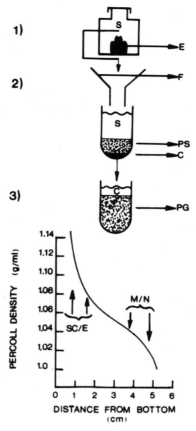

Fig. 1: 1) One ethmoturbinate (E) was shaked in 100 ml of a solution containing 20 mM TRIS/HCl buffer pH 7.8, 0.9% NaCl and 10 mM CaCl$_2$, for 20'. The supernatant was renewed once and further shaking was held for 30'. 2) This 2nd cell suspension (S) was filtered through a nylon gauge (F). Twentyfive ml of it were layered on top of a Percoll (Pharmacia) step (d= 1.06 g/ml) (PS), adjusted with MEM cell culture medium (Sigma). After 10' of a centrifugation at 500xg in a swinging rotor, the upper 25 ml of the cell-suspension were sucked and discarded. Other 25 ml of S were layered, repeating step 2), until completion of the sample. 3) The pellet (C) was then resuspended in 1 ml of MEM medium and layered on top of a pre-formed 10 ml of Percoll gradient (PG) (starting density 1.06 g/ml, centrifuged in a 24 fixed angle rotor for 30' at 40,000xg). The cell suspension on top of the gradient was then centrifuged in a swinging rotor for 15' at 500xg. Mucus, neurons and membranes (M/N) sedimented between 1.02 and 1.04 g/ml; staminal cells and erythrocytes (SC/E) between 1.08 and 1.12 g/ml.

Until now, little has been published on cultures of mammalian olfactory precursor cells. The only remarkable example has been reported by Schubert et al. (1985). These authors detached enzimatically epithelial cells from rat nasal ethmoturbinates and plated all the cells in Petri dishes. They observed that few of them could survive and differentiate into two morphologically and functionally distinct cell types: flat glial lymphoid-like cells (tightly attached) and round neuron-like cells (loosely attached). According to their evidences, flat cells grew first, while round cells appeared five days later from the flat

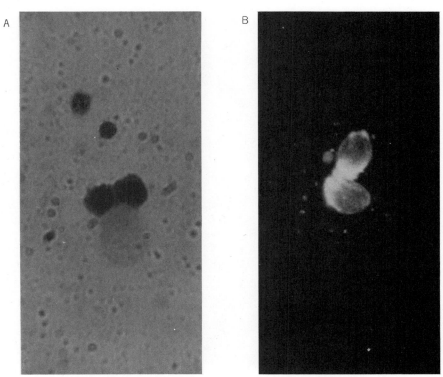

Fig. 2: A) Light microscopy of two bovine nasal staminal prepared as described in Fig. 1 cells stained with haematoxylin-eosin (400x initial magnification). B) Indirect immunofluorescence-microscopy of two bovine nasal staminal cells (500x initial magnification). The cells were isolated as described in Fig. 1. Slides were dried under a fan and incubated with monoclonal antibody-containing ascite fluid $C_{8.235}$, 100-fold diluted in PBS, 0.5 mg/ml BSA and 0.1% NaN3 for 1 hr at RT in a moist chamber. The samples were washed three times, 10' each in PBS, and incubated with a 50-fold diluted anti-mouse FITC-conjugated specific secondary antibody (Sigma), for 1 hr at RT. A final wash was repeated three times. Cells were covered with 1:1 solution of PBS and glycerol.

cells. The round cells showed neither glial nor lymphoid characteristics but rather electrical excitability. After several weeks, the round cells grew short neurite-like bipolar processes. Schubert et al. assumed that glial lymphoid like cells derived from staminal cells and concluded that the neuron-like round cells differentiated from them.

By phase-microscopy inspection, we observed that our preparation of bovine nasal staminal cells (as judged by immunoreaction with Mab C8.235) could differentiate in culture into small flat cells, at first. At 3 days, they gave origin to clumps of round bright-phase cells that started migrating out and proliferating into flat (attached) and round (loosely attached) cells. At 7 days, when flat cells have formed large semi-confluent and orderly aligned monolayers, isolated phase-bright round cells grew out. They were anchored to flat cells simply by means of thin cytoplasmatic projections (see Fig. 3A). By that time, such round cells were collected by gently pipetting and replated. They proliferated and differentiated again into both lineages of flat and round cells. After several weeks, when flat cells were confluent, many round cells grew over bipolar neurite-

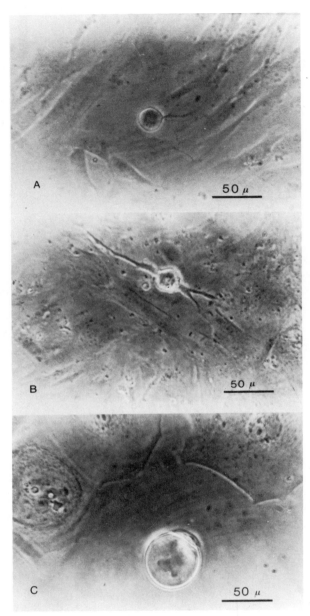

Fig. 3: Phase-microscopy of a preparation of bovine staminal cells differentiated in culture; day 7. A) One small phase-bright round cell is attached to a flat cell with a thin projection. B) After several weeks the type of cells shown in A) grew bipolar processes and tightly attached to the substratum. C) A different type of round non-attached cells appeared early in culture. Compare its gigantic size to cells in A) and B) and note its transparent body and a peculiar cap aside. These cells probably reproduced and partly attached to the dish giving rise to patches of very large polygonal flat cells. Three of them are present in this picture. These cells avoided interactions with cells in A) and B). Conditions: glucose/glutamine-containing DME medium (GIBCO) enriched with 10% cadet calf serum (GIBCO), 5% CO_2, 37^0C. The medium was renewed every week.

like processes (see Fig. 3B). After three months, the plates were still generating such cells. None of the differentiated cells, could be stained anymore by our monoclonal antibody. The existence of two phenotypically distinct but parentally cross-linked cells in our cultures apparently confirmed the data of Schubert et al. The differences seemed to be the reversible pattern of differentiation from round to flat cells and vice-versa, observed by us and excluded by Schubert et al. and the smaller dimensions of our cells. This, however needs a confirmation through the experiments of subcloning we are now carrying with the aid of a micromanipulator.

In addition to what described, we also observed a third phenotipically distinct lineage of extremely large cells (see Fig. 3C). Such newly observed cells, in suspension, have a shape that resembled that of sustentacular cells, as described by Hirsh and Margolis (Hirsch et al., 1981). Once attached, they formed small patches of very large polygonal flat cells. The arrangements in patches of this fourth type of cells resembled a typical cross-section of a Bowman's gland. Many independent patches lately went to confluency without interwining with the other cells previously described. We are now investigating the possibility that the large cells come directly from staminal cell differentiation, or from a completely different population of cells that contaminated our preparation. Preliminary experiments show their positive reactivity to our Mab and to PAS staining. Due to their shape and dimension, these cells are much closer to those described by Schubert et al.

Concluding, we have now four stable cells types in culture that could be representative of all cell members more directly involved in the ontogenesis of olfactory epithelium.

Schubert et al. observed that the glial-type flat cells in culture expressed the T200 antigen. Eventhough it is not uncommon for lymphoid cell-surface molecules to be associated with cells of the nervous system, this finding is appealing to us in view of the model based on the analogy between immune system and olfaction. Due to methodological restrictions, those authors could not apply their investigation to undifferentiated staminal cells. Now, our new method of their isolation, improved in the future with the aid of our Mab, will make it feasible.

ACKNOWLEDGEMENTS

We thank Professor A. Cavaggioni and Professor P. Graziadei for helpfull discussions.

REFERENCES

Amoore, J. E., 1971, in:"Handbook of Sensory Physiology," Vol.4, part 1, Beidler, L.M., ed., pp. 245-256, Springer-Verlag.

Anholt, R. R. H., 1987, *Trends Biochem. Sci.*, 12:58-62.

Bignetti, E., Cattaneo, P., Cavaggioni, A., Damiani,G. and Tirindelli R., 1988, *Comp. Biochem. Physiol.*, 90B:1-5.

Bignetti, E., Cavaggioni, A. and Tirindelli, R., 1989, This book.

Doty, R. L., 1986, *Experientia*, 42:257-271.

Engen, T., 1986, *Experientia*, 42:211-213.

Getchell, T. V., 1986, *Physiol. Rev.*, 66:772-818.

Graziadei, P. P. C. and Monti-Graziadei, G.A., 1979, *J. Neurocyt.*, 8:1-18.

Hildebrand, J. G., 1987, "The 2nd World Congr. of Neurosci.,"(IBRO), abstract 580w, Budapest.

Hirsch, J. D. and Margolis, F. L., 1981, *in*: "Biochemistry of taste and olfaction," Cagan R.H. and Kare M.R., ed., pp 311-332, Academic Press Inc., London.

Kaissling, K. E., 1986, *Ann. Rev. Neurosci.* 9:121-145.

Kleene, S. J., 1986, *Experientia*, 42:241-250.

Koshland, D. E. Jr, 1983, *Trends Neurosci.*, 9:133-137.

Lancet, D., 1986, *Ann. Rev. Neurosci.*, 9:329-355.

Lancet, D. and Pace, U., 1987, *Trends Biochem. Sci.*, 12:63-66.

Nakamura, T. and Gold, G.H., 1987, *Nature*, 325:442-444.

Pace, U., Hanski, E., Salomon, Y. and Lancet, D., 1985, *Nature*, 316:255-258.

Price, S., 1984, *Chem. Senses*, 8:341-345.

Sklar, P. B., Anholt, R. R. H. and Snyder, S. H., 1986, *J. Biol. Chem.*, 261:15538-15543.

Schubert, D., Stallcup W., LaCorbiere, M., Kikidoro, Y. and Orgel, L., 1985, *Proc. Natl. Sci. USA*, 82:7782-7786.

Phototransduction

MICROVILLAR PHOTORECEPTOR CELLS OF INVERTEBRATES

Anatomy and Physiology

H. Stieve

Institut für Biologie II
RWTH Aachen, Germany

ABSTRACT

The photoreceptor cells of invertebrate animals differ from those of vertebrates in morphology and physiology. Our present knowledge of structure and transduction mechanism of the microvillar photoreceptor cell of invertebrates is described. Less than for vertebrates is known about the biochemistry of the enzyme cascade which leads in the invertebrate photoreceptor from the light-activated rhodopsin molecule to the formation of the excitatory intracellular transmitter. A GTP-binding protein is activated. The activity of phospholipase C, which catalyzes the formation of inositol-trisphosphate, is essential. The intracellular transmitter which binds to the cation channel to open it has not yet been identified; cGMP or Ca ions are suspects. The single-photon-evoked events, bumps, are assumed to be based on a light-induced concerted opening of many cation channels. The bumps vary greatly in delay, size and shape. The intensity dependence of the size of the macroscopic receptor current has a region of supralinear slope. The mechanisms of light/dark adaptation are better understood in photoreceptors of invertebrates than in those of vertebrates. Calcium is a desensitizing intracellular transmitter for light adaptation. cAMP is apparently another controller of sensitivity in dark adaptation.

INTRODUCTION

Nature has solved the task of visual transduction differently for invertebrated than for vertebrated animals. The most conspicuous differences in structure and mechanism of the two groups comprise the following:

1. The photosensory cell membrane is greatly enlarged in the rhabdomeric type of photoreceptor of invertebrates by glove finger-like protrusions, the microvilli, in the ciliary type of photoreceptor of vertebrates by flat-spread invaginations of the cell membrane which may part to form layers of flat sacks or disks.

2. The rhodopsin molecules display large rotational and translational mobility within the vertebrate photosensory membrane, whereas in the photoreceptor membrane of invertebrates the rhodopsin molecules are virtually immobile.

Sensory Transduction
Edited by A. Borsellino *et al.*
Plenum Press, New York, 1990

3. In invertebrates the visual pigment is converted by light to a thermostable meta-state which is not, or only very slowly, metabolically regenerated.

4. The electrical signal of visual excitation in invertebrates is based upon increased membrane conductance as opposed to conductance decrease in vertebrates, resulting in membrane voltage signals (receptor potentials) of opposite polarity.

5. Single photon-evoked excitatory events, which are believed to be due to a concerted action (the opening in invertebrates or the closing in vertebrates) of many light-modulated cation channels, are very different in characteristics of size and time course for photoreceptors of invertebrates and those of vertebrates. In invertebrates the single photon events (bumps) evoked under identical conditions vary greatly in delay (latency), time course, and size. The multiphoton response to brighter stimuli is several times as long as a response evoked by a single photon. The single photon responses of vertebrates have a standard size, a standard latency and a standard time course, all three parameters showing relatively small variations. Responses to flashes containing several photons have a shape and time scale that is similar to the single photon-evoked events, varying only by an amplitude scaling factor, but not in latency and time course (Baylor 1984).

These apparent distinctions may indicate fundamental differences in the transduction mechanism of the two groups of animals. At present some aspects of the transduction process are better understood, or are easier to study, in the photoreceptors of vertebrates, others in those of invertebrates.

The photoreceptor cell of vertebrates has an outer segment which is a highly specialized apparatus solely for phototransduction, separated from the other part of the cell which serves for the metabolic cell functions with mitochondria and nucleus, and a part which is typical for a nerve cell with an axon and a synaptic region. This is advantageous for biochemical and certain electrophysiological investigations. The separation of these regions is not as distinct in photoreceptors of invertebrates. The visual cells of some invertebrates are very large and therefore readily accessible to microprobing and microinjections. The microvillar photoreceptors of the fruit fly Drosophila which are so suitable for genetic studies of the transduction process are very small, making both electrophysiological and biochemical studies rather difficult. Until now the molecular mechanism of visual transduction is better understood for photoreceptors of vertebrates whereas the mechanism of adaptation is better understood in invertebrates (Stieve 1985, 1986a; Becker et al., 1988).

STRUCTURE

Visual cells of invertebrates, the microvillar photoreceptors, can differ greatly in form. Figure 1 shows a scheme of the ventral photoreceptor cell of *Limulus* (Calman and Chamberlain 1982). They contain a portion which is enveloped by the photosensory membrane holding the visual pigment. The microvilli, protrusions of the photosensory membrane, effect a ca. 15-fold enlargement of the cell surface. Microvilli have a diameter of 0.1 μm and may be as long as 1-20 μm. They are supported by a cytoskeleton consisting of an axial filament and spike-like processes to the cell membrane. Beneath the microvilli there is a layer of cisternae, derivates of the smooth endoplasmatic reticulum, called subrhabdomeric cisternae SRC. The rest of the visual cell, containing the typical set of cellular organelles, is more or less distinctly separated from the transducing portion and

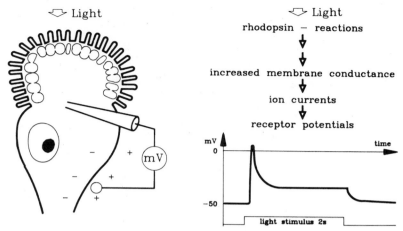

Fig. 1: Functional diagram of the ventral nerve photoreceptor cell of *Limulus*. It has a diameter of 50 to 100 μm. The cell membrane of its distal lobe has glove finger-like protrusions, the microvilli, which contain the visual pigment rhodopsin. Such a cell contains about 10^6 microvilli, ca.1 μm long and 100 nm in diameter; they contain all together a total of about 10^9 rhodopsin molecules (Stieve 1985).

is surrounded by a cell membrane which differs from the photosensory membrane and resembles that of a nerve cell. The photosensory membrane contains besides rhodopsin at least two other integral membrane proteins, the light-modulated cation channel and a sodium/calcium antiporter.

The sensory cell ends in an axon which after a few micrometers (insects) or centimeters (*Limulus*) forms a synapse with the following nerve cell. The photoreceptor cell is usually surrounded by glial or pigment cells which control access and exit of substances from and into the small extracellular compartment surrounding the visual cell membrane.

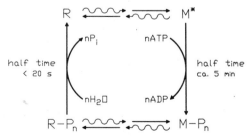

Fig. 2: Light and dark reactions of rhodopsin R of the photoreceptor cell of the fly. M* light-activated metarhodopsin, R-Pn phosphorylated rhodopsin; M-Pn phosphorylated metarhodopsin. The undulating arrows symbolize light reactions, the straight arrows dark reactions (after Paulsen and Bentrup 1986 and Hamdorf 1979).

THE TRANSDUCTION PROCESS

Pigment reactions and successive enzymatic   steps in the transduction process

The molecular processes involved in transduction and amplification in photoreceptors of invertebrates are less well known than in those of vertebrates. The light-induced
rhodopsin reactions, starting with the stereo-isomerization of the 11-cis-retinal chromophore, lead after less than 1 - 5 ms to an activated state metarhodopsin M* which is
inactivated after a certain life time due to multiple enzymatic phosphorylation (Fig. 2,
Paulsen and Bentrup 1986).

This phosphorylated, inactive state which contains the all-trans-retinal chromophore
is stable for hours. After the absorption of a second photon a phosphorylated rhodopsin is
formed (again containing 11-cis-retinal) which is quickly dephosphorylated enzymatically
and thereby reconverted into rhodopsin which is ready to be activated again by light.

Probably by a similar mechanism as in vertebrates a guanyl-nucleotide binding protein (G-protein) binds to the light-activated metarhodopsin M* and is activated itself
while bound.

The degree of amplification involved in this first step in the cascade is not yet known,
but it is probably much smaller than in vertebrates. A number of findings indicate that
artificial (chemical) activation of G-protein in photoreceptor cells of *Limulus* and flies
can evoke a response similar to a light response (Corson and Fein 1983; Corson, Fein
and Walthall 1983).

Activation of G-protein causes the activation of a phosphodiesterase (phospholipase C) which is different from the cGMP-phosphodiesterase of the photoreceptors of
vertebrates. Phospholipase C hydrolyses a membrane lipid, phosphatidyl inositol-4,5-
bisphosphate (PIP$_2$, Fig. 3) into diacylglycerol and inositol-trisphosphate (IP$_3$); so in
the photoreceptor cell of invertebrates, the concentration of IP$_3$ is raised due to illumination (Brown et al., 1984; Brown and Rubin 1986; Fein et al., 1984). Microinjection

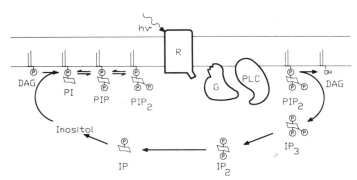

Fig. 3: Light-controlled inositol-phospholipid reaction cycle: Phosphatidyl inositol-4,5-
bisphosphate (PIP$_2$) is hydrolyzed by phospholipase C (PLC) into diacylglycerol
(DAG) and inositol-trisphosphate (IP$_3$). Phospholipase is activated by a G-protein
which is activated via binding to light-activated rhodopsin (R). *IP$_3$* is then step
by step dephosphorylated to inositol. Inositol and DAG are co-substrates for the
synthesis of phosphoinositol (PI), which is phosphorylated step by step to PIP$_2$
(after several authors).

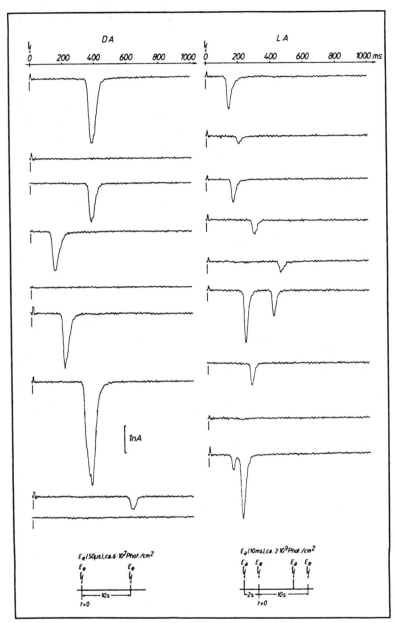

Fig. 4: Current bumps measured from a *Limulus* ventral nerve photoreceptor cell under voltage clamp conditions. Left column dark-adapted photoreceptor (DA);
right column photoreceptor weakly light-adapted (LA) by a conditioning light
flash, 2 s prior to the bump-evoking flash. Bump-evoking flash: E_e ca. 6×10^7
photons/cm^2, duration 50 μs, repetition time 10 s; light-adapting flash: E_a ca.
2×10^9 photons/cm^2, duration 10 ms, both flashes 540 nm, membrane potential
constant - 40 mV; 15°C. Bottom: stimulus regime (Stieve 1985).

of IP$_3$ into the visual cell of *Limulus* induces an electrical response quite similar to the electrical light response. The only known difference between these two responses is that the response to IP$_3$ injection is prevented after prior injection of calcium chelators, such as EGTA, whereas the light response persists under these conditions.

IP$_3$ is not the terminal intracellular transmitter, the binding of which causes the opening of the light-modulated cation channels, but acts earlier in the transduction chain. IP$_3$ causes the intracellular release of calcium ions from intracellular (and extracellular?) compartments into the cytosol. An experimental increase in intracellular calcium ion concentration causes desensitization of the visual cell of *Limulus* in the state of light adaptation (Lisman and Brown 1972, 1975). However, under certain conditions calcium injections into this cell evoke an increase in sodium conductance of the cell membrane like an excitation by light. Payne et al. (1986) therefore postulated that intracellular calcium ions act also as the terminal excitatory transmitter which binds to the channel molecules and causes the light-induced conductance increase. An alternative hypothesis was proposed by Johnson et al. (1986) who found that injection of cGMP into the microvilli-bearing R-lobe of the *Limulus* ventral nerve photoreceptor sometimes (< 10%) produced responses similar to the light-induced depolarization "if the injection was into a cGMP-sensitive region" of the photoreceptor. Our experiments (Heuter and Stieve, submitted for publication, and Stieve et al., 1988) demonstrate a cGMP-induced cation release from vesicles of photosensory membrane of Sepia which were preloaded with lithium or calcium ions. A similar release was evoked by light, but not by cAMP or IP$_3$. Although we are now somewhat more in favour of cGMP being the terminal excitatory transmitter in the microvillar photoreceptor cell, the biochemistry of the transduction process in the microvillar photoreceptor is far from being resolved. Functional phospholipase C and thus the formation of IP$_3$ seems to be essential for visual transduction (Devary et al., 1987; Shortridge et al., 1988). It is still an open question whether the transduction sequence of reactions separates into two branches, one leading via IP$_3$ to Ca^{2+} release, the other one to a transient increase in cGMP or a yet unknown terminal transmitter.

ELECTROPHYSIOLOGY

The absorption of a single photon by a rhodopsin molecule leads in the *Limulus* photoreceptor to the generation of a relatively large elementary excitatory response, the

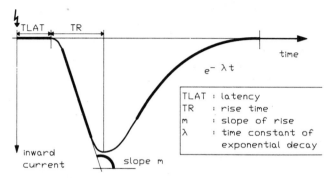

Fig. 5: Primary bump parameters. TLAT latency, TR rise time, m slope of rise, Lambda time constant of the exponential decay.

bump or single photon-evoked event (Stieve et al., 1986b). With stronger stimuli many of these bumps of a visual cell superimpose to a up to 1000-fold larger "macroscopic" response signal, the receptor current or receptor potential. Size and time course of this signal are modulated by adaptation, thus enabling the visual cell to adapt its sensitivity to the ambient illumination in a large range.

If the dark-adapted photoreceptor of *Limulus* is stimulated by light flashes which are so weak that not every flash evokes a light response, one observes bumps, responses of the photoreceptor to the successful absorption of single photons. These bumps can be recorded as membrane voltage signals or as membrane current signals (Fig. 4). A bump is a transient increase of the cation conductance of the visual cell membrane and follows photon absorption after a long, greatly variable delay (latency). Fig. 5 shows a scheme of the generalized bump shape. The latency lasts on the average ca. 200 ms at 15°C. The bump rise lasts about 30 ms and has an almost linear section. After the bump maximum the bump declines for almost 60 ms, for the last two thirds or half of this time the decline is almost exponential (Fig. 5).

Bumps evoked under identical experimental conditions by equal photons (same wave-length) vary greatly in size, shape and latency (Figs. 4 and 6). A bump of a dark-adapted photoreceptor cell of *Limulus* is based on a sodium ion-preferring conductance increase of on the average about 5 nS, occasionally up to 20 nS in the maximum of the bump. Bump size and bump latency are not correlated; they vary independently of each other. The four characteristic *primary shape parameters* (Keiper et al., 1984) of the bumps (Fig. 5), latency TLAT, maximal slope m of the bump rise, duration TR of bump rise and time constant lambda of the exponential phase of the bump decline are not correlated with each other for the dark-adapted photoreceptor under constant experimental conditions.

In contrast to these 4 *primary independent shape parameters* bump amplitude A and bump size (current-time-integral F) are *secondary bump parameters*, which depend mainly on rise time and slope of rise, and F to a lesser degree also on lambda.

The large variability of the latency of the bump indicates that it is determined by the reaction of a small number of molecules. The duration of the latency is not determined by diffusion or by processes of a simple Michaelis-Menten type of kinetics. According to Schnakenberg (1988) this can be concluded from the relatively narrow distribution of bump latencies (Fig. 6). The characteristic shape of the distribution of latencies is in agreement with the assumption that cooperative processes determine the duration of the latency. The average delay of 200 ms is considerably larger than the time needed by rhodopsin after photon absorption to reach the activated state of metarhodopsin M* (<10 ms). It may be due to the time needed for the activation of an enzyme.

A bump-generating mechanism which is consistent with the bump phenomena as yet known is described in Fig. 7. The absorption of a photon by any one rhodopsin molecule in a microvillus of the photosensory membrane causes, after a considerable delay, the activation of a source Q which produces (or releases) many molecules of an internal messenger (intracellular transmitter). A plausible description of the mechanism of bump generation includes the enzymatic production of intracellular transmitter and

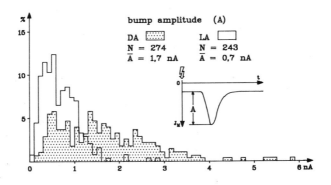

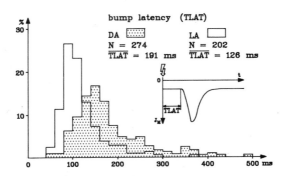

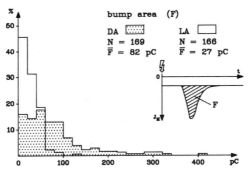

Fig. 6: Frequency distributions of bump amplitudes A (above), bump latencies TLAT (middle), and bump current-time integral F (area, below) of a *Limulus* ventral nerve photoreceptor cell in the dark-adapted (DA) and weakly light-adapted (LA) conditions. The stimulus program is described in Fig. 4. Only latencies of first bumps following the bump-evoking flash are plotted; amplitudes plotted from all single and first bumps, current-time integrals of all single bumps. N is the number of the bumps accounted for. The arithmetical averages of the bump amplitudes, bump current-time integrals and bump latencies, respectively, are given. Bump-evoking flash: E_e ca. 9×10^7 photons/cm^2; conditioning, light-adapting flash: E_c ca. 8×10^9 photons/cm^2; otherwise as in Fig. 4 (Stieve 1986).

transmitter diffusion to the light-controlled ion channels which are distributed over a large area (of up to 2-4 μm in diameter) of the photosensory membrane (Fig. 7). A time-consuming enzyme activation could determine the latency. Transmitter diffusion over the "bump-speck" could be responsible for the linear bump rise, and the stochastic closure of ion channels or the time course of transmitter decay could determine the exponential bump decay.

Bacigalupo and Lisman (1983), using the patch-clamp technique, recorded single channel events in the visual cell membrane of *Limulus* The opening probability of the channels was strongly increased by illumination. The plausible assumption that a bump is based on the superposition of many incompletely synchronized, i.e. concerted, opened ion channels is supported by the registration of these single channel events. One can estimate that in the maximum of a large bump up to 10^3 - 10^4 ion channels with a single channel conductance of 10-20 pS are opened simultaneously. Bump generation thus involves a considerable amplification - as defined by the number of ion channels opened due to the light-activation of a single rhodopsin molecule.

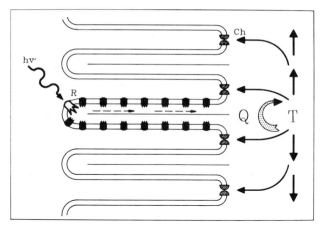

Fig. 7: Schematic diagram of proposed mechanism of bump generation in *Limulus* ventral nerve photoreceptor. A light-activated rhodopsin molecule R in a microvillus starts the activation of an enzyme (cascade) which finally leads to the activation of a transmitter source Q. This may be situated at the basis of this microvillus. The transmitter source thereupon produces, activates or releases many transmitter molecules which are built from precursors. The transmitter T diffuses along the bases of the microvilli and is bound by the cation channel Ch which plausibly may be situated close to the bases of the microvilli. Following transmitter binding the cation channels are transiently open. Thus develops a more or less circular "bump-speck" which in the dark-adapted photoreceptor cell should have a diameter of about up to 4 μm. The average bump-speck of a light-adapted photoreceptor cell is smaller. The bump amplitude is proportional to the number of simultaneously opened ion channels and should be thereby more or less proportional to the area of the bump-speck (Stieve 1985).

BUMP SUMMATION AND "MACROSCOPIC" RESPONSE SIGNALS TO BRIGHTER FLASHES

The number of bumps evoked by a light flash depends linearly on the energy of the light stimulus, i.e. on the number of absorbed photons. With stronger light stimuli bumps superimpose and fuse to the macroscopic receptor current (Fig. 8). Basically the "macroscopic" receptor current evoked by a brighter flash is the summation of a volley of bumps. However, the shape of the signal is secondarily modified by bump interactions and a feed-back process (automatic gain control, Stieve et al., 1986).

Due to the scattering of bump latencies the macroscopic receptor current signal is much (several times) longer than a single photon response. The latency of the first bump occurring after the flash is the latency of the macroscopic light response. With increasing intensity of the light stimulus the latency of the macroscopic response of invertebrates is shortened much more as compared to vertebrates. This drastic shortening of the latency may be sufficiently explained for the visual cell of *Limulus* on the basis of the statistics of the bump latency distribution: With higher stimulus intensities even short latencies occur increasingly frequently.

The stimulus energy versus response characteristics is described in Fig. 9. The light-evoked receptor current signal becomes larger with the stimulus energy and saturates

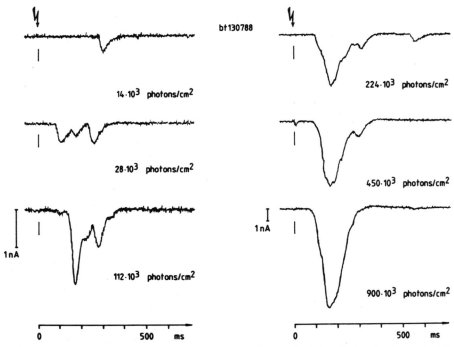

bt 130788

$14 \cdot 10^3$ photons/cm^2

$28 \cdot 10^3$ photons/cm^2

$112 \cdot 10^3$ photons/cm^2

$224 \cdot 10^3$ photons/cm^2

$450 \cdot 10^3$ photons/cm^2

$900 \cdot 10^3$ photons/cm^2

1 nA

1 nA

0   500   ms   0   500   ms

Fig. 8: Light-induced membrane current signals evoked by light flashes of different stimulus energies. *Limulus* ventral nerve photoreceptor, voltage-clamp. Very weak flashes evoke individual bumps, with increasing stimulus energy bumps overlap and fuse to the "macroscopic" receptor current signal where individual bumps are no longer recognizable. Stimulus duration ca. 100 μs, stimulus energy inserted.

at very high energies. The light energy versus response characteristics (current time integral F) shows four regions with different slopes in the double logarithmic plot:

a) a linear section (slope about 1)
b) a supralinear section (slope 2-4)
c) a sublinear section (slope < 0.5)
d) a second sublinear section (slope > 0.5).

Light adaptation shifts the curve to higher light intensities and reduces the steepness of the supralinear region. The point of transition between region a and b ("curve heel") and b and c (curve knee) are at the same ordinate values for the plot of the dark-adapted and the light-adapted state.

A deviation from slope 1 indicates that the single photon-evoked events which constitute the macroscopic receptor current are not independent from each other. The supralinear slope results from some kind of positive cooperativity. The observed slope of up to four could be explained by four ligands which have to be bound by a key molecule. This molecule could be, but need not be, the light-modulated cation channel (Stieve 1986b). The two sublinear slopes are due to response interactions like negative cooperativity or negative feed-back (automatic gain control, Stieve et al., 1986), in effect reducing the amplification (Stieve and Schlösser 1987).

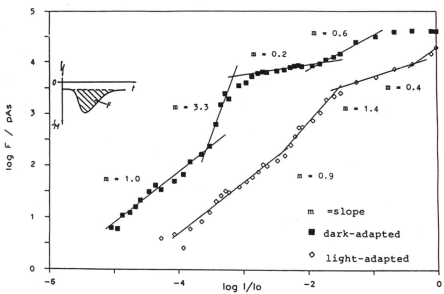

Fig. 9: The intensity dependence of the light-evoked receptor current signal (current-time-integral F) of the *Limulus* ventral nerve photoreceptor under voltage clamp conditions at two defined states of adaptation: moderate light adaptation (LA) and considerable dark adaptation (DA). The light energy vs response characteristics show four regions with different slopes in the double logarithmic plot:a) a linear section (slope about 1) b) a supralinear section (slope 2-4) c) a sublinear section (slope < 0.5) d) a second sublinear section (slope > 0.5). (Stieve and Schloesser, 1987).

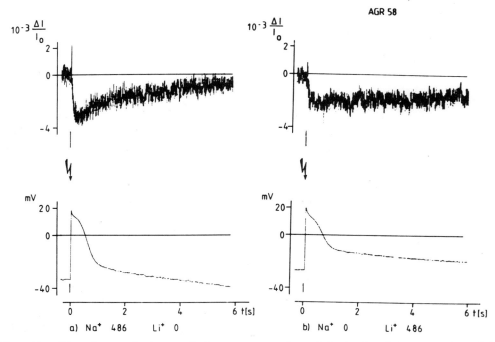

Fig. 10: The change in intracellular concentration of ionized calcium monitored by the calcium indicator arsenazo III. Arsenazo response (upper trace) and receptor potential (lower trace) upon stimulation by a 10 ms light flash. a) physiological saline (PS); b) after 15 min stay in a saline in which all Na^+ was replaced by Li^+. Inset ion concentration in mmol/l. Average of 4 records.

ADAPTATION

Phototransduction involves a considerable amplification. The degree of this amplification is controlled by light adaptation (Stieve, Bruns and Gaube 1983, Claßen-Linke and Stieve 1986).

The visual cell changes its sensitivity according to the ambient illumination. This light adaptation (by background light or pre-illumination) causes the stimulus energy versus response size characteristics (Fig. 9) to be shifted parally to higher stimulus energies. The sensitivity control in dark/light adaptation is caused by several processes (Claßen-Linke and Stieve 1986). Calcium is a transmitter for light adaptation (desensitization in the *Limulus* photoreceptor, Lisman and Brown 1972, 1975). A light-induced transient increase in intracellular Ca^{2+} controls the sensitivity via a yet unknown molecular mechanism. The shift in sensitivity due to light adaptation is the stronger, the higher the external calcium concentration (Stieve, Bruns and Gaube 1984; Stieve, Gaube and Klomfaß 1986). The light stimulus evokes, besides the membrane current signal, a transient increase in intracellular calcium ion concentration which can be monitored by calcium indicators such as aequorin or arsenazo III (Brown 1986, Fig. 10). Amplitude, increase and most of all decline of this calcium signal depend on the membrane voltage to which the cell is clamped (Nagy and Stieve 1983; Ivens and Stieve 1984). Ivens and Stieve (1984) suggested that the rise of the calcium signal is based - to a substantial degree - on a calcium inward current probably through the light-activated sodium channels, whereas the decline of the calcium signal is due to a calcium/sodium exchange transport across the cell membrane. Recently O'Day and Keller (1987) have shown a sodium/calcium exchange transport in the *Limulus* photoreceptor using aequorin. Here we show that the recovery of the light-induced intracellular calcium signal monitored by arsenazo III becomes strongly retarded when the external sodium is replaced by lithium (Fig. 10, Stieve et al., 1988). The calcium signal is greatly enlarged and both rise and decay retarded when the external calcium is replaced by strontium. Both these results are in agreement with an assumed sodium/calcium exchange transport. While strontium is measured by arsenazo and accepted by the sodium/calcium exchanger, lithium is almost not. Strontium has a weaker desensitizing effect than calcium and is probably much less effectively sequestered in the cell.

In addition to the calcium-regulated feed-back desensitization other control mechanisms have been found. Apparently one of them is regulated by cyclic adenosine monophosphate (cAMP) (Stieve et al., 1988). In addition light-induced conversion of visual pigment to inactive states (phosphorylated metarhodopsin) is also involved in desensitization (Paulsen and Bentrop 1986).

Acknowledgements. This work was supported by the Deutsche Forschungsgemeinschaft SFB 160. I wish to thank K. Nagy and J. H. Nuske for helpful comments on the manuscript, H. Gaube for improving the English, H. T. Hennig for drawing the figures, and him and I. Wicke for considerable help with the manuscript.

REFERENCES

Bacigalupo, J. and Lisman, J. E. , 1983, Single-channel currents activated by light in *Limulus* ventral photoreceptors, *Nature*, 304:268-207.

Baylor, D. A., Nunn, B. J. and Schnapf, J. L., 1984, The photocurrent, noise and spectral sensitivity of rods of the monkey Macaca fascicularis, *J. Physiol.*, 357:575-607.

Becker, U. W., Nuske, J. H. and Stieve, H., 1988, Phototransduction in the microvillar visual cell of *Limulus*: Electrophysiology and Biochemistry, *in*: "Progress in Retinal Research," N. Osbarne, J. Chader, ed., Pergamon Press, Vol 8, 229-253

Brown, J. E., 1986, Calcium and light adaptation in invertebrate photoreceptors, *in*: "The Molecular Mechanism of Photoreception," H. Stieve, ed., Dahlem Konferenzen, 231-240, Springer, Berlin.

Brown, J. E., Rubin, L. J., Ghalayini, A. J., Tarver, A. P., Irvine, R. F., Berridge, M. J. and Anderson, R. E., 1984, Myo-inositol polyphosphate may be a messenger for visual excitation in *Limulus* photoreceptors, *Nature*, 311:160-163.

Brown, J. E. and Rubin, L. J., 1986, Signal transduction: The putative participation of inositol trisphosphates in *Limulus* photoreceptors, *Fortschritte der Zoologie*, 33:321-331.

Calman, G. B. and Chamberlain, S. C., 1982, Distinct lobes of *Limulus* ventral photoreceptors, II. Structure and ultrastructure, *J. Gen. Physiol.*, 80:839-862.

Claßen-Linke, I. and Stieve, H., 1986, The sensitivity of the ventral nerve photoreceptor of *Limulus* recovers after light adaptation in two phases of dark-adaptation, *Naturforsch.*, 41c:657-667.

Corson, D. W. and Fein, A., 1983, Chemical excitation of *Limulus* photoreceptors: I. Phosphatase inhibitors induce discrete-wave production in the dark, *J. Gen. Physiol.*, 82:639-657.

Corson, D. W., Fein, A. and Walthall, W. W., 1983, Chemical excitation of *Limulus* photoreceptors: II.Vanadate, GTP-gamma-S, and fluoride prolong excitation evoked by dim flashes of light, *J. Gen. Physiol.*, 82:659-677.

Devary, O., Heichal, O., Blumenfeld, A., Cassel, D., Suss, E., Barash, S., Rubinstein, C. T., Minke , B., and Selinger, Z., 1987, Coupling of photoexcited rhodopsin to inositol phospholipid hydrolysis in fly photoreceptors, *Proc. Natl. Acad. Sci.USA*, 84:3939-3943.

Fein, A., Payne, R., Corson, D. W., Berridge, M. J. and Irvine, R. F., 1984, Photoreceptor excitation and adaptation by inositol 1,4,5-trisphosphate, *Nature*, 311:157-160.

Hamdorf, K., 1979, The physiology of invertebrate visual pigment, *In:* "Handbook of Sensory Physiology," H. Autrum, ed., Vol 7, 145-224, Springer, Berlin.

Ivens, I., and Stieve, H., 1984, Influence of the membrane potential on the intracellular light induced Ca^{2+}-concentration change of the *Limulus* ventral photoreceptor monitored by arsenazo III under voltage clamp conditions, *Z. Naturforsch.*, 39c:986-992.

Johnson, E. C., Robinson, P. R. and Lisman, J. E., 1986, Cyclic GMP is involved in the excitation of invertebrate photoreceptors, *Nature*, 324:468-470.

Keiper, W., Schnakenberg, J. and Stieve, H., 1984, Statistical analysis of quantum bump parameters in *Limulus* ventral photoreceptors, *Z. Naturforsch.*, 39c:781-790.

Lisman, J. E. and Brown, J. E., 1972, The effects of intracellular iontophoretic injection of calcium and sodium ions on the light response of *Limulus* ventral photoreceptors, *J. Gen. Physiol.*, 59:701-719.

Lisman, J. E. and Brown, J. E., 1975, Light-induced changes of sensitivity in *Limulus* ventral photoreceptors, *J. Gen. Physiol.*, 66:473-488.

Nagy, K. and Stieve, H., 1983, Changes in intracellular calcium ion concentration in the course of dark adaptation measured by arsenazo III in the *Limulus* photoreceptor, *Biophys. Struct. Mech.*, 9:207-223.

O'Day, P. M. and Gray-Keller, M. P., 1989, Evidence for electrogenic Na^+/Ca^{2+}-exchange in *Limulus* ventral photoreceptors, *J. Gen. Physiol.*, 93:473-495.

Paulsen, E. and Bentrop, J., 1986, Light-modulated biochemical events in fly photoreceptors, *Fortschritte der Zoologie*, 33:299-319.

Payne, R., Corson, D., Fein, A. and Berridge, M. J., 1986, Excitation and adaptation of *Limulus* ventral photoreceptors by inositol 1,4,5-trisphosphate result from a rise in intracellular calcium, *J. Gen. Physiol.*, 88:127-142.

Schnakenberg, J., 1989, "Amplification and latency in photoreceptors: Integrated or separated phenomena?", Biol. Cybernetics, 60:421-437.

Shortridge, R. D., Bloomquist, B. T., Schneuwly, S., Perdew, M. H. and Pak, W. L., 1988, Molecular isolation and analysis of a photoreceptor-specific phospholipase C gene, norpA, of Drosophila, *in:* "XI Yamada Conference, Molecular Physiology of Retinal Proteins," T. Hara and Mt. Hiei, ed., Kyoto, Japan.

Stieve, H., 1985, Phototransduction in invertebrate visual cells. The present state of research - exemplified and discussed through the *Limulus* photoreceptor cell, *in:* "Neurobiology," R. Gilles and J. Balthazart, ed., 346-362.

Stieve, H., 1986 a, Introduction, *in:* "The Molecular Mechanism of Phototransduction," H. Stieve, ed., Dahlem Konferenzen, 1-10, Springer, Berlin.

Stieve, H., 1986 b, Bumps, the elementary excitatory responses of invertebrates, *in:* "The Molecular Mechanism of Phototransduction," H. Stieve, ed., Dahlem Konferenzen, 199-230, Springer, Berlin.

Stieve, H., Bruns, M., Gaube, H., 1983, The intensity dependence of the receptor potential of the *Limulus* ventral nerve photoreceptor in two defined states of light- and dark adaptation, *Z. Naturforsch.*, 38c:1043-1054.

Stieve, H., Bruns, M., Gaube, H., 1984, The sensitivity shift due to light adaptation depending on the extracellular calcium ion concentration in *Limulus* ventral nerve photoreceptor, *Z. Naturforsch.*, 39c:662-679.

Stieve, H., Gaube, H., Klomfaß, J., 1986, Effect of external calcium concentration on the intensity dependence of light-induced membrane current and voltage signals in two defined states of adaptation in the photoreceptor of *Limulus*, *Z. Naturforsch.*, 41c:1092-1110.

Stieve, H., Heuter, H., Hua, P., Nuske, J. Rüsing, G., Schloesser, B., 1988, Excitation and adaptation in the photoreceptor cell of *Limulus, in:* "XXI Yamada Conference, Molecular Physiology of Retinal Proteins," T. Hara, Mt. Hiei, ed., Kyoto, Japan.

Stieve, H., and Schloesser, B., 1987, The intensity dependence of the *Limulus* photoreceptor current shows consecutive regions of linear, supralinear and sublinear slope, *in:* "New Frontiers in Brain Research. Proc. 15th Goettingen Neurobiology Conference," Elsner N., Creutzfeld, O., ed., Georg Thieme, Stuttgart.

Stieve, H., Schnakenberg, J., Kuhn, A., Reuss, H., 1986, An automatic gain control in the *Limulus* photoreceptor, *Fortschritte der Zoologie*, 33:367-376.

THE RHODOPSIN-TRANSDUCIN-cGMP-PHOSPHODIESTERASE

CASCADE OF VISUAL TRANSDUCTION

F. Bornancin, C. Pfister and M. Chabre

Laboratoire de Biophysique Moleculaire et Cellulaire

(Unité Ass. 520 du CNRS)

Departement de Recherche Fondamentale, CENG, Grenoble, France

The light activated cascade of vision shares many features with transduction pathways in hormonal systems involving G-proteins (L. Stryer, 1986, M. Chabre et al., 1984, L. Stryer and H.R. Bourne, 1986). Indeed rhodopsin can be considered as a particular hormone receptor where the retinal, the equivalent of an hormone, is already bound but will become fully active only after illumination. Apart from this original aspect, rhodopsin clearly appears as a member of the "seven α-helix" receptors family which includes the M1 and M2 muscarinic, the $\alpha 2$, $\beta 1$ and $\beta 2$ adrenergic and the substance K receptors (H. G. Dohlman et al., 1987). In all cases, activation of the receptors triggers an enzymatic cascade involving interactions with G-proteins which are thus allowed to exchange GDP for GTP, become active and interact with effector proteins that control an internal messenger. Many analogies (in the subunit structure of the G-proteins, in the mechanisms of activation at the different steps of the cascades) have been documented: evolution seems to have conserved typical key structures. The visual system is however special in many respects: (i) rhodopsin is exceptionally abundant in the photoreceptor cell, at least thousand times more abundant than receptors in hormone responsive cells. This extremely dense population of receptors is here necessary in order to capture efficiently photons. The total number of rhodopsins is ten times higher than the total number of G proteins, transducins. However, at a physiological level of illumination, the number of transducins involved in transduction will be much higher than that of photoexcited rhodopsins, like in hormone responsive systems. (ii) Transducin (T) and the cGMP phosphodiesterase (PDE, the effector protein of the visual system) are soluble proteins that can be detached from the disc membrane in the absence of detergent. (iii) The system is conveniently excited by light flashes. (iv) Spectral properties of the retinal buried in the receptor allow time-resolved studies of the conformations of the receptor. A simplified sketch of the complete transduction process, from the photon to the closure of the cGMP dependent cationic channels in the rod cell plasma membrane is shown on fig. 1. Here we shall only discuss the amplifying cascade from rhodopsin to the cGMP phosphodiesterase.

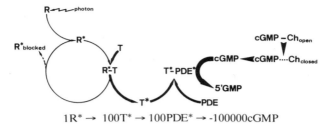

$$1R^* \rightarrow 100T^* \rightarrow 100PDE^* \rightarrow -100000cGMP$$

Fig. 1: The basic scheme of the phototransduction cascade.

THE MAJOR COMPONENTS OF THE cGMP CASCADE

The 5 major proteins involved in the first steps of the cGMP cascade have been characterized and purified.

Rhodopsin (R), the light receptor protein, is an integral membrane protein of 39 kD folded into seven α-helices spanning the disc membrane (E.A. Dratz and P.A. Hargrave, 1983) (fig. 2). The chromophore, linked by a Schiff-base bond to a lysine in the middle of the last helix, is buried in the central hydrophobic core. One half of the protein is within the bilayer membrane, the remaining being divided into two hydrophilic domains protuding in the intradiscal space and in the cytoplasm. The sites of interaction with the proteins of the cascade will be on the cytoplasmic surface. The intradiscal domain, whose role is not well understood, might contain a binding site for calcium.

Transducin (T), as all G proteins involved in signal transduction, is made up of three polypeptide chains: T_α (39 kD), T_β (36 kD) and T_γ (8 kD). T_α has a guanyl nucleotide site which can bind GDP or GTP. T_β and T_γ form an undissociable subunit $T_{\beta\gamma}$. When the nucleotide site contains GDP, T_α GDP and $T_{\beta\gamma}$ are associated and the holoenzyme T_α GDP-$T_{\beta\gamma}$ remains membrane attached at physiological ionic strength (H. Kuhn, 1984), (F. Bruckert et al., in press); it is solubilized only in hypo-ionic conditions. When the GDP is exchanged for GTP, T_α GTP dissociates from $T_{\beta\gamma}$; T_α GTP becomes soluble under physiological conditions whereas $T_{\beta\gamma}$ remains membrane bound. T_α GTP is the activator of the phosphodiesterase. The only known role for $T_{\beta\gamma}$ would be to "present" T_α GDP to photoactivated rhodopsin (R*). Among the three subunits making up all known signal-transducing G-proteins the β polypeptide (36 kD) of Gs and Gi for example is strictly identical to $T\beta$. The sequences of the various α subunits are all slightly different but there is evidence that G_α i can interact with a photoactivated rhodopsin. Moreover, G_α and T_α have the same tryptic proteolytic pattern, thus confirming that these different α subunits of G proteins are structurally very similar. However, differences can be seen at the γ subunit level: $T\gamma$, whose sequence is the only one known, differs immunologically and electrophoretically from the γ subunits of other G-proteins. These differences in γ subunit could account for the large difference in membrane attachment of the various G-proteins.

Cyclic-GMP phosphodiesterase (PDE), the effector protein of the visual system, is also a peripheral protein and is made up of four subunits: two undissociable cyclic-GMP hydrolysing units whose peptide maps are remarkably similar, PDEα(88 kD) and PDEβ(84 kD); two apparently identical inhibitory subunits PDEγ, of 13 kD each, which will be removed by the action of activated transducin.

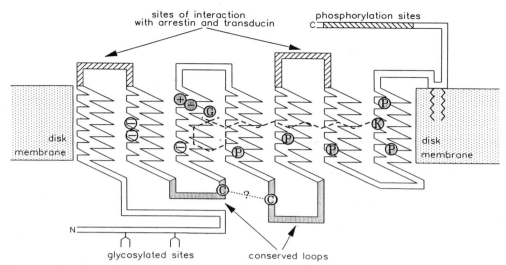

sites of interaction
with arrestin and transducin

phosphorylation sites

disk
membrane

disk
membrane

glycosylated sites

conserved loops

Fig. 2: Structural model of bovine rod rhodopsin. Some residues conserved among
known G-protein related receptors are shown (shadowed circles). Empty circles
indicate the three negative residues specifically mutated in cone pigments. The
C-terminal is anchored in membrane by palmitoylation of two cysteines (Y. A.
Ovchinnikov et al., 1988).

Rhodopsin kinase (K) is a soluble ATP-dependent kinase of 68 kD, whose activation
does not require any cofactor: it is selectively active on photoexcited rhodopsin. It
can phosphorylate as many as 9 serin or threonin residues on the C-terminal end of
photoexcited rhodopsin.

Arrestin (A), also known as 48 kD protein or S-Antigen, is a very soluble protein,
present at a hight concentration in the cytoplasm of rods, even in the inner segment
which appears to store most of the arrestin pool in the dark. Arrestin is a blocker
protein that will bind to phosphorylated photoactivated rhodopsin. This protein is also
named "S-Antigen" due to its involvement in the generation of an autoimmune disease
of retina (C.Pfister et al., 1985). The location of these various proteins with respect to
the disc membrane is shown on fig. 3.

THE FIRST STAGE OF THE CASCADE: R*-T COUPLING

It is essentially a two steps cascade. The central protein of the cascade is transducin
which shuttles between rhodopsin and PDE, carrying the excitation signal.

The triggering event. In the resting state (stage 0 in fig. 4), that is in the dark,
rhodopsin is non-active (the "thermal noise" is very weak) and has no affinity for transducin; it bears a retinal in the 11-cis conformation. Upon a flash (stage 1), the primary
event, occuring in picoseconds, is the photoisomerization of retinal which turns to the all-
trans conformation. This conformational change requires 35 kcal/mol which represents
more than 60 % of the energy of the absorbed photon, a high efficiency for photoen-
ergetic conversion (A. Cooper, 1979). In later dark reactions, this stored energy will
force conformational changes in the protein to relax the strain on the chromophore. The
conformational changes of the protein lead to successive states, characterized by their
absorption spectra. The capital intermediate is Metarhodopsin II also called R* (N. Ben-

nett et al., 1982, D. Emeis et al., 1982). That state is reached in about 1 ms, a long enough delay for long distance conformational changes of the protein to be completed: two sites for protein-protein interactions are created on the cytoplasmic surface: one for transducin, the other one for the ATP-dependent kinase (H. Kuhn, 1981), (M. Chabre, 1983).

The R*-T interaction and its regulation by rhodopsin kinase and arrestin. When no rhodopsin is photoactivated, transducin is in the T_α GDP-$T_{\beta\gamma}$ holoprotein state with an unexchangeable GDP locked in the nucleotide site of T_α , and diffuses on the membrane surface without interacting with rhodopsins. But transducin has a high affinity for photoexcited rhodopsin, to which it binds by collision coupling, as soon as R* is formed (stage 2): this induces the opening of the nucleotide site on T_α ; the intrinsic GDP becomes exchangeable and can be released in the medium (N. Bennett et al., 1985). One must stress here that R* provides no energy and only catalyses the exchange reaction, which is driven by the energy provided by the previous hydrolysis of GTP in GDP: there is a free energy release in the conformational change that accompanies the exchange of GDP for GTP in T_α . Conformational energy had been stored in T-GDP upon the hydrolysis of GTP in a preceding tun of the cycling process. The R*-T complex is transiently emptied of nucleotide (R*-Te, "e" for empty site). GTP can enter the site and binds very quickly (within less than a millisecond) (T.M. Vuong et al., 1984). In the GTP bound state T_α has no longer an affinity for R*. T_α GTP dissociates from R* and from $T_{\beta\gamma}$, and readily solubilizes, leaving $T_{\beta\gamma}$ on the membrane (stage 3). The solubilized T_α GTP, with the GTP locked again in the nucleotide site, will carry the activation signal to the PDE. The liberated R* can catalyse exchanges on other transducins: since R* is long-lived, this allows the successive interaction with many transducins, thus providing a high amplification. But a fast blocking system must exist to shut off the process before the spontaneous decay of Meta II rhodopsin which occurs only after many seconds. This is needed to block the physiological response and also to allow the system to be sensitive to a new flash. R* is first phosphorylated (stage 4) by the rhodopsin kinase which was unable to act on the dark-adapted rhodopsin. Phosphorylated R* is still able to activate transducins (stage 5), although with a lower efficiency, depending on the number of phosphates bound. This ability will be completely abolished by the binding of arrestin, which recognizes specifically the multiphosphorylated R* (stage 6), (U. Wilden et al., 1986). We shall present now some of our studies concerning this first amplification step.

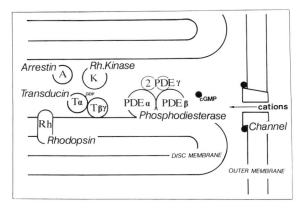

Fig. 3: Intracellular compartments of the rod outer segment and the main molecular effectors.

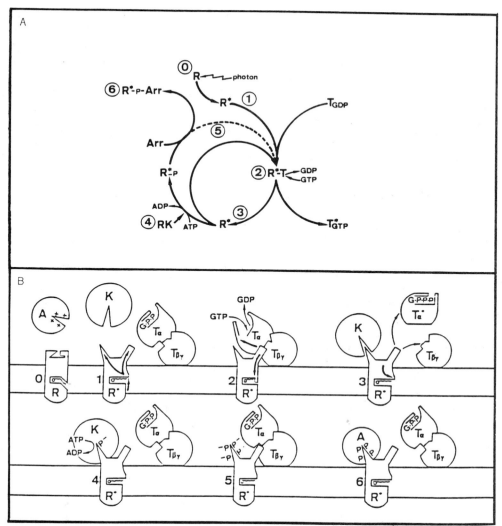

Fig. 4: Regulation by rhodopsin kinase and arrestin of the catalytic action of R* on transducin. The numbers in A and B correspond to the same stage of reaction.

Light scattering study of the kinetics of activation of T by R*. H. Kuhn (1984) had demonstrated that the reactions between R* and transducin could be monitored by measurements of the variations of the near infra-red light scattering from ROS disc membrane suspensions. Extending the method to oriented suspensions of structurally intact rods, Vuong et al (1984) showed that under conditions close to the physiological ones, a photoexcited rhodopsin sequentially activates a few hundreds of transducins at a rate of one transducin per ms. This rate is close to what can be computed for the collision coupling rate of R* with T, assuming that rhodopsin is the only mobile component in the membrane and taking its known lateral diffusion coefficient ; that assumption is often implicitly made, mainly because the lateral diffusion coefficient of rhodopsin is easily measurable and well known while that of transducin is difficult to assess. To check this point, Bruckert et al., (to be published) have devised an experiment based on the analysis of near infra-red light scattering transients observed on oriented rod suspension, combined with photoexcitation by fringe interference patterns. The insensitivity of the kinetics to the fringe pattern suggests that the relative diffusion of rhodopsin with respect to transducin is not a rate limiting factor for the rising phase of the cascade. This seems only compatible with a lateral mobility for transducin that would be higher than that of rhodopsin. This could be expected since the peripherally bound transducin moves in the cytoplasm on the membrane whereas rhodopsin is embedded in the 100 times more viscous lipid bilayer. For other G-protein systems it is likely that, as for the visual transduction cascade, the most mobile component is always the G-protein: even a G-protein that is attached to the membrane by a mirystyl chain, but whose entire polypeptide chain is in the cytoplasm, probably has a faster lateral diffusion than that of a necessarily transmembraneous receptor.

The intermediate complex R*-Te. The key molecular interaction occurs in the transient catalytic complex between R* and T. *In vitro*, due to the large excess of rhodopsin over transducin, after a very strong flash all of the transducin pool will be induced to bind on the R* formed. In the absence of GTP, the exchange reaction cannot proceed and transducin remains quasi-permanently bound to R*. If the previously bound GDP is washed out from the suspension after illumination, the complex cannot even reverse back to R* + T-GDP and the transitory state intermediate R*-Te, with no nucleotide in the site of T_α is fully stabilized and can be studied. We found that this R*-Te complex is very tight and does not dissociate, neither at extreme dilution of the membrane, nor after a long incubation period at 30^0. Both component proteins seem to be "frozen" in the complex : not only is the conformation of T modified by its interaction with R* (its nucleotide site gets opened), but conversely the tight binding of Te acts on R* at the level of the chromophore site, and blocks it in the Metarhodopsin II state. This spectral state is normally transient and evolves in tens of seconds toward another spectral state Meta III (absorbance peak at 470 nm), followed by the release of the chromophore from the protein part, the opsin. We had previously noticed (C. Pfister et al., 1983) that the

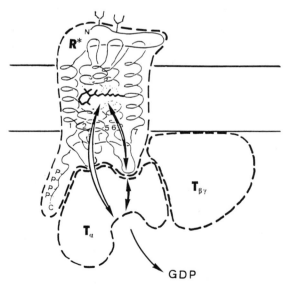

Fig. 5: The R*-Tempty complex. The representation for rhodopsin takes into account what is known about its structure. The thick arrows symbolize interactions between distant sites.

spectral decay was notably slower for transducin-bound R* than for free R*, but these measurements had been done without totally eliminating the endogenous GDP. We have reinvestigated this problem (Bornancin et al., to be published) and studied the spectral evolution of rhodopsin that is blocked in R*-Te complexes: after various incubation periods at 30^0C, R* is liberated from the complex by addition of GTPγS which allows the formation of T_α GTPγS and its dissociation from R*. The spectral decay of the newly liberated R* is measured by difference with that of an aliquot sample in which no GTPγS is added. One sees (fig. 6) that even after an incubation of 30 minutes at 30^0C, which would allow the total decay of all free R*, the decay pattern of the R* that had been maintained in R*-Te complexes during the incubation is identical to that observed before the incubation. This proves that the T-bound R* has been quantitatively maintained in the spectral state Meta II. The retinal binding site, located in the middle of the transmembrane core of rhodopsin, and the transducin binding site on the cytoplasmic surface therefore remain allosterically coupled after photoexcitation. Initially it is the isomerization of retinal that induces the formation of the transducin binding site, but later the binding of transducin can block the release of the chromophore from its internal site: the chromophore in its binding site in the hydrophobic core of R*, is sensitive to the presence of transducin on the cytoplasmic surface, which on turn depends on the occupancy of the nucleotide site (fig. 5).

The same kind of retroaction of the G-protein on the receptor can be seen in hormonal systems too (R.J. Lefkowitz et al., 1983): it results in an hindrance of the release of the hormone from the receptor, hence an increased affinity for the hormone in the absence of GTP.

We further demonstrated that the stabilized R* remains functional with respect to its catalytic capacity: after incubation of the complex R*-Te at 30^0C and subsequent release of transducin upon addition of GTPγS, the "liberated R*" was tested for its ability to bind a newly added T-G9P and to catalyze in it a nucleotide exchange.(see fig. 6).

Artificial activation of T-GDP by fluoroaluminates (AlF$_4^-$). The physiological activation of transducin requires its complexation with a photoactivated rhodopsin for a nucleotide exchange to occur. Fluoroaluminates provide an interesting mechanism that allows the activation of transducin to bypass this need for a receptor. The long known activating effect of fluorides on G-proteins had been shown by P.C. Sternweis and A.G. Gilman (1982) to depend on the presence of traces of Aluminum which is commonly etched from the glassware by NaF or KF solutions. The effective entity would be the complex AlF$_4^-$. We observed (J.Bigay et al., 1985; J.Bigay et al., 1987) that AlF$_4^-$ activates transducin stoichiometrically, but only when a GDP is bound in the nucleotide site of T_α : transducin on ROS membrane was activated in the dark upon addition of AlF$_4^-$, but the activation could be reversed by a strong illumination, that is when transducin bound to R* and lost its GDP. Structural analogies between the two complexes AlF$_4^-$ and PO$_4^-$ led us to propose a model where the fluoroaluminate complex would enter the nucleotide site, directly bind to the free oxygen of the β-phosphate of GDP and then simulate the presence of the of a GTP, hence turning T_α into the active form. The GDP is directly "complemented" in the non exchangeable site. Therefore, by contrast with receptor-catalysed permanent activation by GTPγS, activation by fluoroaluminates, which does not require the release of the guanine part of the nucleotide, is reversible. This model has been shown now to be valid also in other systems, involving proteins interacting with phosphates or nucleoside polyphosphates.

131

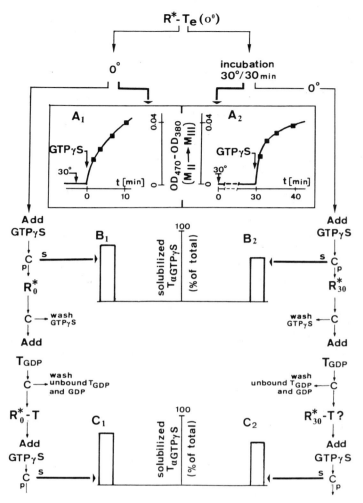

Fig. 6: Perturbation of spectral and functional evolution of R* by its interaction with
nucleotide-free transducin. The R*-Tempty complexes are obtained by illuminat-
ing ROS membranes to create R* in excess over the pool of T and by washing
the membrane suspension extensively in low ionic strength buffer at 0°C. The
pellets are resuspended in a medium ionic strength buffer (20 mM Hepes, pH
7.4, 120 mM NaCl, 0.1 mM MgCl$_2$). (A) Spectral study. The suspension (0.5
mg/ml rhodopsin) is brought to 30°C and poured into two identical temperature-
controlled cuvettes set in two beams of a Perkin-Elmer λ 5 dual beam spectrophotometer. A first fast scan of the difference spectrum determines that the absorption of the two samples is identical from 250 to 650 nm. After 1 min of incubation
at 30°C (A$_1$), GTPγS (10μM) is added to one of the cuvettes to liberate R* from
its complex with T. Fast scanned difference spectra then show a decrease in absorption at 380 nm compensated by an increase at 470 nm. This is characteristic
of the spectral transition Metarhodopsin II $\longrightarrow$ Metarhodopsin III, which occurs
in the cuvette where GTPγS has been added. The time course is a few minutes
for the decay of free R* from Metarhodopsin II to Metarhodopsin III at 30°C. The
amplitude of the observed spectral difference is the same if GTPγS is added after a 30-min incubation at 30°C (A$_2$). This demonstrates that transducin-bound
R* was maintained in the Metarhodopsin II state over this incubation period.

(B) Biochemical control of the release of transducin. Aliquots of the same ROS suspension are sedimented after GTPγS has been added after 1 or 30 min of incubation at 30^0C. The released transducin is titrated in the supernatant by the technique of Bradford and by Coomassie blue staining on gels and is normalized to the total transducin content measured by further extraction of the membranes at low ionic strength. The GTPγS-dependent release is the same in both samples. *(C) Test of functionality of R* that has been "liberated" after incubation.* R* that has been released from R*-Tempty complexes after an incubation of 30 min at 30^0C is still able to specifically bind to newly added T-GDP and to catalyse a GDP/GTP exchange on it.

THE SECOND STAGE OF THE CASCADE: ACTIVATION OF PDE BY T_α GTP

The cGMP specific phosphodiesterase is an hydrolytic enzyme inhibited in the resting state by the constraint imposed by the binding of an inhibitor. In the absence of precise data on the inhibitor abundance, a 1:1:1 stoichiometry had been usually assumed for PDEα, PDEβand I (PDEγ) subunits in the inactive holoenzyme. N. Miki et al., (1975) first observed that PDE activity could be elicited by tryptic treatment of ROS membrane, suggesting that PDE is maintained in its basal state by an easily proteolysable inhibitor. A 13 kD subunit isolated from the purified enzyme by Hurley and Stryer (1982) was shown to inhibit, with high affinity, the 85+88 kD catalytic complex PDEαβ of the trypsin-activated enzyme. Therefore, PDE activation *in vivo* was expected to result from an interaction between TαGTP and the inhibitor. Starting from a mixture of purified TαGTPγS and native PDE, analysed by ion exchange chromatography, Ph. Deterre et al., (1987) provided evidence for a scheme where TαGTP binds to I and forms with it a stoichiometric complex that dissociates from the catalytic units in the conditions used in the experiments. Further analysis of the T-PDE interaction demonstrated the existence of two inhibitory subunits per native PDE (Ph. Deterre et al., 1988): the chromatography of an activated ROS extract resolved two different peaks of activated PDE beside the excess native (inactive) PDE peak. The last peak is totally devoid of nucleotide and, in the intermediate peak, the I/PDEαβ ratio is close to 50 % of that found for native PDE. The simplest explanation is to assume two I in the inactive PDE (PDEαβ -I_2), and the existence of two different active states with one I (PDEαβ -I) and no I (PDEαβ). In fact, these two activated states can be reached upon progressive mild trypsin proteolysis of native PDE, which degrades and successively eliminates the two inhibitors. In preliminary experiments, the specific activity of PDEαβ -I appears to be about half that measured on totally stripped PDEαβ . However, an exchange of inhibitory subunits can be observed from PDEαβ -I_2 to PDEαβ when these two species are mixed together. As nothing is known yet regarding the affinities of I for the different states of PDE, it is still impossible to estimate unambiguously the enzymatic properties of the intermediate complex PDEαβ -I. Specific activities of trypsin-activated PDEαβ and T_α GTP-activated PDEαβ are the same, when tested in solution: about 1000 cGMP hydrolysed per sec. However, they differ by their ability to bind to disc membranes: the transducin-activated PDE remains peripherally membrane bound, whereas the trypsin-activated species no longer interacts with the membranes. In fact, before digesting the inhibitors, trypsin cleaves a very short terminal peptide which seems to be necessary for membrane attachment of the catalytic units (T. G. Wensel and L. Stryer, 1986), (Catty and Deterre, in preparation). Removing that small part could interfere with the catalytic efficiency. Thus, tryptic treatment is not a good model for a quantitative analysis of the "natural" activation process by T_α GTP.

The existence of two inhibitory subunits per PDE complex and their exchangeability between inactive PDE$\alpha\beta$ -I$_2$ and fully activated PDE$\alpha\beta$ could allow a diffusion-controlled fast inactivation of PDE: a transducin-activated PDE$\alpha\beta$, diffusing away from the proximity of the R* that creates the pool of TαGTP, gets into a membrane area where it encounters excess native PDE$\alpha\beta$ -I$_2$, from which it might regain on inhibitor, forming then two PDE$\alpha\beta$ -I. If PDE$\alpha\beta$ -I has a lower Vmax or a higher Km for cGMP than PDE$\alpha\beta$, this process could provide a rapid quenching of the cGMP hydrolysis that would precede the permanent inactivation of transducin resulting from the GTP hydrolysis.

TERMINATION OF THE CASCADE: THE LIFETIME OF TαGTP

We have already discussed the fast blocking of R* by the combined action of rhodopsin kinase and arrestin. But in order to terminate the signal, a fast deactivation of the PDE is also required. The activation of PDE lasts as long as the complex T$_\alpha$ GTP-PDEγ is stable, that is until the GTP in T$_\alpha$ is hydrolysed. Then T$_\alpha$ GDP dissociates from PDEγ which binds back onto PDE$\alpha\beta$ and inhibits it. The lifetime of the GTP in T$_\alpha$, limited by its spontaneous hydrolysis in GDP, should then be the limiting factor. This causes a problem, since the GTPase activity of transducin, as measured *in vitro*, seems too slow: the turn over rate of phosphate release at saturating illumination is on ROS membrane suspension of 1 to 5 Pi/min/Transducin, depending on the medium and on the preparation. This seems to imply a lifetime of at least 10 seconds for the active state T$_\alpha$ GTP, that is much too long to account for the termination of the physiological signal in less than one second. The problem might arise from the fact that in these *in vitro* studies, the rods are broken and T$_\alpha$ GTP, which is soluble, diffuses away from the disc membrane stack; it might take a substantial time for T$_\alpha$ GDP, after the hydrolysis, to come back and reassociate with T$_{\beta\gamma}$ which remained membrane bound, before it can have its GDP exchanged for a new GTP loaded by interacting with R*. The real lifetime of the T$_\alpha$ GTP state might then be much shorter than what is estimated from the turn over rate of the GTPase.

Microcalorimetric measurements of the lifetime of T$_\alpha$ GTP. An original experimental approach of this problem is to try to analyse the time course of enthalpy release in a population of T$_\alpha$ GTP formed by a short pulse of R* activation. For each GTP hydrolysed in GDP + P, 8 kilocalories are released. We have built a fast time-resolved micro-

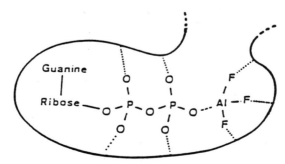

Fig. 7: The reversible binding of AlF$_4^-$ to GDP in the nucleotide-binding site of transducin.

calorimeter based on the use of pyroelectric PVDF film (M. Chabre et al., 1988). Each sample contains the rods extractable from 2 to 4 cattle retinas (4 to 8 mg/ml rhodopsin). The sensitivity of the device is of the order of one μcal/sec and its impulse response, as measured by flashing an absorbing dye, is 0.5 sec (full width at half maximum). When GTP is added to a rod suspension, illumination elicits a heat pulse that far exceeds the heat from the flash light itself (which can be measured accurately and subtracted). The time course of the heat pulse depends not only on the hydrolysis lifetime of the T_α GTP produced, but also on the lifetime of the photoexcited rhodopsins: as long as the R* are active they will reload GTP on the transducins that have already hydrolyzed their first loaded GTP. To reach the real lifetime of T_α GTP, that of R* must be reduced to a neglegible value. This is probably insured *in vivo* by the combined action of rhodopsin kinase and arrestin. These soluble enzymes are however very diluted in the broken rod preparations, making their action too slow. To hasten the decay of R*, we therefore added in the suspension mM amounts of hydroxylamine which does not act on rhodopsin in the dark but inactivate photoexcited rhodopsins. Experiments conducted at 20^0C showed that the duration of the heat pulse resulting from the hydrolyse of GTP can be reduced to 3 seconds when sufficient hydroxylamine is present. Increasing the hydroxylamine to higher concentrations (up to 50 mM) then reduced only the amplitude of the heat pulse, probably by destroying R* too fast to allow the loading of GTP on all the transducins, but did not modify its time course. This 3 seconds limit must therefore represent the lifetime of hydrolysis of GTP in T_α GTP at 23^0C. This is one order of magnitude faster than what one would expect from the known turn-over rate of the GTPase which, one should notice, has alxays been measured in diluted membrane suspension. If one extrapolates these results to the physiological temperature of 37^0C, the lifetime of T_α GTP might be short enough to insure the turn off of the cGMP PDE within less than a second.

CONCLUSION

The cycle of transducin shuttling between rhodopsin and PDE is known in more details than for any other G-protein systems. One is still however far from understanding the interactions between the three proteins at the molecular level. This would first require the determination of their tridimensional structures. The analysis of the two key protein-protein interactions would also demand cristallographic studies of the two corresponding complexes: R*-T_α empty-$T_{\beta\gamma}$ and T_α GTP-I. The mechanism of action of the guanine nucleotide in the system needs to be further analysed, perhaps by drawing on possible analogies not only with other GTP-binding proteins such as elongation factors or tubulin, but even with ATP-binding proteins like actin or other contractile proteins. Notwithstanding some peculiar characteristics related to the specific nature of the visual signal, the transducin cascade has proven, and will still prove to be a very instructive model for G-protein transduction processes.

REFERENCES

Bennett, N., Michel-Villaz, M. and Kuhn, H., 1982, Light-induced interaction between rhodopsin and the GTP-binding protein, Metarhodopsin II is the major photoproduct involved, *Eur. J. Biochem.*, 127:97-103.

Bennett, N, and Dupont, Y., 1985, The G-protein of retinal rod outer segments (Transducin): Mechanism of interaction with rhodopsin and nucleotides, *J. Biol. Chem.*, 260:4156-4168.

Bigay, J., Deterre, P., Pfister, C. and Chabre, M., 1985, Fluoroaluminates activate transducin-GDP by mimicking the g-phosphate of GTP in its binding site, *FEBS Lett.*, 191:181-185.

Bigay, J., Deterre, P., Pfister, C. and Chabre, M., Fluoride, 1987, complexes of aluminium or beryllium act on G-proteins as reversibly bound analogues of the g-phosphate of GTP, *EMBO J.*, 6:2907-2913.

Bruckert F., Vuong, T. M., and Chabre, M., Light and GTP dependence of transducin solubility in retinal rods, Further analysis by near infra-red light scattering, *Eur. Biophys. J.*, (in press).

Chabre M., 1983, Conformational and functional change induced in vertebrate rhodopsin by photon capture, *in*: "The Biology of Photoreception", D.J. Cosens and D. Vince Prue S.E.B, ed., Symposia XXXVI, Cambridge University Press.

Chabre, M., Pfister, C., Deterre, P. and Kuhn H., 1984, The mechanism of control of cGMP phosphodiesterase by photoexcited rhodopsin, Analogies with hormone controlled systems, *in*: "Hormone and Cell regulation," vol. 8, 87-98, J. Dumont and J. Nunez, ed., Elsevier.

Chabre, M., Bigay, J., Bruckert, F., Bornancin, F., Deterre, P., Pfister, C. and Vuong, T. M., 1988, Visual Signal Transduction: The cycle of Transducin Shuttling between Rhodopsin and cGMP Phosphodiesterase, Cold Spring Harbor Symposia on quantitative biology, 53: "Molecular Biologie of Signal Transduction", 313-323.

Cooper, A., 1979, Energy uptake in the first step on visual excitation, *Nature*, 282:531-533.

Deterre, P., Bigay, J., Robert, M., Pfister, C., Kuhn, H. and Chabre, M., 1987, Activation of retinal rod cyclic-GMP phosphodiesterase by transducin: Characterization of the complex formed by phosphodiesterase inhibitor and transducin α-subunit, *Prot. Struct. Funct. Genet.*, 1:188-193.

Deterre, P., Bigay, J., Forquet, F., Robert, M. and Chabre, M., 1988, The cGMP phosphodiesterase of retinal rods is regulated by two inhibitory subunits, *Proc. Natl. Acad. Sci. USA*, 85:2424-2428.

Dohlman, H. G., Caron, M. G. and Lefkowitz, R. J., 1987, A family of receptors coupled to guanine nucleotide regulatory proteins. *Biochem.*, 26:2657-2664.

Dratz, E. A. and Hargrave, P. A., 1983, The structure of rhodopsin and the rod outer segment disc membrane, *TIBS* 8:128-131.

Emeis, D., Kuhn, H., Reichert, J. and Hofmann, K. P., 1982, Complex formation between metarhodopsin II and GTP-binding protein in bovine photoreceptor membranes leads to a shift of the photoproduct equilibrium, *FEBS Lett.*, 143:29-34.

Hurley, J. B. and Stryer, L., 1982, Purification and characterization of the g regulatory subunit of the cyclic-GMP phosphodiesterase from retinal rod outer segments, *J. Biol. Chem.*, 257:11094-11099.

Kuhn, H., 1980, Light- and GTP-regulated interaction of GTPase and other proteins with bovine photoreceptor membranes, *Nature* 283:587-589.

Kuhn, H., 1984, Interactions between photoexcited rhodopsin and light-activated enzymes in rods, *in*: "Progress in Retinal Research", N. Osborne and J. Chader, ed., vol. 3, 123-156, Oxford: Pergamon Press.

Lefkowitz, R. J., Stadel, J. M and Caron, M. G., 1983, Adenylate cyclase-coupled beta-adrenergic receptors: Structure and Mechanisms of Activation and Desensitization, *Ann. Rev. Biochem.*, 52:159-186.

Miki, N., Baraban, J. M., Keirns, J. J. Boyce, J. J. and Bitensky, M. W., 1975, Purification and properties of the light-activated cyclic nucleotide phosphodiesterase of rod outer segments, *J. Biol. Chem.*, 250:6320-6328.

Ovchinnikov, Y.A., Abdulaev, N. G. and Bogachuk, A. S., 1988, Two adjacent cysteine

residues in the C-terminal cytoplasmic fragment of bovine rhodopsin are palmity-
lated, *FEBS Lett.*, 230:1-5.

Pfister, C., Kuhn, H., and Chabre, M., 1983, Interaction between photoexcited rhodopsin
and peripheral enzymes in frog retinal rod, Influence on the post-metarhodopsin II
decay and phosphorylation rate of rhodopsin, *Eur. J. Biochem.*, 136:489-499.

Pfister, C., Chabre, M., Plouet, J., Tuyen, V. V., De Kozak, Y., Faure, J. P. and Kuhn,
H., 1985, Retinal S antigen identified as the 48K protein regulating light dependent
phosphodiesterase in rods, *Science*, 228:891-893.

Sternweis, P.C., and Gilman, A. G., 1982, Aluminium: A requirement for activation of
the regulatory component of adenylate cyclase by fluoride, *Proc. Natl. Acad. Sci.
USA*, 79:4888-4891

Stryer, L., 1986, Cyclic GMP cascade of vision, *Ann. Rev. Neurosci.*, 9:87-119.

Stryer, L. and Bourne, H. R., 1986, G Proteins: a family of signal transducers, *Ann.
Rev. Cell Biol.*, 2:391-419.

Vuong, T. M., Chabre, M. and Stryer, L., 1984, Millisecond activation of transducin in
the cyclic nucleotide cascade of vision, *Nature*, 311:659-661.

Wensel, T. G. and Stryer, L., 1986, Reciprocal control of retinal rod cyclic GMP phosphodiesterase by its g subunit and transducin, *Proteins Struc. Funct. Genet.*,
1:90-99.

Wilden, U., Hall, S. W. and Kuhn, H., 1986, Phosphodiesterase activation by photoexcited rhodopsin is quenched when rhodopsin is phosphorylated and binds the 48K
protein of rod outer segments, *Proc. Natl. Acad. Sci. USA*, 83:1174-1178.

GUANYLATE CYCLASE OF RETINAL ROD OUTER SEGMENT

I.M. Pepe, I. Panfoli and C. Cugnoli

Istituto di Cibernetica e Biofisica del C.N.R.
Genova, Italy

ABSTRACT

Guanylate cyclase activity was studied in the rod outer segments of the toad retina. The enzyme becomes sensitive to calcium ions after illumination, showing an enhancement of its activity when Ca^{2+} concentration is lowered from 10^{-5} M to less than 10^{-8} M. A possible pathway of cyclase activation by light was also investigated. When Zaprinast was used as inhibitor of phosphodiesterase, the light-activation of cyclase and of phosphodiesterase were simultaneously suppressed, suggesting a coupling of the two antagonist enzymes.

INTRODUCTION

Cyclic GMP metabolism plays a central role in phototransduction (Lamb, 1986; Pugh and Cobbs, 1986; Stryer, 1986). Cation-specific channels in the plasma membranes of vertebrate retinal rods are activated by molecules of cyclic GMP (cGMP) (Caretta et al., 1979; Fesenko et al., 1985). Light triggers an enzymatic cascade that leads to the activation of a phosphodiesterase (PDE) that hydrolyzes cGMP. The resulting hyperpolarization of the plasma membrane constitutes the rising phase of the neuronal signal. The falling phase which restores the normal dark potential requires the resynthesis of cGMP, catalysed by guanylate cyclase. The drop of internal calcium concentration following a flash of light (Yau and Nakatani, 1985; McNaughton et al., 1986; Lamb et al., 1986) could activate cyclase, as suggested by experiments on rod outer segments kept in low calcium. Pepe et al. (1986) reported that the guanylate cyclase of the rod outer segments of toad retina was strongly stimulated at 10^{-8} M Ca^{2+} only after a flash of light. In contrast Koch and Stryer showed recently (1988) that guanylate cyclase activity of unilluminated bovine rod outer segments was increased by 5-fold when calcium level was lowered from 10^{-7} to 10^{-8} M.

In this paper we report cyclase activity measurements on toad rod outer segments carried out both in darkness and in dim red light which could conciliate the different observations of the two groups.

Moreover we show that cyclase activation by light follows the same enzymatic cascade that binds excited rhodopsin with cGMP hydrolysis.

METHODS

Preparations. Rod outer segments were obtained from dark adapted eyes of toads Bufo-Bufo, enucleated under dim red light. The eyes were hemisected under infrared light and retinae were removed and gently shaken in 35 % of sucrose (w/w) in Ringer solution which did not contain $CaCl_2$ (115 mM NaCl, 2.5 mM KCl, 1 mM $MgCl_2$, buffered at pH 7.5 with 10 mM Hepes and tetramethylammonium hydroxyde). Rod outer segment suspension was centrifuged at 6000 rpm (3000 g). The pellet was discarded and the supernatant diluted with Ringer solution and centrifuged at 4000 rpm for 10 min. The resulting supernatant was discarded and the pellet, which contained rod outer segments from 4 retinae, was hypotonically shocked in 100μl of 5 mM Mops pH 7.1 containing 5 mM dithiolthreitol. The obtained disks membranes suspension was homogenized with a 100μl pipette with a plastic tip (0.2 mm diameter orifice) to prevent disks from aggregating.

Guanylate cyclase assay. Guanylate cyclase was assayed following the methods of Pannbacker (1973) and Fleischman (1982) with minor modifications. The reaction mixture contained 100 mM Mops pH 7.1, 140 mM KCl, 20 mM NaCl, 5 mM $MgCl_2$, 2 mM GTP, 10 mM phosphocreatine, 1 mg/ml creatine phosphokinase,4mM IBMX or 0.4 mM Zaprinast (2-o-Propoxyphenyl-8-azapurin-6-one : a specific inhibitor of PDE) , 4 mM cGMP, 3μCi of [^{14}C] GTP (450 mCi/mmol) and 10μCi of [^{3}H] cGMP (15 Ci/mmol). The reaction was initiated by adding a volume of rod outer segment homogenate to an equal volume of reaction mixture with a final protein concentration of about 1-2 mg/ml. The mixture was incubated at room temperature (about 24^0C) and aliquots of 20μl were collected at different times. Reactions were stopped by adding 5μl of 100 mM EDTA and boling for 2 min. After centrifuging, 10μl of the supernatants were applied to a polyethyleneimine-cellulose thin layer plate (20x20 cm) which was then developed in one dimension, first with distilled water and, after drying, with 0.2 M LiCl for 7 cm and then with 1 M LiCl. 5'-GMP, GDP and Guanosine were added as internal standard to the samples. Nucleotides and nucleosides were separated and visualized under short wave UV light. The spots were cut out and transferred to scintillation vials; 0.5 ml of 0.7 M $MgCl_2$ /1 M Tris-HCl pH 7.4 (10/2, v/v) were added to each vial to elute the compounds; 3 ml of Instagel was added and samples were counted in a LKB liquid scintillation spectrometer. Counting efficiency was about 34% and 20% for [^{14}C] and [^{3}H] respectively.

Different concentrations of Ca^{2+} were obtained adding 2 mM EGTA to samples containing: 1.8 mM $CaCl_2$ (1.6 10^{-6}M Ca^{2+}), 1.0 mM $CaCl_2$ (1.8 10^{-7} M Ca^{2+}), 0.2 mM $CaCl_2$ (2.1 10^{-8} M Ca^{2+}).

Protein concentration was determined with a method based on Coomassie blue staining (Asim Esen, 1978).

RESULTS

Guanylate cyclase activity in ROS homogenates was measured from the formation of [^{14}C] cGMP from [^{14}C] GTP, taking into account cGMP hydrolysis due to PDE activity which was monitored following the decrease of tritium-labeled cGMP.

In the first experiments complete darkness was not achieved because some sources of dim red light were still present in the dark room where the ROS preparations and cyclase assays were carried out. In these conditions we obtained results (Fig. 1, open circles) indi-

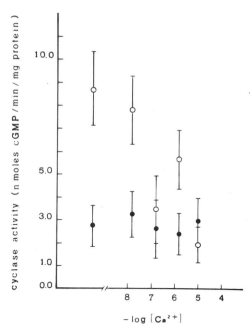

Fig. 1: Effect of calcium on cyclase activity: in the dark (●) each value represents the mean ± SD of 5 - 10 determinations; in dim red light (○) each value represents the mean ± SD of 3 determinations. Samples were incubated 4 min for cyclase assay.

cating that cyclase activity increased by decreasing calcium ion concentration from 10^{-5}M to 10^{-8}M. When darkness was improved by eliminating every light sources, we were no longer able to repeat those initial observations. More than 30 experiments were done in complete darkness, using only infrared light, and the value of 3 ± 1 nmol cGMP / min x mg protein was obtained for cyclase activity independently of free Ca^{2+} concentration (Fig. 1, filled circles).

Supposing that light would affect cGMP synthesis, we decided to measure cyclase activity after illumination with a flash of light. We found that cGMP synthesis was enhanced by light up to about 30-fold upon lowering Ca^{2+} concentration from 10^{-6}M to less than 10^{-8} M (Pepe et al., 1986). Some of those results are reported in Table 1.

Cyclase activity measured at intervals longer than 1 min after the flash of light did not give reliable values because 90% or more of the cGMP formed was degraded by PDE, despite the presence of 2mM IBMX. In order to be able to follow the kinetics of cyclase activation by light, some experiments were carried out in the presence of Zaprinast, a stronger inhibitor of PDE. The results are shown in Fig 2 where cyclase activity could be monitored up to 4 min after a flash of light, thank to the slowing down of PDE activity in 0.2 mM Zaprinast. Following a flash of light, cyclase activity was increased by about 5 times at about 10^{-8}M Ca^{2+}, and by 3 times at about 10^{-7}M Ca^{2+}, while it did not change appreciably when calcium concentration was raised up to 10^{-4}M.

Table 1: Effect of light and calcium on cyclase activity

Ca^{2+} conc.	Cyclase activity (nm min $^{-1}mg^{-1}$)	
	dark	light
$1.0 \times 10^{-5}M$	3.0 ± 1.0 (5)	3.8 ± 0.8 (3)
$1.6 \times 10^{-6}M$	3.2 ± 0.9 (8)	3.1 ± 0.9 (4)
$1.8 \times 10^{-7}M$	2.8 ± 0.9 (8)	8.7 ± 2.0 (4)
$2.0 \times 10^{-8}M$	3.2 ± 0.8 (10)	33.1 ± 8.0 (4)
$< 10^{-8}M$ (2mM EGTA only)	2.9 ± 0.9 (9)	103 ± 30.0 (3)

Dark values were obtained from samples incubated 4 min in darkness. Values in the light were obtained from samples incubated 4 min in darkness and illuminated with a flash bleaching about 10^{-3} % of rhodopsin, 1 min before the reaction was stopped. The values given are the mean $\pm$ S.D. for the numbers of determinations shown in parentheses. Cyclase assay was carried out in the presence of 2 mM IBMX.

In the presence of 2mM IBMX and in the same condition of illumination and calcium concentration, light-activation of PDE as well as of cyclase was higher (about 10-fold for cyclase, at $2\ 10^{-8}$M Ca^{2+}, as shown in Table I) , suggesting that cyclase activation by light should increase as the inhibition of PDE was less effective.

In order to examine this hypothesis, experiments were carried out in the presence of different concentrations of Zaprinast. The results are shown in Table II. In the dark, different concentrations of Zaprinast had no effect on cyclase activity. On the contrary after a flash of light, cyclase activation was enhanced upon lowering the concentration of Zaprinast. From Table II it can be easily noticed that maximal activation of cyclase by light corresponded to minimal inhibition of PDE. Moreover when PDE was completely inhibited by 1 mM Zaprinast, light-activation of cyclase was suppressed.

Table 2: Effect of Zaprinast on the light-activation
of cyclase and phosphodiesterase

Zaprinast conc. (mM)	cyclase activity (nm min^{-1}mg^{-1})		PDE activity (nm min^{-1}mg^{-1})	
	dark	light	dark	light
0.05	2.6	46.8	187	750
0.20	3.0	15.1	225	450
1.00	2.8	2.7	75	85

Samples-from the same ROS preparation-were incubated for 5 min in the dark, then illuminated with a flash of light bleaching about 10^{-3} % of rhodopsin. Ca^{2+} concentration was about 2×10^{-8} M. Values are the means of duplicate determinations.

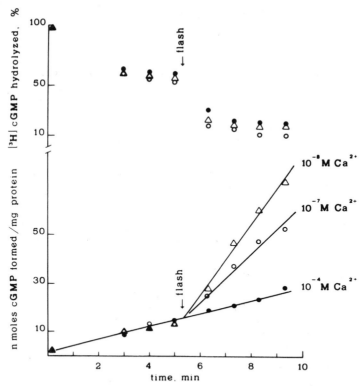

Fig. 2: Effect of light on cyclase activity at different calcium concentrations. Samples - from the same ROS preparation - were assayed for cyclase in the dark and illuminated with a flash (arrow) bleaching about $10^{-3}\%$ of rhodopsin, in the presence of 0.2 mM Zaprinast. In the upper part of the figure, cGMP hydrolysis owing to PDE activity is shown.

DISCUSSION

Our measurements on toad ROS show that a large stimulation of guanylate cyclase activity by lowering cytosolic calcium is observed only following a flash of light and this stimulation is strongly dependent on PDE inhibition. A relevant consequence of these results is that light-activation of cyclase may follow the same enzymatic cascade that links photoexcited rhodopsin to PDE activation, via the G-protein transducin. Consistent with this hypothesis is the observation that, when antibody 4A against transducin was added to cyclase assay, PDE activity after a flash of light was inhibited, and so was the light-activation of cyclase (Pepe et al., 1988). The same effect was obtained in the presence of 1 mM Zaprinast (see Table II). Furthermore, when the concentration of Zaprinast was decreased, both PDE and cyclase activation by light increased. These results suggest a coupling between the two antagonist enzymes: PDE activated by light, would in turn be able to activate cyclase, in low calcium ions concentration.

One simple mechanism for such coupling could be the enhancement of cyclase activity by light as a consequence of the rapid removal of its product - cGMP - by light-activated PDE. Nevertheless, experiments carried out with different concentrations of cGMP - from 20 mM to 0.1 mM - a range of concentrations wider than that occurring in the usual assay, gave almost identical values for cyclase activity (Pepe and Panfoli, 1988).

Cyclase activity was always measured in the presence of some PDE inhibitor. The values for this activity in the dark were independent on PDE-inhibitor concentration, while those after a flash of light depended strongly on PDE inhibition as shown in Table II. This leads to the conclusion that cyclase activity in the light was always underestimated. At calcium concentration less than $10^{-8}M$, for instance, the extent of light-activation of cyclase was 30-fold its basal activity in the presence of IBMX (see Table 1). Without IBMX this value would be certainly higher. Such an enhancement of cyclase activity would be necessary to antagonize the tremendous activity of PDE which, in similar condition of illumination, gave values of about 950 nm cGMP hydrolyzed/min x mg protein (Keirns et al., 1975).

Our observations about the independence of cyclase basal activity from calcium concentration in darkness, might seem in contrast with the large increase of the dark-current observed in rods after removal of the extracellular Ca^{2+} (Rispoli et al., 1988; Lamb and Matthews, 1988). Given the highly cooperative effect of cGMP on the dark-current (Zimmermann and Baylor, 1986; Haynes et al., 1986), the increase of this latter can be explained by assuming a simple two-fold increase of the level of free cGMP that our technique would not be able to reliably detect, as it cannot follow variations in cyclase activity smaller than a two-fold change.

Similarly our results seem to be in desagreement with those of Koch and Stryer (1988) where they report that cyclase activity of unilluminated bovine ROS increased 5-fold when calcium level was lowered from $2\ 10^{-7}M$ to $5\ 10^{-8}M$. One possible reason for the discrepancy could be the fact that Koch and Stryer did retina dissection and ROS preparation in dim red light, which led them to results for cyclase activity that are very similar to the results of our measurements carried out in dim red light, showed in Fig. 1.

REFERENCES

Asim, Esen, 1978, *Analyt. Biochem*, 89:264.

Caretta, A., Cavaggioni, A. and Sorbi, R. T., 1979, *J.Physiol. Lond.*, 295:171.

Fesenko, E. E., Kolesnikov, S.S. and Lyubarsky, A.L., 1985, *Nature*, 313:310.

Fleishman, D., 1982, *Methods Enzymol.*, 81:522.

Haynes, L. W., Kay, A.R. and Yau, K.W., 1986, *Nature*, 321:66-70.

Keirns, J. J., Miki, N., Bitensky, M.W. and Keirns, M., 1975, *Biochemistry*, 14:2760.

Koch, K. W. and Stryer, L., 1988, *Nature*, 334:64.

Lamb, T. D., 1986, *Trends Neurosci.*, 9:224.

Lamb, T. D., Matthews, H. R. and Torre, V., 1986, *J. Physiol. Lond.*, 372:315.

Lamb, T. D. and Matthews, H. R., 1988, *J.Physiol* 403:473-494.

McNaughton, P. A., Cervetto, L. and Nunn, B. J., 1986, *Nature*, 322:261.

Pannbaker, R. G., 1973, *Science*, 182:1138.

Pepe, I. M., Panfoli, I. and Cugnoli, C., 1986, *FEBS Lett.*, 203:73.

Pepe, I. M. and Panfoli, I., 1988, *In*: "Light in Biology and Medecine," vol 1, pp. 357-361, R. H., Douglas, J., Moan and F., Dall'Acqua, ed., Plenum Press, New York.

Pepe, I. M., Panfoli, I. and Hamm, H. E., 1989, *Cell Biophysics* (in the press).

Pugh Jr, E. N. and Cobbs, W. H., 1986, *Vision Res.*, 26:1613.

Rispoli, G., Sather, W. A. and Detwiler, P. B., 1988, *Biophys. J.*, 53:388a.

Stryer, L., 1986, *Ann. Rev. Neurosci.*, 9:87.

Yau, K. -W. and Nakatani, K., 1985, *Nature*, 313:579.

Zimmerman, A. L. and Baylor, D. A., 1986, *Nature*, 321:70-72.

PHYSICAL STUDIES OF α-$\beta\gamma$ SUBUNIT INTERACTIONS OF THE ROD OUTER SEGMENT G PROTEIN, G_t: EFFECTS OF MONOCLONAL ANTIBODY BINDING

M. R. Mazzoni and H. E. Hamm

Department of Physiology and Biophysics
University of Illinois, College of Medicine at Chicago, U.S.A.

INTRODUCTION

A guanine nucleotide-binding regulatory protein, called transducin or G_t, couples light-activated rhodopsin with the cGMP phosphodiesterase (Hurley, 1980). Like all G proteins, transducin is a heterotrimer composed of two distinct subunits: α_t (39 kDa) and $\beta\gamma_t$ (β_t: 36 kDa, γ_t: 8 kDa). The α_t subunit binds GDP and GTP (Fung and Stryer, 1980), and in its GTP-bound form activates the cGMP phosphodiesterase (Fung et al., 1981). The activation is terminated when the bound GTP is hydrolyzed to GDP by an intrinsic GTPase activity (Wheeler and Bitensky, 1977, Fung et al., 1981). Although the $\beta\gamma_t$ subunit does not directly participate in either GTP hydrolysis or phosphodiesterase activation, its presence is important for effective binding of α_t to photolyzed rhodopsin and GTP-GDP exchange (Fung, 1983). Transducin binds tightly to photolyzed rhodopsin in intact rod outer segment membranes (Kühn, 1980). It also binds selectively to unphotolyzed rhodopsin in rod outer segment membranes (Hamm et al., 1987) and in reconstituted phospholipid vesicles (Fung, 1983).

Several monoclonal antibodies (mAb) were generated to test the functional roles of transducin in phototransduction (Hamm and Bownds, 1984; Witt et al., 1984). One of these antibodies, mAb 4A, was found to block light-activated GTP-GDP exchange and cGMP phosphodiesterase, whereas other antibodies, including mAb 4H, did not have this effect (Hamm and Bownds, 1984). When the mechanism of action of mAb 4A was examined, this antibody appeared to block a site on transducin that interacts with rhodopsin (Hamm et al., 1987; Hamm et al., 1988). Using the combination of enzyme digestion and antibody binding of transducin, Deretic and Hamm (1987) determined the antigenic sites of five monoclonal antibodies which are able to inhibit the association of transducin with rhodopsin in the dark (Hamm et al., 1987). Particularly, the major binding site of mAb 4A appears to be within region Arg^{310} to Lys^{329} at the COOH-terminus of the α_t subunit. These results, obtained by tryptic digestion studies, were also confirmed by ELISA competition experiments using synthetic peptides of known amino acid sequences (Hamm et al., 1988).

Recently, the three-dimensional structure of an antigen-antibody complex was resolved (Amit et al., 1986). It was shown that the surface of this antigen-antibody interaction is rather flat, with a diameter of approximately 20 Å. This allows different segments of the polypeptide chain to form the antigen residues which are contacted by the antibody. However, tryptic digestion studies can detect only the center or the most tightly binding region of the antigen-antibody binding area. Thus the participation of surrounding amino acids to the binding site cannot be proven using these methods. For example, according to the model proposed by Deretic and Hamm (1987), the amino-terminal region of α_t would be relatively close to the COOH-terminus and would contain a part of the antigenic site. On the other hand, this possibility cannot be demonstrated by tryptic digestion methods since the amino-terminal 2-kDa tryptic fragment is lost in the first minutes of trypsinization and is too small to bind to nitrocellulose.

The exact location on α_t of the binding domain for $\beta\gamma_t$ is not known, although the involvement of the amino-terminal region has been implicated (Fung and Nash, 1983; Watkins et al., 1985). More recently, Navon and Fung (1987) showed by immunoprecipitation using an anti-α_t monoclonal antibody, that removal of the amino-terminal region of α_t by *Staphylococcus aureus* V8 protease impairs $\beta\gamma_t$ binding.

In this study, the hydrodynamic properties of the transducin-mAb 4A complex were examined in order to test whether the binding of this antibody to α_t interferes with subunit interaction. This anti-α_t antibody appears to affect the binding of α_t to $\beta\gamma_t$, while control IgG and another monoclonal antibody (4H), which does not inhibit transducin interaction with rhodopsin, do not dissociate the $\alpha\beta\gamma_t$ complex.

MATERIALS AND METHODS

Preparation of Protein

Bovine ROS were prepared according to Papermaster and Dreyer (1974). Bovine transducin was extracted from rod outer segments (ROS) as described by Kühn (1980), with some modifications. Pure α_t and $\beta\gamma_t$ subunits were prepared by chromatography of * GTPγS-activated a$\beta\gamma_t$ on Blue Shepharose CL-6B, essentially as described by Kleuss et al. (1987). Protein concentrations were determined by the Coomassie blue binding method (Bradford, 1976) using IgG as a standard. Protein concentration and purity were also quantified by SDS-polyacrylamide gel (12.5%) electrophoresis followed by Coomassie blue staining and densitometric scanning (Ephortec, Joyce Loebl densitometer), using bovine serum albumin as a standard.

Sucrose Density Gradient Centrifugation

To activate G_t, protein samples (30 μg) were incubated in buffer A with dark adapted ROS membranes and GTPγS (100μM) for 20 min at 4^0C in light. Samples (150-170 μl) containing purified $\alpha\beta\gamma_t$, activated $\alpha\beta\gamma_t$, α_t or $\beta\gamma_t$ (30-60 μg) and marker proteins were layered on the top of 3.2 ml gradients of 5-20% sucrose prepared in buffer A (10 mM MOPS, 60 mM KCl, 30 mM NaCl, 2 mM MgCl$_2$, 1 mM DTT, 0.1 mM PMSF, pH 7.5) or

---

* The abbreviations used are: GTPγS, guanosine-5'-O-(3- thiotriphosphate); SDS, sodium dodecyl sulphate; MOPS, 3-(N- morfolino)propanesulfonic acid; DTT, dithiothreitol; PMSF, phenylmethyl-sulfonyl fluoride.

buffer B (10 mM MOPS, 60 mM KCl, 30 mM NaCl, 2 mM MgCl$_2$, 0.1 mM DTT, 0.1 mM PMSF, pH 7.5). Samples containing activated G$_t$ were layered on gradients prepared in buffer A containing 100 μM GTPγS. Purified G$_t$ was incubated for 30 min at room temperature in buffer B in the presence either of monoclonal antibodies or nonimmune rabbit IgG. The marker protein mix consisted of 14 μg of bovine serum albumin, 24 μg of carbonic anhydrase and 5 μg of cytochrome c. The gradients were centrifuged at 41,000 rpm for 15 h in a Beckman SW 50.1 rotor at 4^0 C and fractionated into approximately 25 fractions. Aliquots of these fractions were subjected to precipitation by acetone and the position of the proteins was determined by SDS-polyacrylamide gel (12.5%) electrophoresis, according to Laemmli (1970), followed by Coomassie blue staining and densitometric scanning.

Table I Hydrodynamic properties of α_t, $\beta\gamma_t$ and $\alpha\beta\gamma_t$

	Sedimentation coefficient $s_{20,w}$(S)	
Buffer B		
$\alpha\beta\gamma_t$	4.10 ± 0.07	(5)
Buffer A		
$\alpha\beta\gamma_t$	4.00 ± 0.07	(4)
α_t	3.48 ± 0.34	(3)
$\beta\gamma_t$	3.75 ± 0.13	(3)
α_t	3.4	
$\alpha\beta\gamma_t$ + GTPγS		(2)
$\beta\gamma_t$	3.6	

Purified transducin or subunits were applied to 5 - 20% linear sucrose density gradients prepared in either buffer A or B and centrifuged as described in Methods. Following the centrifugation, each gradient was fractionated, and the sedimentational profiles of transducin, α_t and $\beta\gamma_t$ were determined by electrophoresis on 12.5% SDS-polyacrylamide gels, Coomassie blue staining and densitometric scanning. Sedimentational mobility was defined as follows : mobility = [(total no. of fractions) - (peak fraction)]/(total fractions). The sedimentation coefficients [$s_{20,w}$ in Svedberg units (S)] were determined from the calibration curves, as shown in panel A of Fig. 1. The values are the averages ± S.E. The numbers in parentheses are the number of experiments.

RESULTS

Table I shows the values obtained for sedimentation coefficients ($s_{20,w}$) of free α_t, free $\beta\gamma_t$, and the associated complex. Under our experimental conditions, standard proteins sedimented in a linear relationship to their $s_{20,w}$ values (panel A, Fig. 1). The sedimentation rate of $\alpha\beta\gamma_t$ did not change when the sucrose density gradient experiments were carried out either in buffer A or buffer B. In fact, similar sedimentation coefficient values were obtained (Table I). This indicates the formation of $\alpha\beta\gamma_t$ aggregates is not made easier by lower concentrations of DTT (0.1 mM).

Therefore, sucrose density gradients prepared in buffer B were used to study the physical properties of antibody-transducin complex because in this low concentration of reducing agent the disulfide bonds of antibody are maintained. However, sometimes the

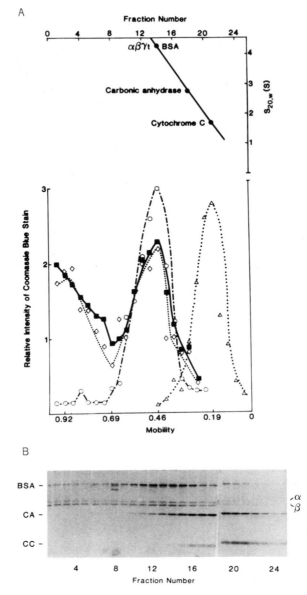

Fig. 1: Hydrodynamic behavior of transducin. Transducin (60 μg) was centrifuged on 5-20% linear sucrose density gradients prepared in buffer B as described under Methods. The gradients were recovered as 125 μl fractions with 1 representing the highest density fraction. The results were obtained from one experiment. A, The sedimentation mobilities of each standard protein and transducin were determined as described in Table I and expressed in Svedberg units (S). The gradient was calibrated by BSA, carbonic anhydrase, and cytochrome c. The profile of the distribution of α_t ($\diamond$•••••$\diamond$), $\beta\gamma_t$ (■————■), BSA (o•—•—•o), and cytochrome c ($\triangle$•••••$\triangle$) are shown at the bottom, as determined by gel densitometric scanning. B, Photograph of the Coomassie blue-stained polyacrylamide gel.

formation of transducin oligomers was noted in sucrose gradients prepared in both buffer A or buffer B, as shown by the presence of α_t and $\beta\gamma_t$ in high density fractions (panel B, Fig. 1). A direct relationship exists between the concentration of transducin used in the gradient and formation of oligomers. Holotransducin as a monomer migrates at a faster rate relative to the individual subunits, α_t and the $\beta\gamma_t$ complex. The $s_{20,w}$ values determined from at least three separate experiments, which were performed using 5-20% linear sucrose density gradients prepared in buffer A, were as follows: transducin, 4.0 ± 0.07; α_t, 3.48 ± 0.34; $\beta\gamma_t$, 3.75 ± 0.13.

In physiological conditions, the light activated rhodopsin tightly binds transducin and promotes its activation by GDP-GTP exchange and subsequent subunit dissociation. This activation is studied *in vitro* using a nonhydrolyzable GTP analog (GTPγS). When we incubated transducin with ROS membranes and GTPγS (100 μM) before performing the linear sucrose density gradient, we observed an alteration in the sedimentation behavior of transducin as compared to control experiments. We found that transducin, under these conditions, displays a shift in sedimentation, indicating dissociation of the $\alpha\beta\gamma_t$ complex. In fact, α_t and $\beta\gamma_t$ showed different sedimentation rates. The sedimentation coefficient values were similar (Table I) to those obtained from the sucrose density gradient centrifugations either of pure α_t or $\beta\gamma_t$. Before performing sucrose density gradient centrifugations of G_t in the presence of mAb 4A, we studied the sedimentation behavior of mAb 4A alone. In this case, a single peak was found in the high density fractions and a sedimentation coefficient of 5.7 S was estimated (Table II). No other minor peaks were found in fractions of higher density. This suggests that mAb 4A migrates in our linear sucrose density gradient as a monomer.

When the sedimentation behavior of transducin was examined in the presence of mAb 4A, a large change in the sedimentation rate was observed (panel B, Fig. 2). In normal conditions the $\alpha\beta\gamma_t$ complex comigrates with bovine serum albumin (BSA) in the sucrose density gradient (panel A of Fig. 1). In the presence of mAb 4A, the $\beta\gamma_t$ subunit migrates within one fraction of BSA, with an apparent $s_{20,w}$ value of 3.8 S (3.5 S and 4.2 S in duplicate experiments; panel A of Fig. 2), while the α_t peak is found in fractions of much higher density, even higher than mAb 4A alone. This shift in the sedimentation of transducin can only be the result of subunit dissociation by antibody. The main mAb 4A peak is not in the same position of α_t peak (panel A, Fig. 2), but it is found in a lower density fraction. This means that the α_t peak results from the formation of a stable, high density complex between α_t and mAb 4A, and there is some unbound antibody. This is because there is an excess of mAb 4A with respect to transducin (5:1). An easier formation of transducin aggregates in the presence of mAb 4A cannot explain the observed shift in the sedimentation of α_t, because $\beta\gamma_t$ shows a sedimentation behavior similar to the free monomeric subunit, and the amount which is found in high density fractions is low.

As a control experiment transducin was incubated in the presence of nonimmune rabbit IgG before performing the sucrose density gradient centrifugation. In this case, holotransducin co-migrates with BSA. Moreover, some preliminary experiments with another monoclonal antibody (mAb 4H), which has no effect on any studied function of α_t, show no dissociation of $\alpha\beta\gamma_t$ complex. Therefore, the effect of mAb 4A appears to be specific and is the result of subunit dissociation. Table II shows the sedimentation values which were estimated for mAb 4A alone, G_t plus mAb 4A, G_t plus IgG, and G_t plus mAb 4H.

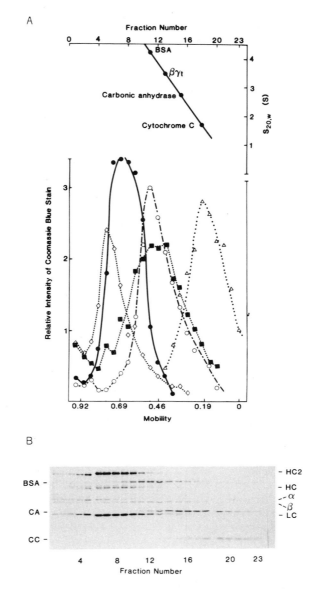

Fig. 2: Hydrodynamic behavior of transducin in the presence of monoclonal antibody 4A. The sample, containing transducin and mAb 4A (1:5), was layered on the top of 5-20% linear sucrose density gradient prepared in buffer B, as described under Methods. The results were obtained from one experiment. A, The sedimentation mobilities of each standard protein and $\beta\gamma_t$ were determined as described in Table I and expressed in Svedberg units (S). The gradient was calibrated by BSA, carbonic anhydrase, and cytochrome c. The profile of distribution of α_t, (◇••••••◇), $\beta\gamma_t$ (■•••••■), mAb 4A (●————●), BSA (○•—•—○), and cytochrome c (△•••••△) are shown at the bottom, as determined by gel densitometric scanning. B, Photograph of the Coomassie blue-stained polyacrylamide gel.

Table II Hydrodynamic properties of G_t and antibodies

		Sedimentation coefficient $s_{20,w}$ (S)	
mAb 4A		5.7	(1)
	IgG	5.1	
IgG + G_t			(1)
	$\alpha\beta\gamma_t$	4.1	
	mAb 4A	6.1	
mAb 4A + G_t	α_t	6.7	(2)
	$\beta\gamma_t$	3.85	
	mAb 4H	4.6	
mAb 4H + G_t			(1)
	$\alpha\beta\gamma_t$	6.8	

The sedimentation coefficients [$s_{20,w}$ in Svedberg units (S)] were estimated from the calibration curves. The values are the averages. The numbers in parentheses are the number of experiments.

DISCUSSION

Our sucrose density gradient experiments show that holotransducin migrates at a faster rate than α_t or $\beta\gamma_t$ subunits. The sedimentation coefficient which we obtained for G_t either in buffer A or B (Table I), is close to those reported for G_o (Sternweis, 1986), G_i, G_s (Codina et al., 1984; Sternweis, 1986) and G_t (Wessling-Resnick and Johnson, 1987). It is interesting to note that all previous sedimentation studies of G-proteins except G_t were carried out in the presence of detergent, because G_s, G_i and G_o can be extracted from membranes and mantained in solution only with detergent, while G_t does not need this agent. Sternweis (1986) showed that G_s, G_i and G_o remained soluble in sucrose gradients only in the presence of detergent while they aggregated extensively if detergent was not present. This aggregation was reversed by inclusion of Al^{3+}, Mg^{2+} and F^- in the gradients, which causes dissociation of the subunits. The $\beta\gamma$ subunits are responsible for this aggregation of G-proteins. In fact α subunits migrate as monomers even in the absence of detergent, while $\beta\gamma$ subunits aggregate very easily. The only exception to this rule, according to Sternweis (1986), is $\beta\gamma_t$ that migrated in sucrose density gradients as a monodisperse peak either in the presence or absence of detergent. We found that $\beta\gamma_t$ shows a single sharp peak at low concentrations, and it displays a tendency to form oligomers at higher protein concentrations. This was also true for holotransducin (Fig. 1). The sedimentation coefficient which we calculated for the $\beta\gamma_t$ monomer is very similar to that reported by Sternweis (1986). In our sucrose density gradients α_t migrated as a single peak, without formation of oligomers, and at a slower sedimentation rate than $\beta\gamma_t$. This is in agreement with the observation that α_o, α_i, and α_s exist as monomeric species either in the presence or absence of detergent (Sternweis, 1986). The sedimentation coefficient which we obtained is similar to those reported for α_o, α_i (Sternweis, 1986) and α_t (Wessling-Resnick and Johnson).

It has been shown that monoclonal antibody 4A disrupts transducin interaction with rhodopsin (Hamm et al., 1987), and the major binding site of this antibody is located within the region Arg^{310} and Lys^{329} at the carboxyl terminus (Deretic and Hamm, 1987). This result was confirmed using synthetic peptides corresponding to this region. These peptides completely block mAb 4A binding to transducin (Hamm et al., 1988).

Furthermore, this antibody completely blocks access of pertussis toxin to Cys^{347} of α_t and subsequent ADP-ribosylation (Deretic and Hamm, 1987). This could be the result of a direct effect of antibody 4A to sterically hinder pertussis toxin binding, because it is known that amino acids 340-350 take part in the antigenic site of mAb 4A from the ability of α_t-340-350 to block antibody binding (Hamm et al., 1988). On the other hand, pertussis toxin-catalyzed ADP-ribosylation requires the presence of the $\beta\gamma_t$ subunit (Watkins et al., 1985) and the site of interaction between subunits is thought to be in the amino-terminal region of α_t (Fung and Nash, 1983; Watkins et al., 1985; Navon and Fung, 1987). In the three-dimensional model proposed by Deretic and Hamm (1987), the amino-terminal region is also very close to the carboxyl-terminal region. In the previous epitope mapping studies, antibody binding could detect only the center or the most tightly binding region of the epitope area. Participation of surrounding amino acids, such as those of the amino-terminal region, could not be shown, because the amino-terminal 2-KDa tryptic fragment is lost in the first minutes of trypsinization and is too small to bind to nitrocellulose. However, synthetic peptides from the amino-terminal region have no effect on antibody binding, suggesting that it is not a major part of the antigenic site (Hamm et al., 1988).

The main purpose of this work was to study the effect of monoclonal antibody 4A on the $\alpha\beta\gamma_t$ complex using another approach under physiological conditions of α_t folding and subunit binding. The results, which we obtained using linear sucrose density gradients, clearly suggest that mAb 4A causes dissociation of $\alpha\beta\gamma_t$ complex. It is interesting to note that some immunoprecipitation experiments (Hamm et al., 1987) have shown that monoclonal antibody 4A co-precipitates both subunits, suggesting no dissociation of the complex. This discrepancy of results can perhaps be explained by the different interaction times between monoclonal antibody 4A and transducin in the two experiments. In the case of immmunoprecipitation the incubation period is 1h, while in the case of the sucrose density gradient the interaction time is much longer, 30 min of incubation and then 15 h of centrifugation. Recently, Hamm et al. (1988) showed, using synthetic peptides of known amino acid sequence, that the site of interaction between transducin and rhodopsin is located in the COOH-terminal region of α_t, and is close to the antigenic site of monoclonal antibody 4A. This antibody directly inhibits G-protein-rhodopsin interaction, but also disrupts the binding between subunits. At this point we suggest three possibilities to explain our present results:

1. The antigenic site of monoclonal antibody 4A is completely within the region Arg^{310} to Phe^{350} at the COOH terminus of α_t, but the binding of antibody to this region modifies the binding affinity between subunits, because the site of subunit interaction, the amino-terminal region, is very close to the antigenic site of this antibody (Deretic and Hamm, 1987).

2. The region Arg^{310} to Lys^{329} of α_t is the center of the antigen-antibody binding area, but surrounding amino acids of the amino terminus take part in the formation of the antigenic site. Therefore, the interaction between both transducin and rhodopsin and the transducin subunits are directly inhibited by monoclonal antibody 4A.

3. The COOH-terminal region of α_t is involved, together with the amino-terminal region, to form the site of interaction with $\beta\gamma_t$.

Actually, we prefer the first hypothesis for three reasons. Panel A of Fig. 2 shows the $\beta\gamma_t$ peak tails into the heavier densities, is broadened, and partly overlaps the α_t peak. This indicates the sequential formation, during centrifugation, of two types of complexes between antibody and protein. First, monoclonal antibody 4A binds to the holoprotein and forms a transducin-antibody complex which sterically hinders $\beta\gamma_t$ binding to α_t resulting in the formation a stable α_t-antibody complex and free $\beta\gamma_t$. This hindrance may be relatively slow, explaining the different results obtained with immunoprecipitation and sucrose density gradient experiments. Finally, we have evidence that synthetic peptides corresponding to the amino-terminal sequence of α_t (Hamm et al., 1988) do not compete with monoclonal antibody 4A binding to G_t in ELISA competition experiments.

Therefore, we conclude that dissociation of the $\alpha\beta\gamma_t$ complex by monoclonal antibody 4A is probably the result of a steric effect of antibody. Other experiments will be useful to prove this hypothesis. We plan to perform faster sucrose density gradient experiments that will be run in one hour instead of 15 hours. Moreover, we will carry out some gradients using the Fab fragment of mAb 4A. In this case the antigen binding site of antibody is still intact, and the lack of the Fc fragment could eliminate the steric hindrance. Finally, proteolytic fragmentation of α_t by different proteases may elucidate whether the amino-terminal region participates in the formation of the antigenic site of monoclonal antibody 4A.

REFERENCES

Amit, A. G., Mariuzza, R. A., Phillips, S. E. V., and Poljak, R. J., 1986, Three-dimensional structure of an antigen-antibody complex at 2.8 Å resolution, *Science*, 233:747.

Bradford, M. M., 1976, A rapid and sensitive method for the quantitation of microgram quantities of protein utilizing the principle of protein-dye binding, *Anal. Biochem.*, 72:248.

Codina, J., Hildebrandt, J. D., Sekura, R. D., Birnbaumer, M., Bryan, J., Manclark, C. R., Iyengar, R., and Birnbaumer, L., 1984, N_s and N_i, the stimulatory and inhibitory regulatory components of adenylyl cyclase, *J. Biol. Chem.*, 259:5871.

Deretic, D., and Hamm, H. E., 1987, Topographic analysis of antigenic determinants recognized by monoclonal antibody to the photoreceptor guanyl nucleotide-binding protein, transducin, *J. Biol. Chem.*, 262:10831.

Fung, B. K.-K., and Stryer L., 1980, Photolyzed rhodopsin catalyzes the exchange of GTP for bound GDP in retinal rod outer segments, *Proc. Natl. Acad. Sci.*, 77:2500.

Fung, B. K.-K., Hurley, J. B., and Stryer, L., 1981, Flow of information in the light-triggered cyclic nucleotide cascade of vision, *Proc. Natl. Acad. Sci.*, 78:152.

Fung, B. K.-K., 1983, Characterization of transducin from bovine rod outer segments. I. Separation and reconstitution of the subunits. *J. Biol. Chem.*, 258:10495.

Fung, B. K.-K., and Nash, C. R., 1983, Characterization of transducin from bovine rod outer segments. II. Evidence for distinct binding sites and conformational changes revealed by limited proteolysis with trypsin, *J. Biol. Chem.*, 258:10503.

Hamm, H. E., and Bownds, M., 1984, A monoclonal antibody to guanine nucleotide binding protein inhibits the light-activated cyclic GMP pathway in frog rod outer segment, *J. Gen. Physiol.*, 84:265.

Hamm, H. E., Deretic, D., Hofmann, K. P., Schleicher, A., and Kohl, B., 1987 Mechanism of action of monoclonal antibodies that block the light activation of the guanyl nucleotide-binding protein, transducin, *J. Biol. Chem.*, 262:10831.

Hamm, H. E., Deretic D., Arendt, A., Hargrave P. A., Koening B., Hofman K. P., 1988, Site of G-protein binding to rhodopsin mapped with synthetic peptides from the α subunit, *Science*, 241:832.

Hurley, J. B., 1980, Isolation and recombination of bovine rod outer segment cGMP phosphodiesterase and its regulators, *Biochem. Biophys. Res. Commun.*, 92:505.

Kleuss, C., Pallat, M., Brendel, S., Rosenthal, W., and Schultz, G., 1987, Resolution of transducin subunits by chromatography on blue sepharose, *J. Chrom.*, 407:281.

Kühn, H., 1980, Light and GTP-regulated interaction of GTPase and other proteins with bovine photoreceptor membranes, *Nature*, 283:587.

Laemmli, U. K., 1970, Cleavage of structural proteins during the assembly of the head of bacteriophage T4, *Nature*, 227:680.

Navon, S. E., and Fung, B. K.-K., 1987, Characterization of transducin from bovine retinal rod outer segments, Participation of the amino-terminal region of Ta in subunit interaction, *J. Biol. Chem.*, 262:15746.

Papermaster, D. S., and Dreyer, W, J., 1974, Rhodopsin content in the outer segment membranes of bovine and frog retinal rods, *Biochemistry*, 13:2438.

Sternweis, P. C., 1986, The purified a subunit of Go and Gi from bovine require $\beta\gamma$ for association with phospholipid vesicles, *J. Biol. Chem.*, 261:631.

Watkins, P. A., Burns, D. L., Kanaho, Y., Liu, T.-Y., Hewelett, H. L., and Moss, J., 1985, ADP-ribosylation of transducin by pertussis toxin, *J. Biol. Chem.*, 260:13478.

Wheeler, G. L. and Bitensky, M. W., 1977, A light-activated GTPase in vertebrate photoreceptors: regulation of light-activated cyclic GMP phosphodiesterase, *Proc. Natl. Acad. Sci.*, 74:4238.

Witt, P. L., Hamm, H. E., and Bownds, M.D., 1984, Preparation and characterization of monoclonal antibodies to several frog rod outer segment proteins, *J. Gen. Physiol.*, 84:251.

COMPARATIVE EFFECTS OF PHOSPHODIESTERASE INHIBITORS

ON DETACHED ROD OUTER SEGMENT FUNCTION

G. Rispoli, *P. G. Gillespie and P. B. Detwiler

Department of Physiology and Biophysics and
*Department of Pharmacology
University of Washington School of Medicine, Seattle, U.S.A.

INTRODUCTION

In vertebrate retinal rods light reduces a standing inward current that flows into the outer segment in darkness. Current is carried primarily by sodium (70%) and calcium (15%) ions which pass through channels in the outer segment surface membrane that are opened by cGMP (Yau and Baylor, 1989). A photoproduct (Rh*) formed by isomerization of rhodopsin catalyzes GTP-GDP exchange to activate a G protein (transducin, T) which stimulates cGMP-specific phosphodiesterase (PDE) causing cGMP hydrolysis, channel closure and a decrease in dark current. The recovery of current following light exposure involves inactivation of the light stimulated PDE and an increase in cGMP synthesis by activation of guanylate cyclase (Hodgkin and Nunn, 1988; Detwiler et al., 1989). The sequential activation and inactivation of PDE in the onset and recovery of rod light responses has been probed using IBMX, a membrane permeant phosphodiesterase inhibitor (Beavo et al., 1970; Capovilla et al., 1983; Hodgkin and Nunn, 1988). While such studies have provided useful information about the orchestration of transduction IBMX is not an ideal inhibitor. It is not very potent ($K_i = 1$ x 10^{-5}M) which makes it practically impossible to use to fully inhibit light-stimulated PDE. It is also non-specific in that it inhibits many kinds of cyclic nucleotide phosphodiesterases which could potentially complicate the interpretation of results. Assays on isolated enzymes have shown two compounds (dipyridamole and M&B:22,948) selectively inhibit cGMP phosphodiesterases (Beavo, 1988; Gillespie and Beavo, 1989) and are 20 to 100 times more potent than IBMX ($K_i = 3.8$ x 10^{-7}M and 1 x 10^{-7}M respectively). This short paper compares the effects of IBMX, dipyridamole (FitzGerald, 1987) and M&B:22,948 (Broughton et al., 1974) on dark current and light responses recorded from detached rod outer segments during whole-cell voltage clamp. Our results show that there are significant differences in the action of the three inhibitors, which suggests that they may affect more than PDE. In the course of these experiments we also discovered that in nucleoside triphosphate depleted outer segments steps of bright light cause an increase rather than a decrease in dark current. Since these cells contain little or no GTP and dark current is maintained by an exogenous source of cGMP, the production of inward or inverted light responses suggests that steps of strong light can inhibit PDE possibly through a G protein independent pathway.

Sensory Transduction
Edited by A. Borsellino *et al.*
Plenum Press, New York, 1990

METHODS

Experiments were done on retinal rods from dark-adapted (4-16 hours) Gekko lizards. All manipulations were done in darkness using infrared illumination and image converters. The animal was decapitated and pithed. Both eyes were removed: one was stored whole in Ringer on ice for latter use and the other was hemisected. The back half of the eyeball was cut into four pieces which were stored in oxygenated Ringer on ice and used when needed. Detached outer segments were obtained from a single piece by gently stretching the isolated retina in 200 μl of Ringer solution (mM): 160 Na, 3.3 K, 1 Ca, 1.7 Mg, 168 Cl, 1.7 SO$_4$, 2.8 HEPES, 10 Dextrose, pH 7.4.

The fluid drop containing outer segments was held by surface tension between two cover slips that were supported on one side and formed the floor and ceiling of a recording chamber that was open on three sides. The chamber was mounted on a modified inverted microscope and the cell was viewed with an infrared-sensitive T.V. camera using long wavelength illumination >850 nm. The recording and perfusion pipets entered the chamber from opposite sides. Patch pipets were made from VWR 100 μl micropipet glass (Cat. No. 53432-921) in the usual way (Sakmann and Neher, 1983) and were filled with a standard internal solution supplemented with nucleotides and in some cases calcium buffers (Table 1).

A firepolished pipet (R_e = 8-12 MΩ) was pressed against the surface of a detached outer segment to obtain a cell-attached gigaseal (typically 10-30 GΩ) recording. A brief pulse of suction was applied to the pipet to breakthrough to whole-cell recording. The pipet voltage was set at -20 mV which when corrected for liquid junction potential of 9 mV gave a holding potential of -29 mV. The access resistance of the recording, estimated from the initial current transient evoked by a brief voltage step, was typically 25-50 MΩ.

Table 1. Composition of dialysis solutions (mM)

	High Triphosphate Solutions		Zero Triphosphate Solutions	
	I 0 Chelators	II + Chelators	III 0 Chelators	IV + Chelators
ATP	5	5	0	0
GTP	1	1	0	0
cGMP	0	0	0.05	0.05
BAPTA	0	10	0	10
EDTA	0	1	0	1
K	138	138	139	138
Na	10	10	9	7
Mg	6.05	6.75	0.05	0.75
Ca	0	0	0	0
Aspartate	111	67	124	79
Cl	24	26	22	22
HEPES	5	5	5	5
pH	7.4	7.4	7.4	7.4

Osmolality adjusted with sucrose to ~310 mOs/Kg.

Outer segments were stimulated with uniform 520 nm illumination having an unattenuated intensity of 3.4×10^3 photons/μm^2/flash. Neutral density filters were used to reduce the intensity of the unattenuated light. For convenience stimulus strength is reported in figures as the nominal log attentuation of the filters used. The actual attenuation, however, measured with a calibrated photodiode differed from one filter combination to another. These values were used to calculate the intensities listed in the figure legends.

The external solution was changed rapidly during whole-cell recording by moving the detached outer segment between fluid streams that flowed from adjacent barrels of a theta tube pipet containing either Ringer or Ringer plus inhibitor. M&B:22,948 (trade name: Zaprinast; a gift from the May & Baker Co.) was dissolved in Ringer containing 0.2% dimethylsulfoxide (DMSO). In these experiments control responses were recorded in Ringer containing an equal amount of DMSO. Dipyridamole (trade name: Persantine) and IBMX were obtained from Sigma Chemical Company, St. Louis, MO and dissolved in Ringer without DMSO. Other reagents include: guanosine 3':5'-cyclic monophosphate, sodium salt; guanosine 5'-triphosphate, sodium salt semicolon adenosine 5'-triphosphate, dipotassium salt; ethylenediaminetetraacetic acid (EDTA) all from Sigma Chemical Co. and BAPTA tetrapotassium salt from Molecular Probes, Eugene, OR.

RESULTS

Detached outer segments dialyzed with millimolar amounts of ATP and GTP but no cGMP develop stable light-sensitive inward dark current and support light responses having normal kinetics, sensitivity and reversal potential (Sather, 1988). The development of dark current and response recovery in the absence of an exogenous source of cGMP shows that the outer segment contains guanylate cyclase which uses GTP provided by the dialysis solution to synthesize cGMP. Although guanylate cyclase is active in darkness (Rispoli and Detwiler, 1989) cytoplasmic cGMP does not increase steadily because basal PDE activity maintains an equivalent rate of cGMP hydrolysis (Sather and Detwiler, 1987). The balance between the dark activities of the two enzymes maintains a stable dark current and produces a steady turnover or flux of cGMP (Goldberg et al., 1983; Sather, 1988) as it circulates from synthesis to hydrolysis. A decrease in the activity of either enzyme will unbalance synthesis and hydrolysis and cause a change in dark current. The effect of the inhibitors on dark current and light responses was studied in outer segments under two fundamentally different conditions. In one case the outer segment was dialyzed with a high nucleoside triphosphate solution that contained no cGMP (solution I, Table 1). Under these conditions the outer segment uses endogenous guanylate cyclase to synthesize cGMP from the nucleotides provided by the dialysis solution. With this arrangement the outer segment is required to sort out its own cGMP economy. In the second case the outer segment was dialyzed with a solution that contained cGMP but no nucleoside triphosphates (solution III, Table 1). Under these conditions the outer segment is depleted of nucleoside triphosphates and guanylate cyclase is effectively disabled by removing its substrate (GTP). The influx of cGMP by diffusion from the dialysis solution replaces its synthesis by guanylate cyclase. Cytoplasmic [cGMP], and hence dark current, is determined by the level of basal PDE activity. Both the high triphosphate and zero triphosphate solutions were used with or without chelators (10 mM BAPTA, 1 mM EDTA; see Table 1). Solutions with chelators provide indirect information about changes in intracellular calcium. The addition of BAPTA increases the calcium buffer capacity of the cell which slows changes in calcium and alters the timing of events that are controlled by it.

159

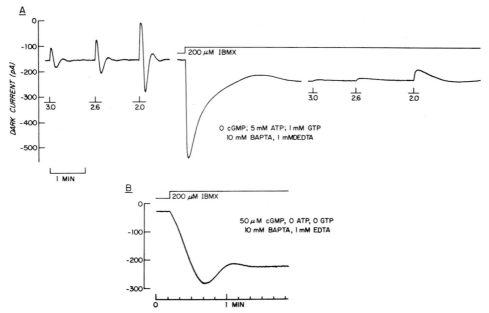

Fig. 1: IBMX induced changes in dark current in the presence and absence of nucleoside triphosphates. A. Recording from a detached outer segment dialyzed for > 14 minutes with solution containing nucleoside triphosphates. Chart records shows the response to 3 test flashes of progressively brighter intensity (3.0 = 1.27, 2.6 = 3.07, 2.0 = 13.9 photons/μm^2/flash) before and after exposure to 200 μM IBMX. B. Response of triphosphate depleted cell to 200 μM IBMX. Composition of dialysis solution shown in figure. Dialysis time 23 minutes. V_{hold} = -29 mV; temperature 17-17.8^{0}C.

The principal finding of our experiments is that IBMX, dipyridamole and M&B:22,948 have the same effect on dark current when recorded from triphosphate depleted outer segments but not when recorded from outer segments with nucleoside triphosphates.

Dialysis with 0 Triphosphate Solutions. In cells dialyzed with 50 μM cGMP, 0 ATP and 0 GTP, all three inhibitors caused an abrupt and sustained increase in dark current. The level of the current increased with the concentration of IBMX and M&B:22,948. Experiments with dipyridamole were limited by its aqueous solubility to 20 μM which was the highest concentration that would dissolve in Ringer without any sign of precipitation. At this concentration dipyridamole increased dark current by about the same amount (200 - 300 pA) as 200 μM IBMX and 8μM M&B:22,948 (Figs. 1B, 2B & 3C). All the inhibitors caused the dark current to reach a maximum level in about 60 seconds followed by a slow fall to a maintained level that was approximately 10% smaller than the maximum. The mean steady-state dark current in 8 μM M&B:22,948 was 210 pA (n = 2) versus 280 pA in 20 μM dipyridamole (n = 2) and 219 pA in 200 μM IBMX (n = 1).

The lack of GTP in triphosphate depleted outer segments interferes with the activation of the transduction cascade by affecting GTP-GDP exchange and in so doing reduces light sensitivity. In such cells bright flashes produced no light response. Although still brighter lights often evoke a small decrease in dark current in some cells steps of unattenuated 520 nm or white light produced a long latency slow increase in

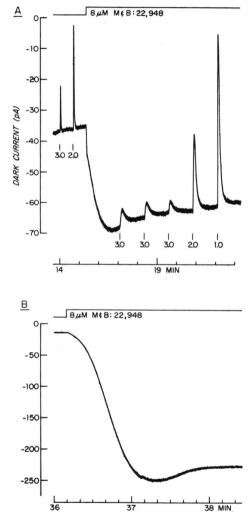

Fig. 2: Effect of M&B:22,948 on dark current and light response in outer segments with (A) and without (B) nucleoside triphosphates. A. Chart record shows responses to test flashes (photocell below) in normal Ringer +0.02% DMSO and in Ringer +8 μM M&B:22,948 +0.02% DMSO. Dialysis solution: 0 cGMP, 5 mM ATP, 1 mM GTP, no chelators. Light intensities: 3.0 = 1.27, 2.0 = 13.9, and 1.0 = 119 photons/μm^2/flash. Temperature 17°C. V_{hold} = -29 mV. B. Response of triphosphate depleted cell to 8 μM M&B:22,948. Dialysis solution: 50 μM cGMP, 0 ATP, 0 GTP. Temperature 17.5°C. V_{hold} = -29 mV. In both A and B dialysis time is shown on the time axis.

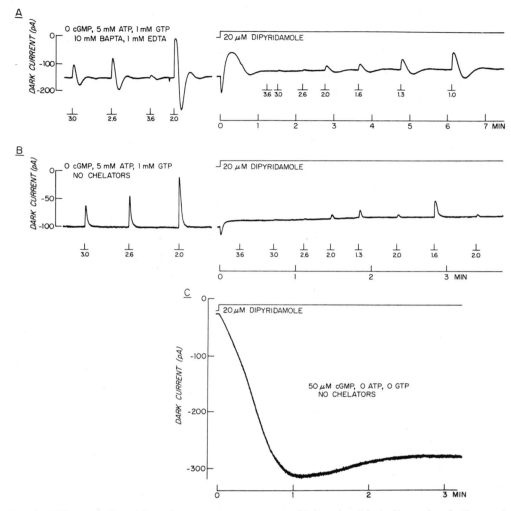

Fig. 3: Effect of dipyridamole on outer segments dialyzed with indicated solutions. A
& B. Chart record showing responses to test flashes before and after exposing the
outer segment to 20 μM dipyridamole. Flash intensities: $1.0 = 119$; $1.3 = 60$; 1.6
$= 25$; $2.0 = 13.9$; $2.6 = 3.07$; $3.0 = 1.27$; $3.6 = 0.28$ photons/μm^2/ flash. Dialysis
time: (A) 11 minutes; (B) 17 minutes. C. Response of triphosphate depleted cell
to 20 μM dipyridamole. Dialysis time 23 minutes, $V_{hold} = -29$ mV, Temperature
17^0C.

dark current that was graded with light exposure (Fig. 4). In outer segments depleted
of triphosphates by more than 20 min of dialysis with 50 mM cGMP, 0 ATP and 0 GTP
bright steps evoked inward or "inverted" light responses in 40% of the cells bathed in
normal Ringer (n=5). Twice as many cells produced inward responses when exposed to
Ringer containing M & B:22,948 or dipyridamole (IBMX was not tested). In the presence
of PDE inhibitors the amplitude of the inward response increased from < 10 pA to as
much as 100 pA (Fig. 5).

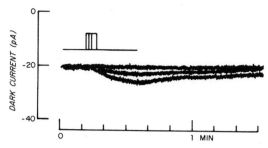

Fig. 4: Inward light responses in a triphosphate depleted outer segment. Superimposed traces show response to 1, 2, and 4 second steps of unattenuated 520 nm light (photocell trace above) recorded from an outer segment bathed in normal Ringer solution and dialyzed for 23 minutes with 50 μM cGMP, 0 ATP and 0 GTP. Temperature 17^{0}C. V_{hold} = -29 mV.

Dialysis with High Triphosphate Solution. In cells dialyzed with 0 cGMP, 5 mM ATP and 1 mM GTP steady-state dark current was increased by IBMX and M&B:22,948 and decreased by dipyridamole. Switching to Ringer solution containing 200 μM IBMX or 8 μM M&B:22,948 (Figs. 1A & 2A) caused an initial rapid increase in current followed by slower phase of growth that stabilized at a level that was 58% (mean 2 cells, IBMX) and 55% (mean 5 cells, M&B:22,948) greater than the dark current before. Switching to Ringer containing 20 μM dipyridamole caused a transient increase in current followed by a rapid decline and final stabilization at a level that was 38% (mean 3 cells) less than the dark current before exposure to inhibitor (Fig. 3B). With all inhibitors the transition to the new steady-state level of dark current was more oscillatory when the internal solution contained chelators in addition to ATP and GTP (solution II, Table 1).

All three PDE inhibitors reduced light sensitivity and slowed the kinetics of the light response (Figs. 1, 2 & 3). The mean reduction of sensitivity in 20 μM dipyridamole, 8 μM M&B:22,948 and 200 μM IBMX was 22x, 7x and 4x respectively. The same concentrations of dipyridamole, M&B:22,948 and IBMX increased the duration of equivalent responses, i.e. responses that caused similar fractional reductions in dark current, by 1.5x, 3x and 6x respectively. It is noteworthy that the inhibitor that caused the greatest reduction in light sensitivity caused the least slowing of kinetics and vice versa.

DISCUSSION

Our results are based on a relatively small number of experiments and for this reason should be considered preliminary. In spite of this caveat there are two unexpected findings that are well established and warrant further discussion. One is the discovery that dark current is increased by bright steady light in triphosphate depleted cells. The other is that dipyridamole decreases dark current in the presence but not the absence of nucleoside triphosphates.

The Inward Light Response. Dark current is determined by the cytoplasmic cGMP concentration (Karpen et al., 1988). In outer segments depleted of nucleoside triphosphates current is maintained by cGMP that diffuses into the cell from the pipet dialysis solution (50 μM cGMP, 0 ATP, 0 GTP). Under these conditions the most likely explanation for a light-induced increase in dark current, i.e. an inward light response is a decrease in cGMP hydrolysis by basal PDE activity. While other explanations are possible our

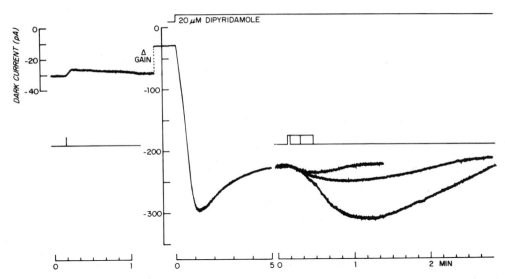

Fig. 5: Inward light responses are enhanced in the presence of a PDE inhibitor. A 20 msec flash of unattenuated 520 nm light produced a small outward light response in normal Ringer. After exposure to 20 μM dipyridamole 2, 10, and 20 second steps of unattenuated light (photocell trace above) caused inward light responses graded with light exposure. Responses are shown superimposed after correction for an outward drift. Note changes in gain and timescale. Dialysis solution: 50 μM cGMP, 0 ATP, 0 GTP. Dialysis time: 21 minutes. Temperature 17.5^{0}C. V_{hold} = -29 mV.

working hypothesis is that the generation of Rh* during intense illumination produces a messenger with little to no amplification that is independent of transducin activation and causes inhibition of PDE activity. Such a signal would represent a divergent pathway in the transduction cascade. It is consistent with other studies that suggest a stage preceding the activation of transducin produces a signal in steady light that accelerates the recovery of the light response (Rispoli and Detwiler, 1989). This raises the possibility that there are two opposing light- activated pathways. One short latency and high gain that is routed through transducin and stimulates PDE. The other long latency and low gain that is not mediated by transducin and inhibits PDE. The latter may provide a secondary mechanism for recovery from strong stimuli. The two pathways are antagonistic such that suppression of one may unmask the other. This may explain why PDE inhibitors, which would be expected to interfere more with the activation than the inhibition of PDE, enhance the inward light response.

The Dipyridamole Paradox. In triphosphate depleted cells all three inhibitors caused a maintained increase in dark current consistent with inhibition of basal PDE activity. Based on the concentration of the different inhibitors needed to produce similar increases in current, the approximate order of relative effectiveness is: 1 (M & B:22,948): 0.4 (dipyridamole): 0.04 (IBMX). This roughly agrees with their relative differences in Ki using trypsin activated bovine rod PDE, i.e. 1 (M & B:22,948):0.33 (dipyridamole): 0.01 (IBMX).

In the presence of nucleoside triphosphates all three inhibitors cause an initial transient increase in current that decays to a lower maintained level. The fact that the transient change in current is observed only in outer segments that contain nucleoside triphosphates suggests that it involves changes in guanylate cyclase activity, as well as inhibition of PDE. The initial change in current is larger and the transition to steady-state slower and more oscillatory when chelators are included in the dialysis solution. This suggests that the underlying events are influenced by changes in intracellular calcium. In detached outer segments guanylate cyclase is regulated by calcium (Rispoli et al., 1988). With increasing intracellular calcium guanylate cyclase activity decreases to a minimum but finite level of activity which persists even in the presence of mM Ca in the dialysis solution (Sather et al., 1988; Koch and Stryer, 1988).

Thus it is reasonable to assume that the resting level of intracellular calcium sets the basal dark rate of cGMP production. The stable maintenance of dark current requires that the rate of cGMP-synthesis be matched by an equivalent rate of hydrolysis. The PDE inhibitors reduce basal PDE activity and upset the balance between synthesis and hydrolysis. Cytoplasmic cGMP increases, causing an increase in dark current and an accompanying increase in calcium influx. The resulting rise in intracellular calcium reduces cyclase activity and cGMP synthesis decreases. Dark current falls, and ultimately stabilizes when the rate of cGMP synthesis matches the new inhibited rate of hydrolysis.

This sequence of events can explain how PDE inhibitors cause an initial increase in current and a subsequent decline to a steady level. What is not clear, however, is why equivalent concentrations of the three inhibitors cause the current to stabilize at different levels. Equivalent concentrations refers to inhibitor concentrations that cause similar increases in dark current in triphosphate depleted cells - which we assume indicates equivalent inhibition of basal PDE activity. If the three inhibitors, at these concentrations, affect PDE equally and that is all they do, then dark current should finally stabilize at roughly the same steady level for all three inhibitors.

That this does not happen suggests inhibitors either do not inhibit PDE equally, which makes it difficult to explain the results in triphosphate depleted cells, or do more than inhibit PDE. While it is possible that the presence of nucleoside triphosphate alters the relative potency of the inhibitors, such an action can not account for the decrease in steady-state dark current with dipyridamole. A sustained decrease in current suggests that the drug causes either a delayed stimulation of PDE or a partial inhibition of guanylate cyclase. The former alternative is difficult to reconcile with dipyridamole causing a sustained increase in dark current in triphosphate depleted cells and with it causing an initial increase in cells containing triphosphates.

If PDE activity is reduced during the dipyridamole induced decrease in dark current then there must also be an equivalent reduction in cGMP production, or the current will not settle at a steady level. A decrease in dark current is associated, however, with a decrease in calcium influx and presumably a decrease in intracellular calcium which should stimulate rather than inhibit guanylate cyclase. A possible way around this objection is to postulate that dipyridamole increases the sensitivity of guanylate cylase to inhibition by calcium. In other words, it shifts the cyclase activity versus [Ca] curve to the left, i.e. toward lower [Ca]. This explanation is consistent with chelators causing an exaggerated shut down of current during the "decline" phase of the initial dipyridamole induced change in current (Fig. 3A).

Changes in the Light Response. All three inhibitors reduced light sensitivity and slowed response kinetics. It is not surprising that inhibition of PDE would reduce the amplitude and slow the time-to-peak of the light response; in the presence of the inhibitors light stimulated PDE will hydrolyze less cGMP per unit time. It is not clear, however, why there is also a slowing of the recovery phase of the light response. The recovery of current following a flash depends on the termination of the transduction cascade, i.e. inactivation of Rh*, T*, and PDE*, and on cGMP production by stimulation of guanylate cyclase (Hodgkin and Nunn, 1988; Sather et al., 1988; Rispoli et al., 1988; Rispoli and Detwiler, 1989).

Of these alternatives it would seem most likely that the inhibitors would act via PDE or guanylate cyclase, as suggested in the preceding section. The presence of PDE inhibitors might prolong a low level of PDE activity by interfering with the reassociation of the inhibitory γ subunit and hence the full inhibition of enzyme activity.

Another possibility is that the drugs affect response recovery by suppressing the activation of guanylate cyclase. Since guanylate cyclase activation will reduce the amplitude of the light response by hastening its recovery, one would expect that the drug that inhibited guanylate cyclase to the greatest extent would reduce the sensitivity the least and slow kinetics the most. In fact, such a reciprocal relationship between the effects of the inhibitors on sensitivity and kinetics does exist. For example, IBMX reduced sensitivity the least of the three inhibitors and slowed kinetics the most. According to the hypothesis under consideration this result would suggest that IBMX inhibited guanylate cyclase activation more than the other drugs. However, this conclusion does not explain why the dark current stabilize at a higher level in IBMX than in dipyridamole or M&B:22,948.

One would expect dark current to settle at lower and lower levels as the extent of cyclase inhibition was increased. This point could be countered by postulating that the phosphodiesterase and guanylate cyclase that are basally active and set the resting dark current are distinct from the phosphodiesterase and guanylate cyclase that participate in turning the light response on and off. If this is the case our results could be explained by differentially weighting the inhibitory effect of the three drugs on the four different enzymes.

While our experiments allow little more than speculation about the explanation for the relative difference in IBMX, M&B:22,948 and dipyridamole they clearly show that not all PDE inhibitors are the same and suggest that their actions may arise from effects on more than one type of enzyme.

REFERENCES

Beavo, J. A., 1988, Multiple isotymes of cyclic nucleotide phosphodiesterase, *in*:"Advances in Second Messenger and Phosphoprotein Research," 22:1-38.

Beavo, J. A., Rogers, N. L., Crofford, O. B., Hardman, J. G., Sutherland, E. V., and Newman, E. V., 1970, Effects of xanthine derivatives on lipolysis and on adenosine 3',5' monophosphate activity, *Mol. Pharm.*, 6:597-603.

Broughton, B. J., Chaplen, P., Knowles, P., Lunt, E., Pain, D. L., Wooldridge, K. R. H., Ford, R., Marshall, S., Walker, J. L., and Maxwell, D. R., 1974, New inhibitor of reagin-mediated anaphylaxis, *Nature*, 251:650-652.

Capovilla, M., Cervetto, L., and Torre, V., 1983, The effects of phosphodiesterase inhibitors on the electrical activity of toad rods, *J. Physiol.*, 343:277-295.

Detwiler, P. B., Rispoli, G., and Sather, W.A., 1989, Manuscript in preparation.

Fitzgerald, G. A., 1987, Medical intelligence. Drug therapy:dipyridamole, *N. Engl. J. Med.*, 316:1247-1256.

Gillespie, P. G., and Beavo, J. A., 1989, (Manuscript submitted).

Goldberg, N. D., Ames, A., III, Grander, J. E., and Walseth, T. F., 1983, Magnitude of increase in retinal cGMP metabolic flux determined by ^{18}O incorporation into nucleotide a-phosphoryls corresponds with intensity of photic stimulation, *J. Biol. Chem.*, 258:9213-9219.

Hodgkin, A. L., and Nunn, B. J., 1988, Control of light-sensitive current in salamander rods, *J. Physiol.*, London, 403:439-471.

Karpen, J. W., Zimmerman, A. L., Stryer, L., and Baylor, D. A., 1988, Gating kinetics of the cyclic GMP-activated channel of retinal rods: flash photolysis and voltage-jump studies, *Proc. Natl. Acad. Sci.*, USA, 85:1287-1291.

Koch, K. W., and Stryer, L., 1988, Highly cooperative feedback control of retinal rod guanylate cyclase by calcium ion, *Nature*, 334:64-66.

Rispoli, G., and Detwiler, P. B., 1989, Light adaptation in Gecko rods may involve changes in both the initial and terminal stage of the transduction cascade, *Biophys. J.*, 55:380P.

Rispoli, G., Sather, W. A., and Detwiler, P. B., 1988, Effects of triphosphate nucleotides on the response of detached outer segments to low external Ca^{++}, *Biophys. J.*, 53:390P

Sakmann, B., and Neher, E. ,eds., 1983, "Single channel recording", Plenum Press, New York.

Sather, W. A., 1988, Phototransduction in detached rod outer segments, *Doctoral thesis*, University of Washington, Seattle, WA.

Sather, W. A., and Detwiler, P. B., 1987, Intracellular biochemical manipulation of phototransduction in detached outer segments, *Proc. Natl. Acad. Sci.*, USA, 84:9290-9294.

Sather, W. A., Rispoli, G., and Detwiler, P.B., 1988, Effect of calcium on light adaptation in detached Gecko rod outer segments, *Biophys. J.*, 53:390P.

Yau, K. W. and Baylor, D. A., 1989, Cyclic GMP-activated conductance of retinal photoreceptor cells, *Ann. Rev. Neurosci.*, 12:289-327.

CATION SELECTIVITY OF THE cGMP-GATED CHANNEL OF

MAMMALIAN ROD PHOTORECEPTORS

H. Lühring, W. Hanke, R. Simmoteit and U.B. Kaupp

Institut für Biologische Informationsverarbeitung
Kernforschungsanlage Jülich, Germany

INTRODUCTION

Biochemical and electrophysiological attempts to elucidate the mechanism of phototransduction in vertebrate rod cells have identified a cation permeant channel in the plasma membrane of rod outer segments being responsible for the inward current in the dark which is directly and reversibly activated by cGMP. The light-induced rapid decrease in inward current and concomitant hyperpolarisation are caused by the closure of these channels due to hydrolysis of the agonist cGMP and subsequent desorption from the channel protein (for reviews see Stryer, 1986; Kaupp and Koch, 1986; Lamb, 1986; Pugh, 1987). To date, most biochemical investigations of the cGMP-gated channel were carried out on bovine rod outer segments, whereas electrical characteristics of the channel were almost exclusively determined on amphibian rod cells (Yau and Baylor, 1989). To fill this gap, we examined the electrical properties of the cGMP-gated channel of bovine rod photoreceptors by employing the patch clamp method. The selectivity of the cation channel for monovalent ions was obtained by exposing excised patches to solutions containing different cation species and determining reversal voltages (V_{rev}) and macroscopic conductances (g). The dependence of macroscopic current upon cGMP concentration yielded EC_{50} values.

MATERIALS AND METHODS

Preparation. Bovine rod outer segments (ROS) have been isolated in dim red light, by gently shaking a retina in 1ml of ice-cold solution containing (in mM): 120NaCl, 5Tris-HCl at pH 7.4. The suspension was passed through a nylon cloth (50μm mesh), and samples of cell suspension were kept on ice in the dark. Best results have been obtained using ROS within 4 hours after preparation. Thereafter, membranes started to disintegrate and cell debris impeded the formation of gigaohm seals between glass-micropipette and plasma membrane.

Electrical recordings. Borosilicate glass capillaries with an internally fused filament, providing quick-filling of the pipette, were used throughout all experiments. They were pulled in two steps, fire-polishing was omitted. A micropipette was placed at the ROS membrane surface by means of a hydraulic manipulator, the driving head attached to

Sensory Transduction
Edited by A. Borsellino *et al.*
Plenum Press, New York, 1990

a micromanipulator. Formation of gigaohm seals could be monitored on an oscilloscope by feeding rectangular current pulses through the pipette to ground. When a tight seal between pipette rim and membrane had formed by establishing a negative pressure at the pipette interior, i.e., by gentle suction, the current pulse disappeared and capacitive spikes were left. Capacitive currents could be abolished by the amplifier's internal compensation circuitry. The patch clamp amplifier was commercially obtained (List L/M-EPC-7, Darmstadt FRG). After seal formation ($> 8 G\Omega$), an excised patch was obtained either by moving the pipette with the attached cell through the air-water interface or by detaching the ROS in the solution jet in front of the perfusion pipette. Inside-out configuration was established when a cGMP-induced current increase across the excised patch could be detected. Control solution and the solution used for preparation of the ROS were identical. Typically, pipette resistances were of the range of $25 M\Omega$. Macroscopic currents across excised patches were recorded under voltage clamp conditions, where voltage was applied as a triangle with $\pm 80 mV$ peak and a $0.25 Hz$ cycle. Membrane current was low-pass filtered ($50 Hz$, 8-pole Bessel) and together with the voltage continuously recorded on-line controlled by a commercially available program (pCLAMP 4.0, Axon Instruments, CA).

Superfusion system. Using the inside-out configuration, the cytoplasmic face of the excised plasma membrane could be exposed to various solutions at the outlet of a double-barreled pipette. One of the barrels permanently supplied control solution, e.g., without cGMP, the other barrel was connected to the output of an 8-port valve, thus delivering up to 8 different test solutions with a delay of about 20s after changing the ports. The patch electrode could be displaced in less than a second from control to test solution. IV curves were recorded in control and varying test solutions, intermittently.

RESULTS AND DISCUSSION

cGMP-dependent currents. In the presence of cGMP, the macroscopic current across excised patches increases markedly (Fig. 1a). Maximum current amplitudes determined at $V_m = +60 mV$ varied between 80 and 200pA in the absence of divalent cations. Currents of similar amplitude have been recorded from amphibian ROS membranes (Fesenko et al., 1985; Stern et al., 1986; Haynes et al., 1986; Zimmerman and Baylor, 1986). The activation of the current by cGMP was analysed in a Hill-plot according to:

$$\lg \{I/(I_{max}-I)\} = f(\lg[cGMP]), \qquad (1)$$

where I indicates the macroscopic current recorded at the actual cGMP concentration, and I_{max} the maximum current in the presence of saturating cGMP concentrations ($500 \mu M$). Current increased within a smaller range of cGMP concentrations than could be expected from simple Michaelis-Menten kinetics. Fig. 1b shows as a typical result a half-saturating cGMP concentration of $EC_{50} = 48 \mu M$, and a Hill coefficient n=2.3, suggesting that at least 3 cGMP molecules bind cooperatively to open the channel. In the absence of divalent cations, the IV curve displayed non-linearity with a slight rectification at positive voltages (cf., Fig. 1c), similar to that reported for the amphibian cGMP-gated channel (Yau et al., 1986).

Ion selectivity. The ion selectivity of the cGMP-gated channel was determined from relative ion permeabilities and conductances, respectively, recorded under biionic conditions. The pipette contained 120mM NaCl, whereas the ionic composition of the solutions at the cytoplasmic face of the excised membrane was changed to isoosmolar

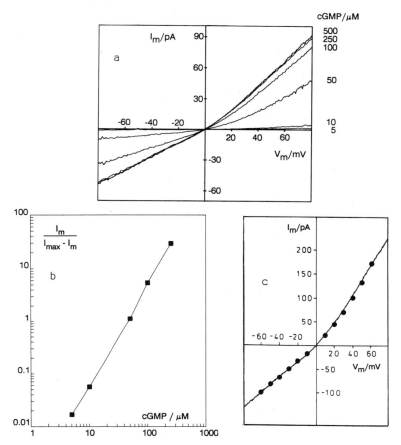

Fig. 1: Dependence of the macroscopic current on cGMP concentration. a) Typical set of current-voltage recordings obtained from a single excised membrane patch at different cGMP concentrations, leak currents have been subtracted. b) Hill plot of the cGMP-activated currents constructed from membrane currents at $+80$ mV clamp voltage, $EC_{50} = 48$ μM, n = 2.3. c) IV characteristics from an excised patch of amphibian rod cell (*filled circles*, data from Yau et al., 1986) and bovine ROS (*continuous line*) in the absence of divalent cations; the IV curve recorded on bovine ROS was expanded by a factor of 1.48 to account for the different current amplitudes.

concentrations of the respective monovalent cations. Fig. 2 shows a series of typical IV recordings from one membrane patch under biionic conditions with different monovalent cations on the cytoplasmic site. All currents have been corrected for the leak current. In the legend of Fig. 2 the mean of reversal voltages, for the different tested cations under biionic conditions are given. According to the Goldman-Hodgkin-Katz equation (Goldman, 1943, Hodgkin and Katz, 1949), the following sequence of permeability ratios P_x/P_{Na} resulted:

$$NH_4^+ > Li^+ > Na^+ > K^+ > Rb^+ > Cs^+ = 1.96 : 1.25 : 1 : 0.89 : 0.73 : 0.67$$

Alternatively, calculating relative currents I_x/I_{Na} at $+50$mV and assuming proportionality to conductances within this small range of membrane voltage of ± 50mV, i.e.,

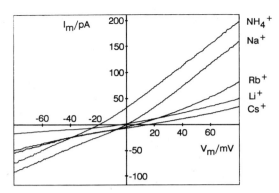

Fig. 2: Ion selectivity of the cGMP-activated current. The set of IV curves was measured under symmetrical X^+ / Na^+ biionic conditions on the same excised membrane patch. The cGMP-activated current (100 μM cGMP) has been corrected for the leak current in the absence of cGMP. The mean reversal voltages $V_{rev} = -RT/F \ln (P_x[X^+]/P_{Na}[Na^+])$ determined in the presence of the different cations were NH_4^+ : -17.2 mV; Li^+ : -5.5 mV; Na^+ : 0 mV; K^+ : $+3.2$ mV; Rb^+ : $+8.7$ mV; Cs^+ : $+11.3$ mV.

$I_x/I_{Na} \approx g_x/g_{Na}$, we obtained the following sequence of relative conductances:

$$NH_4^+ > Na^+ = K^+ > Rb^+ > Li^+ > Cs^+ = 1.46 : 1 : 1 : 0.65 : 0.34 : 0.25$$

Similar sequences are obtained for either determination, except for Li^+ ion. The cGMP-gated channel is much less conductive for Li^+ than for Na^+, whereas its permeability is greater for Li^+ than for Na^+.

The macroscopic current (I) is the product of current through a single channel (i), the number of channel copies within the membrane patch (n), and the fraction of time a channel will remain in a conducting state (P_o):

$$I = n \cdot i \cdot P_o, \tag{2}$$

The reduction in macroscopic outward current across the excised membrane patch, when it is carried by Li^+, may be explained either by a decrease in unit conductance g_i, or a decrease in the open probability P_o, or both. A smaller single-channel current i_{Li} could be explained by stronger binding of Li^+ to a cation binding site and hence a longer dwell time within the pore. Alternatively, a decrease in channel open probability by Li^+ suggests that Li^+ ions may affect the gating, i.e., the kinetics of the open-closed transitions of the channel.

CONCLUSIONS

The cGMP-gated channel of bovine rod photoreceptors exhibits properties similar to those observed in amphibian rod photoreceptors. It is directly and cooperatively activated by at least 3, probably more cGMP molecules (Zimmerman and Baylor, 1986; Haynes et al., 1986). It is selective for cations, but does not discriminate very well between alkali cations (Fesenko et al., 1985; Nunn, 1987; Menini and Torre, 1989; Furman and Tanaka, 1989).

REFERENCES

Fesenko, E. E., Kolesnikov, S. S. and Lyubarsky, A. L., 1985, *Nature*, 313:310-313.

Furman, R. E. and Tanaka, J. C., 1989, *Biophys. J.*, 455a.

Goldman, D. E., 1943, *J. Gen. Physiol.*, 27:37-60.

Haynes, L. W., Kay, A. R. and Yau, K. -W., 1986, *Nature*, 321:66-70.

Hodgkin, A. L. and Katz, B., 1949, *J. Physiol.*, London, 108:37-77.

Kaupp, U. B. and Koch, K. -W., 1986, *Trends Biochem. Sci.*, 11:43-47.

Lamb, T. D., 1986, *Trends Neurosci.*, 9:224-228.

Menini, A. and Torre, V., 1989, *Biophys. J.*, 55:61a.

Nunn, B., 1987, *J. Physiol.*, London, 394, 17P.

Pugh, E. N. Jr., 1987, *Ann. Rev. Physiol.*, 49:715-741.

Stern, J. H., Kaupp, U. B. and MacLeish, P. R., 1986, *Proc. Natl. Acad. Sci. USA*, 83:1163-1167.

Stryer, L., 1986, *Ann. Rev. Neurosci.*, 9:87-119.

Yau, K. -W. and Baylor, D., 1989, *Ann. Rev. Neurosci.*, 12:289-327.

Yau, K. -W., Haynes, L. W. and Nakatani, K., 1986, *In*: "Membrane control of cellular activity," Lüttgau, ed., A.C., Gustav Fischer, Stuttgart, pp. 343-366.

Zimmerman, A. L. and Baylor, D. A., 1986, *Nature*, 321:70-72.

THE EFFECT OF pH ON THE CYCLIC GMP-ACTIVATED

CONDUCTANCE IN RETINAL RODS

A. Menini and B. J. Nunn[*]

Department of Physiology and Department of Neurobiology
Duke University Medical Center, Durham, NC 27710 USA
Istituto di Cibernetica e Biofisica, CNR, Genova, Italy

ABSTRACT

Cytoplasmic pH effect on the Na^+ current through cyclic GMP-activated channels in excised patches from retinal rods of the salamander was studied using rapid solution changes. The current has a plateau from pH 6.7 to 8.2, increases at pH 9.0 and decreases at acidic pH. The shape of the current-voltage relation is unaltered in the range of pH from 5.4 to 9.0 and voltage between -70 and + 70 mV, showing that the effect of cytoplasmic pH is not voltage-dependent. The results are well described assuming that hydrogen ions bind to two charged groups, with apparent pKs of 5.4 and 9.8. Voltage independence of the pH effect suggests that these hydrogen binding sites lie near the inner mouth of the channel.

INTRODUCTION

Fesenko et al. (1985) first demonstrated the existence of cyclic-GMP activated channels in the plasma membrane of retinal rod outer segments. Their patch electrode recordings together with experiments in other laboratories have demonstrated that these cyclic GMP-activated channels are the light-sensitive channels (see Yau and Baylor, 1989 for review). These are closed when light produces the activation of a powerful phosphodiesterase, so reducing the levels of cyclic GMP. Recordings of current from intact cells have shown that although most of the light-sensitive current is carried by Na^+, the channels differ from Na channels in that Ca^{2+} carries a significant fraction of the current (Hodgkin et al., 1985; Nakatani and Yau, 1988; Menini et al., 1988). Furthermore divalent ions such as Ca^{2+} and Mg^{2+} block the passage of monovalent ions so that the average current carried by a channel in Ringer solution is only about 3 fA (see Yau and Baylor, 1989 for review).

---

[*] All the experiments described in this paper were obtained in the laboratory of Dr. B.J. Nunn, who suddenly died September 18, 1987, before completion of this project. The experiments were continued and the paper was written after his tragic death.

Sensory Transduction
Edited by A. Borsellino *et al.*
Plenum Press, New York, 1990

Liebman et al. (1984) studied the sensitivity of intact rods to protons by extracellular perfusion with Ringer solution of altered pH. They found that the dark current was reversibly suppressed by low pH and that perfusion at pH 10.5 caused a slight increase in the dark current. In Ringer solution dark current suppression obeyed a hyperbolic saturation law with apparent dissociation constant $pK = 4.8$.

The effect of pH on the dark current is not only due to a direct action on the channel. In fact Hodgkin and Nunn (1987) showed that external pH has an influence on the Na-Ca exchange: acid pH decreased the saturated pumping current and alkaline pH increased it. The effects of external pH were also rapidly reversible in acid solution if applied for short periods. The outer segment was less tolerant of alkaline pH and seemed to be damaged irreversibly by solutions of pH 10 or more.

By using the pH-sensitive fluorescent dye carboxyfluorescein, Yoshikami and Hagins (1985) demonstrated that the cytoplasm does not show pH changes exceeding 0.002 unit during the light response, suggesting that internal pH does not vary appreciably during the phototransduction process.

We studied the effect of changing cytoplasmic pH on the cyclic GMP-activated channel in excised membrane patches (Hamill et al., 1981) in order to find information about charged groups involved in the ion transport process. Preliminary results have been presented in abstract form (Nunn et al., 1988).

METHODS

Preparation: Rods were mechanically isolated by chopping with a razor blade, without the use of enzymes, the retina of the tiger salamander (Ambystoma tigrinum). All the procedures were the same as described in Menini et al. (1988). In some of the earlier experiments the salamander was dark adapted and the dissection was done in the dark, then the experiments were performed in normal light. The same experiments were repeated with not dark adapted salamander and dissecting the retina in the light. The same results were obtained.

Recording procedures: Inside out membrane patches were excised from the outer segment of isolated rod. Patch pipettes were made from borosilicate glass micro pipettes (Brandt, Germany). A seal between 1 and 10 GΩ was usually obtained by simply touching the membrane of the outer segment and excised inside-out patches were obtained by withdrawing the patch pipette. Currents were activated by the addition of 0.1 mM cyclic GMP sodium disalt (G6129 Sigma) to the bathing solution. Current recordings were made with an Axopatch 1-B amplifier (Axon Instruments).

Solutions: The patch pipette solution, at the extracellular side of the membrane patch contained: 110 mM NaCl, no added divalent salts, 0.1 mM EDTA/TMAOH, 10 mM HEPES/TMAOH, pH 7.6. The bathing solution, at the intracellular side of the patch, contained: 110 mM NaCl, 0.1 mM EDTA/TMAOH and 10 mM of the pH buffer: HEPES, PIPES, MES, CHES or CAPS. The pH was adjusted to the desired value by adding TMAOH and its value was also checked at the end of each experiment. All experiments were performed at room temperature (23-25 °C).

Perfusion system: Solutions at the cytoplasmic side of the membrane patch were rapidly changed by using the method of Nunn (1987). This method is shown in Fig. 1.

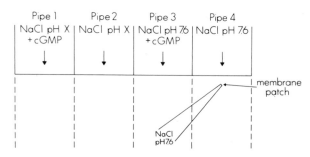

Fig. 1: An excised inside-out patch from a salamander rod outer segment is held in front
of pipe 4 in a flowing solution of 110 mM NaCl, 0.1 mM EDTA, pH 7.6 to measure
the current in the absence of cyclic GMP. 0.1 mM cyclic GMP is rapidly added
to the cytoplasmic face of the membrane by transferring the electrode in front of
pipe 3. Then the electrode is transferred in front of pipe 2 and pipe 1, where a
solution with different pH is flowing. The recovery of the patch current is tested
going back to pipes 3 and 4.

The patch pipette, with the excised inside-out patch, so that the cytoplasmic side
of the membrane was facing the bathing solution, was held in front of pipe 4 were a
solution containing 110 mM NaCl, pH 7.6 was flowing. Voltage steps of ± 10 mV were
given from a holding potential of 0 mV and the leakage current was measured. Then
the stage of the microscope was moved in order to have the patch in front of pipe 3
where the same solution flowing in pipe 4 with the addition of 0.1 mM cyclic GMP was
flowing. The currents, due to the leakage and to the channels activated by cyclic GMP,
were measured in this way as a function of voltage. To obtain the cyclic GMP-activated
current the recordings in the absence of cyclic GMP (pipette 4) were subtracted from
those in the presence of cyclic GMP (pipe 3).

In order to study the effect of different values of pH on the Na^+ current through
the channel the test solutions were flown through pipe 2 (without cyclic GMP) and pipe
1 (with cyclic GMP) and the same technique was applied.

The patch pipette was then positioned again in front of pipes 3 and 4 where the
control solution at pH 7.6 was flowing. In this way the recovery of the response after
each test pH could be checked.

RESULTS AND DISCUSSION

The effect of changing cytoplasmic pH on the cyclic GMP-activated channel was
studied in the presence of a solution containing 110 mM NaCl and low divalents at both
sides of the membrane patch. Solutions at the cytoplasmic side were changed and cyclic
GMP-activated currents were measured as explained in the Methods and in Fig. 1.

Fig.2 shows the effect of cytoplasmic pH 9.0, 5.7 and 5.4 on the Na^+ current through
channels activated by 0.1 mM cyclic GMP.

Traces are the cyclic GMP-activated currents obtained with voltage steps of ± 20 mV
from a holding potential of 0 mV. Outward currents are plotted upward. Traces on the

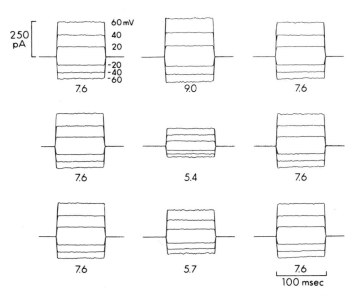

Fig. 2: Effect of changing cytoplasmic pH on the cyclic GMP-activated current. Current is the difference between currents measured in the presence of 0.1 mM cyclic GMP and in the absence of cyclic GMP. Patch voltage was held at 0 mV and switched between -60 and +60 mV in 20 mV steps. Control and recovery in pH 7.6 are shown.

left are the control before testing a pH, traces on the right are the recovery and in the middle the effect of changing pH on the Na^+ current through the cyclic GMP-activated channel is shown. The current at pH 9.0 was sligthly higher than at pH 7.6; pH 5.7 caused a reduction in the current and at pH 5.4 the current was about 50 % of its value at pH 7.6. Both outward and inward Na^+ currents were affected by changing pH at the cytoplasmic side. In this range of pH (from 5.4 to 9.0) the recovery was always complete as shown by the comparison between traces on the left and on the right of Fig. 2.

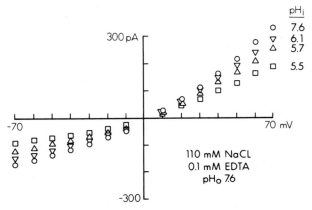

Fig. 3: Current-voltage relations in different pH solutions. Current flowing into the patch pipette is positive. Voltage is V(bath)-V(electrode).

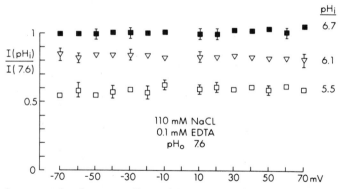

Fig. 4: Ratio of current in the test pH to the average of control and recovery currents in pH 7.6 versus voltage. Collected results from 15 patches.

However at pH below 5.4 the cyclic GMP-activated conductance does not recover completely but, if the exposure time in the low pH solution is sufficiently short (≈ 500 msec), the recovery is complete in a few minutes (data not shown). In this way the pH could be lowered to the value of 5.0. Unfortunately, solutions of pH 9.5 or more caused the irreversible damage of the membrane patch. Low external pH reduces channel current and does not modify the shape of the I-V, as shown in Fig.3 and Fig. 4. Current-voltage relations obtained at different pH from the average current measured at the steady-state in the voltage range from -70 to +70 mV are shown in Fig. 3. The current decreases lowering the pH but the shape of the I - V was not altered. This is better seen in Fig. 4 where the currents at different pH normalized to the control current in pH 7.6 are plotted versus the applied voltage. The average, taken from many patches, is plotted with the standard error. The reduction in current at low pH does not depend on the membrane potential.

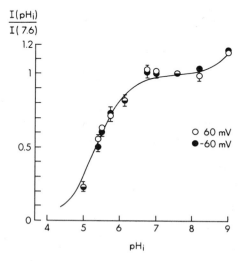

Fig. 5: Cytoplasmic pH dependence of normalized current at two different voltages. Collected results from 20 patches. Continuous line was obtained by considering two hydrogen binding sites with apparent pK of 5.4 and 9.8.

In Fig. 5 the pH dependence of normalized currents, calculated at two different voltages, is plotted. Open circles are data at -60 mV, filled circles are data at +60 mV. The continuous line is the best fit obtained considering the presence of two hydrogen binding sites: one with an apparent pK of 5.4 and the other one with an apparent pK of 9.8. These values are independent from the membrane potential.

In conclusion, the results of these experiments suggest that the conductance of the cyclic GMP-activated channel is controlled by two titratable ionic groups with apparent pKs of 5.4 and 9.8 and the absence of voltage-dependence of the pH effect suggests that these groups are located near the inner mouth of the channel. A comparison with other channels shows that similar values were found for the acetylcholine receptor channel in *Rana pipiens* (Landau et al., 1981) where the presence of two titratable ionic groups with pKs of 4.8 and 9.8 was suggested. In sodium channels (Woodhull, 1973) a titratable group with apparent pK of 5.4 was found and, from the voltage-dependence of the block, it was suggested that this acid site is inside the channel.

ACKNOWLEDGMENTS

Professor John W. Moore is gratefully acknowledged for his continuous encouragement and for many helpful discussions. Mr. Clive Prestt checked the English. This research was supported by grant EY 07106 from the National Institute of Health.

REFERENCES

Fesenko, E. E., Kolesnikov, S. S. and Lyubarsky, A. L., 1985, Induction by cyclic GMP of cationic conductance in plasma membrane of retinal outer segment, *Nature, Lond.,* 313:310-313.

Hamill, O. P., Marty, A., Neher, E., Sakmann, B. and Sigworth, F. J., 1981, Improved patch-clamp techniques for high resolution current recording from cells and cell-free membrane patches, *Pflügers Archiv.,* 391:85-100.

Hodgkin, A. L., McNaughton, P. A. and Nunn, B. J. 1985, The ionic selectivity and calcium dependence of the light-sensitive pathway in toad rods, *J. Physiol.,* 358:447-468.

Hodgkin, A. L. and Nunn, B. J., 1987, The effect of ions on sodium-calcium exchange in salamander rods, *J. Physiol.,* 391:371-398.

Landau, E.M., Gavish, B., Nachshen, D.A. and Lotan, I., 1981, pH dependence of the Acetylcholine receptor channel. *J. Gen. Physiol.,* 77:647-666.

Liebman, P. A., Müller, P. and Pugh, E. N., 1984, Protons suppress the dark current of frog retinal rods, *J. Physiol.,* 347:85-110.

Menini, A., Rispoli, G. and Torre, V., 1988, The ionic selectivity of the light-sensitive current in isolated rods of the tiger salamander, *J. Physiol.,* 402:279-300.

Nakatani, K. and Yau, K. W., 1988, Calcium and magnesium fluxes across the plasma membrane of the toad rod outer segment, *J. Physiol.,* 395:695-729.

Nunn, B. J., 1987, Precise measurement of cyclic GMP-activated currents in membrane patches from salamander rod outer segments, *J. Physiol.,* 394:8P.

Nunn, B. J., Menini, A. and Torre, V., 1988, pH dependence of cyclic GMP-activated channel conductance in patches of rod outer segments from the salamander, *Biophys. J.,* 53:472a

Woodhull, A. M., 1973, Ionic blockage of sodium channels in nerve, *J. Gen. Physiol.,* 61:687-708.

Yau, K. W. and Baylor, D., 1989, Cyclic GMP-activated conductance of retinal photore-ceptor cells, *Ann. Rev. Neurosci.*, 12:289-327.

Yoshikami, S. and Hagins, W. A., 1985, Cytoplasmic pH in rod outer segments and high-energy phosphate metabolism during phototransduction, *Biophys. J.*, 47:101a.

THE CYTOSOLIC CONCENTRATION OF FREE Ca^{2+} IN VERTEBRATE ROD OUTER SEGMENTS

† W. G. Owen and *G. M. Ratto

† Department of Molecular and Cell Biology
   University of California, Berkeley, U.S.A.
* Istituto di Neurofisiologia del CNR, Pisa, Italy

THE ROLE OF Ca IN PHOTOTRANSDUCTION

The role of calcium in the mechanism of phototransduction has yet to be defined. The observation of Yoshikami and Hagins (1971), that an application of calcium to the outside of the rod caused a reduction in the dark-current, and thus mimicked the effect of light, led to the well-known "calcium hypothesis" according to which, calcium ions, released by light from an intracellular store, were internal messengers that diffused to the plasma membrane and caused the closure of "light-sensitive" channels. This hypothesis stimulated an enormous interest in phototransduction during the subsequent decade and a half. During that time, however, several lines of evidence began to point to cyclic GMP as the internal transmitter, the recent demonstration, by Fesenko, Kolesnikov and Lyubarsky (1985, 1986), that the channels in patches of plasma membrane excised from rod outer segments (ROS) are activated by cGMP and are not blocked by calcium ions, being, perhaps, the most direct (see also Yau and Nakatani, 1985b). These channels have properties identical to those of the light-sensitive channels (Matthews, 1986).

Studies of the permeability of the light-sensitive channels, revealed that they are highly permeable to calcium and, indeed, the inward flux of calcium through them was shown to carry about 15% of the dark-current (Yau and Nakatani, 1984; Hodgkin, McNaughton and Nunn, 1985; Cervetto and McNaughton, 1986). This influx of calcium is opposed by a light-insensitive extrusion mechanism which pumps calcium out of the ROS in exchange for sodium (Yau and Nakatani, 1984; McNaughton, Cervetto and Nunn, 1986; Hodgkin, McNaughton and Nunn, 1987). Recent studies (Cervetto, L. Lagnado, Perry, Robinson and McNaughton, 1989) show that potassium ions are also transported by this mechanism, one K$^+$ and one Ca^{2+} being exchanged for four Na$^+$. Yau and Nakatani (1985a) pointed out that since the light-sensitive channels provide the only pathway by which calcium can enter the ROS, their closure by light must reduce the normal influx of calcium and in consequence, since the exchange mechanism is light-insensitive, the concentration of calcium in the cytosol of the ROS should fall, not rise as required by the calcium hypothesis. This prediction was confirmed by McNaughton, Cervetto and Nunn (1986), who loaded single rods with the photoprotein aequorin, elevated internal calcium levels with IBMX to induce measurable luminescence and observed that light caused a drop in that luminescence. The time-resolution of their technique was good enough to show that this light-induced fall in cytosolic $[Ca^{2+}]$ was not preceded be any transient rise, indicating that light does not trigger a measurable release of calcium from internal stores. Today, there are few who would disagree with the view that calcium plays, at most, an indirect role in the generation of the rod's light response.

At the same time that calcium was losing its position as most-favoured candidate for the internal transmitter, new evidence was being developed which pointed to its likely role in light-adaptation. A key finding was that the guanylate cyclase, which catalyses the conversion of GTP to cGMP, is regulated by calcium (Fleischman and Denisevich, 1979; Lolley and Racz, 1982; Pepe, Panfoli and Cugnoli, 1986; Koch and Stryer, 1988). Other studies suggested that, either directly or indirectly, calcium may activate the phosphodiesterase which catalyses the hydrolysis of cGMP to 5' GMP (Bownds, 1980; Robinson, Kawamura, Abramson and Bownds, 1980; Lamb, 1986; Torre, Matthews and Lamb, 1986), though the evidence for this is much less direct. If light induces a fall in cytosolic free calcium, however, either of these mechanisms would tend to restore [cGMP] toward its dark level, thereby light adapting the rod. Recent work by Matthews, Murphy, Fain and Lamb (1988), Nakatani and Yau (1988), and others lends strong support for this idea.

A question that must be answered before any proposed cellular mechanism involving calcium can be assessed concerns the concentration of calcium in the cytosol under normal conditions. In the case of the rod, one must ask whether the normal, dark level of free calcium in the ROS is within the range of activation of the guanylate cyclase and whether light-induced changes in free calcium are large enough to account for the marked changes in response properties that accompany light adaptation.

The technique used by McNaughton, Cervetto and Nunn (1986) yielded an upper limit of 600 nM on the normal cytosolic concentration of free calcium. An exact value could not be obtained, however, because signal-to-noise limitations precluded reliable measurement of Ca concentrations below about 1 μM. Nor could light-induced changes in cytosolic free $[Ca^{2+}]$ be detected without first elevating the dark value of $[Ca^{2+}]_{in}$ into the micromolar range. In collaboration with Roger Tsien and Richard Payne, we undertook to measure cytosolic free calcium in the outer segments of frog rods using the fluorescent calcium chelator, Fura2. This has several advantages over most indicators. One is that by comparing the intensities of the fluorescence excited by light of two wavelengths, 340 nm and 380 nm, a direct measure of the free $[Ca^{2+}]$ can be obtained. Another important advantage is that, in its esterified form, Fura2/AM, it can be loaded into cells by incubation. Once inside, natural esterases cleave the ester groups from the molecule, leaving the indicator trapped within the cell. Thus, we were able to load whole retinae with this indicator and integrate the fluorescence emanating from an area containing about 100,000 rods. We chose to use the retina of the bullfrog, *Rana catesbeiana*, because of the extreme length of its rod outer segments (50 μm). The optical density of rhodopsin is such that whereas about 37% of 360 nm light travelling axially through these outer segments will be absorbed, 90% of 500 nm light will be absorbed. The measurement of fluorescence was optimised by using an optical system of high numerical aperture and a cooled photomultiplier having a low dark noise. The apparatus is illustrated schematically in figure 1. The retina was dark adapted overnight before being isolated and mounted, receptor-side up in a perfusion cell (Sickel, 1965). By both stimulating and collecting fluorescence from above, we ensured that a mere 6%

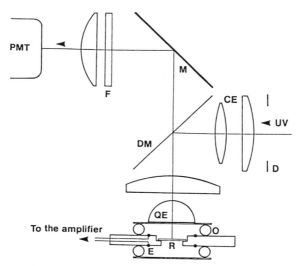

Fig. 1: Schematic diagram of the apparatus. UV light enters the cage from the optical
bench, is collimated and then reflected by a dichroic mirror, DM, onto a quartz
hemisphere of 7.5 mm radius, QE, which produces a uniform, circular patch of
light, 5.5 mm in diameter on the retina, R. The retina is mounted receptor-side
up in a perfusion cell. Transretinal potentials can be measured with the aid of
Ag/AgCl electrodes, E, built into the cell. 500 nm fluorescence is collected by
the hemisphere and transmitted by the dichroic mirror to the cathode of a cooled
photomultiplier, PMT. (Reproduced, with permission of Journal of Neuroscience,
from Ratto, Payne, Owen and Tsien, 1988).

of any fluorescence emitted in the direction of the detector from points proximal to the
rod outer segments would contribute to our measurements.

Dark-adaptation was preserved throughout the experiment by allowing sufficient
time between stimulations for recovery to be complete.

Stimuli were 5.7 second steps of light which alternated between 340 nm and 380
nm every 110 msec. The signal-to-noise ratio was low enough to allow measurement of
fluorescence evoked by stimuli that elicited half-maximal responses, or greater, from the
rods, as determined by monitoring the aspartate-isolated pIII response of the retina; i.e.,
well within the physiological response range of the rods. The limitation on measurements
with stimulus intensities lower than this was set by the shot noise of the system, not by
instrumental noise.

To measure background fluorescence, we made use of the observation (Capovilla,
Caretta, Cervetto and Torre, 1983), that the phosphodiesterase inhibitor, IBMX, renders
the ROS membrane highly permeable to manganese which efficiently quenches Fura2 fluorescence (Grynkiewicz, Poenie and Tsien, 1985). Thus, at the end of each experiment
the retina was perfused with IBMX and Mn^{2+} and background fluorescence was measured. The background counts were subtracted from all experimental counts to yield the
Fura2 fluorescence. $[Ca^{2+}]_{in}$ was calculated as described by Grynkiewicz et al. (1985).
The resulting values should be regarded as "space-averaged" values, since we did not dis-

tinguish between fluorescence emanating from the cytosol and from other dye-containing compartments. Of these, the most important was probably the intradiskal space, as we explain below.

A series of control experiments was performed to determine how much Fura2 was loaded into the retina under our protocol and how it was distributed within the retina. The concentration of Fura2 in the outer segments was found to be in the range 40 - $80 \mu M$. By two different methods we established that 2/3 of the loaded Fura2 was in the rod outer segments, of which 87% was in the cytosol and 13% in the disks. Most of the remaining 1/3 was in the rod inner segments, only negligible amounts being found in the rest of the retina. The finding that most of the Fura2 had loaded into the rods is not surprising when one realises that the total membrane surface area of rod outer segments is about 25 times the planar surface area of the retina.

From these experiments we estimated that better than 90% of the measured fluorescence emanated from Fura2 trapped in the rod outer segments. The 10% of fluorescence that emanated from compartments proximal to the outer segments was a complicating factor but, provided the free Ca^{2+} in those compartments was of the same order of magnitude as that in the outer segments, the error that was introduced by ignoring it was probably less than 5%. A larger potential error is introduced by the Fura2 in the intradiskal space. The free $[Ca^{2+}]$ in the disks is almost certainly high enough to keep the Fura2 saturated throughout our experiments. Given that 13% of the dye was in the disks, we estimate that our values for cytosolic free $[Ca^{2+}]$ could be higher than the true values by up to 30%. Because of the small size of the intradiskal space, however, much of the dye may be sufficiently close to the rhodopsin chromophore for dye fluorescence to be quenched by radiationless fluorescence energy transfer (Steinemann and Stryer, 1973), so that the error in our measurements is probably less than this. It should be noted that even though our measured values of $[Ca^{2+}]_{in}$ could be somewhat higher than the true values, errors in the measured kinetics of light-induced changes in $[Ca^{2+}]_{in}$ are likely to be small. A detailed account of these procedures and experiments is published elsewhere (Ratto, Payne, Owen and Tsien, 1988).

Our principal findings are illustrated in figure 2. The rods were initially dark-adapted. Figure 2A shows the raw fluorescence, in counts per 100 msec, evoked by 340 nm and 380 stimuli which bleached about 3400 rhodopsin molecules per second in each rod, sufficient to saturate the rod response. During illumination, the fluorescence evoked by 380 nm light increased while that evoked by 340 nm light decreased. This is due to a shift in the absorbance spectrum of Fura2 towards longer wavelengths, indicative of a fall in internal free $[Ca^{2+}]$. After correction for background fluorescence, the corrected counts were used to calculate $[Ca^{2+}]_{in}$. The resulting values are plotted in figure 2B. It can be seen that, in this experiment, $[Ca^{2+}]_{in}$ declined roughly exponentially, from an initial value of about 240 nM to about 170 nM. The apparent time constant of this decline was ~1.6 seconds. In 10 such experiments, the average value of $[Ca^{2+}]_{in}$ at t = 0 was 221 nM $\pm$ 9.5 nM (S.E.M.) while the average value at the end of the light step was 140 nM $\pm$ 14.5 nM (S.E.M.).

These values were not significantly different for stimulus intensities in the range 70 Rh* sec^{-1} rod^{-1} and 7500 Rh* sec^{-1} rod−1 though the time-constant of decline was shorter at the higher intensities. The finding that the dark level of $[Ca^{2+}]_{in}$ (at t = 0) is independent of light intensity over a large range argues against the notion that light causes an instantaneous release of Ca from internal stores since if that were the case we would expect the "dark-level" to increase with stimulus intensity. Moreover, we found

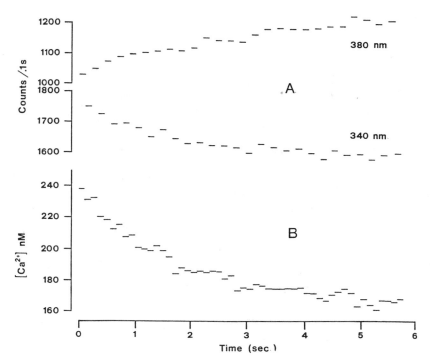

Fig. 2: Effect of light on cytosolic free $[Ca^{2+}]$ under normal conditions. Stimulus intensity was sufficient to bleach 3400 rhodopsin molecules per rod per second. A. The raw fluorescent count evoked by 340/380 nm light as a function of time during the light response. B. Light-induced change in cytosolic free $[Ca^{2+}]$, calculated from the data in A after correction for background. (Reproduced with permission of Journal of Neuroscience from Ratto, Payne, Owen and Tsien, 1988).

that, in darkness, following stimulation, $[Ca^{2+}]_{in}$ recovered to its initial, (t = 0), value. Thus the exponential decline we observe is a true fall from the dark level, not the restoration of a level that had been perturbed by light-induced release of Ca into the cytosol.

CALCIUM REGULATION IN THE ROD OUTER SEGMENT

The normal value of $[Ca^{2+}]_{in}$ in rod outer segments in darkness is about 220 nM, a value similar to that normally found in those parts of nerve cells where there is known to be Ca influx (Connor, 1986). Values of 220 nM and 271 nM were obtained by Miller and Korenbrot (1986) and Korenbrot and Miller (1989), respectively, who restricted their measurements of the fluorescence of the calcium chelator, Quin2, to that which emanated from the outer segments of rods in preloaded retinal slices. All these values are in good agreement with what one would predict on the basis of the known properties of the Na/Ca,K exchanger.

Cervetto, Lagnado and McNaughton (1988) demonstrated that the Na/Ca,K exchanger normally behaves as a first order system having a K_m of $2.3 \mu M$ and thus obeys the relation:

$$\frac{E}{E_{max}} = \frac{[Ca^{2+}]_{in}^{free}}{K_m + [Ca^{2+}]_{in}^{free}} \qquad (1)$$

where E is the rate of $[Ca^{2+}]$ extrusion (maximal rate E_{max}) and K_m is the half-saturating concentration of $[Ca^{2+}]$. Under normal conditions, in darkness, the efflux of $[Ca^{2+}]$ via the exchanger carries a current of about 3 pA (Yau and Nakatani, 1985). This value, which was estimated from the magnitude of the Na/Ca,K exchange current immediately after all light-sensitive channels had been closed by a saturating light, is equivalent to a flux of about 30μM/sec. The value of E_{max}, derived from experiments on toad (Yau and Nakatani, 1984) and salamander (Lagnado, Cervetto and McNaughton, 1988) is about 300 - 400 μM/sec at the normal dark potential of about -40 mV. Substituting these values in equation 1, we estimate that the dark value of $[Ca^{2+}]_{in}$ should be between 190 and 254 nM, in good agreement with our measurement.

Since, in our experiments, $[Ca^{2+}]_{out}$ was 1.5 mM, the rod outer segment must contain mechanisms that tightly regulate $[Ca^{2+}]_{in}$. The primary regulatory mechanism in the plasma membrane is believed to be the Na/Ca,K exchanger (Lagnado, Cervetto and McNaughton, 1988). In addition, there is known to be a powerful Ca buffering system, one consequence of which is the slowness of the light-induced decline in internal free $[Ca^{2+}]$.

At the onset of a saturating light, all the light-sensitive channels are rapidly closed and $[Ca^{2+}]_{in}$ declines because of the continued removal of Ca from the cytosol by light-independent mechanisms such as the Na/Ca,K exchanger. In the aequorin experiments of McNaughton, Cervetto and Nunn (1986), in which $[Ca^{2+}]_{in}$ was elevated by the application of IBMX, the time-constant of the $[Ca^{2+}]$ decline induced by a saturating light was about 0.4 seconds which is close to the time constant of decline of the residual Na/Ca,K exchange current seen in suction pipette recordings. The time-constant of 1.6 seconds observed in our experiments could be due to the additional buffering effect of Fura2 and of a high-affinity buffer which is saturated at the [Ca] levels of the aequorin experiments (Hodgkin, McNaughton and Nunn, 1987). Equally, it could be due to the action of a mechanism (or mechanisms) other than the Na/Ca,K exchanger that contributes significantly to the regulation of cytosolic calcium levels in the physiological range. Significantly, in more recent experiments, we have found that light-induced changes in $[Ca^{2+}]_{in}$ are not eliminated by procedures known to minimise the activity of the exchanger (W. G. Owen and J. P. Younger - personal communication).

We were surprised to observe that the level to which $[Ca^{2+}]_{in}$ declined was approximately the same over the more than two log-unit range of intensities we used. If light adaptation is controlled by $[Ca^{2+}]_{in}$, therefore, this result implies that it must be graded over a range of intensities below those used in our experiments. Moreover, $[Ca^{2+}]_{in}$ did not decline below 140 nM, even in the presence of saturating steps of light which suggests that relatively modest changes in $[Ca^{2+}]_{in}$ must be sufficient to regulate the postulated light adaptation mechanism. This is consistent with the recent finding that guanylate cyclase activity varies inversely as the fourth power of $[Ca^{2+}]_{in}$ below about 200 nM (Koch and Stryer, 1988), and, indeed, recent work by Forti, Menini, Rispoli and Torre (1989) confirms that quite modest changes in $[Ca^{2+}]_{in}$ are sufficient to account for the changes in sensitivity and kinetics of rod responses that are observed when background illumination is switched on.

Initially, we were not surprised to find that $[Ca^{2+}]_{in}$ declined only to 140 nM since a Na/Ca exchanger with a stoichiometry of 3 Na : 1 Ca would be expected to come into equilibrium at about this concentration. With the recent finding that the exchanger also transports K^+, however, this explanation can no longer be regarded as tenable. Such a mechanism, with a stoichiometry of 1 Ca, 1K : 4 Na, should come into equilibrium only when $[Ca^{2+}]_{in}$ has fallen to about 2.5 nM. The excellent agreement between our measured value of the dark concentration of cytosolic free [Ca] and those of Miller and Korenbrot (1986) and Korenbrot and Miller (1989), which were obtained from measurements restricted to the outer segments of rods in retinal slices, argues that it is not an artifact of our method. Korenbrot and Miller (1989) noted that if $[Ca^{2+}]_{in}$ had fallen as low as 50 nM, the change would have been detected by their method. It is also worth noting that, on the basis of their measurements of calcium movements across the outer segment membranes of isolated rods, Miller and Korenbrot (1987) concluded that $[Ca^{2+}]_{in}$ does not fall below 80 nM following exposure to bright light. Clearly, there is a contradiction here for which, at present, we have no explanation.

It remains possible that another mechanism besides the Na/Ca,K exchanger may be involved in the regulation of cytosolic Ca. If so, it is likely to be one which is internal to the rod since the Na/Ca,K exchanger appears to be the only Ca-regulatory mechanism in the plasma membrane. The most obvious site for such a mechanism is the disk membrane which is reported to contain active Ca uptake mechanisms and also cGMP-sensitive pathways by which Ca can leak into the cytosol (Cavaggioni and Sorbi, 1981; George and Hagins, 1983; Koch and Kaupp, 1985). We are currently investigating the possibility that Ca movement between the cytosol and the disks may also contribute significantly to the "buffering" of cytosolic calcium.

ACKNOWLEDGEMENTS

The experiments described in this paper were carried out in collaboration with Richard Payne and Roger Tsien. We thank Drs. P. A. McNaughton and V. Torre for helpful discussions. The work was supported in part by grants EY04372 and EY03785 from the National Eye Institute.

REFERENCES

Bownds, M. D. 1980, Biochemical steps in visual transduction: Roles for nucleotides and calcium ions, *Photochem. Photobiol.*, 32:487-490.

Capovilla, M., Caretta, A., Cervetto, L. and Torre, V., 1983, Ionic movements through light-sensitive channels of toad rods, *J. Physiol.*, 343:295-310.

Cavaggioni, A. and Sorbi, R.T., 1981, Cyclic GMP releases calcium from disk membranes of vertebrate photoreceptors, *Proc. Natl. Acad. Sci. U.S.A.*, 78:3964-3968.

Cervetto, L., Lagnado, L. and McNaughton, P. A., 1988, Ion transport by the Na-Ca exchange in isolated rod outer segments, *Proc. Natl. Acad. Sci. U.S.A.*, 85:4548-4552.

Cervetto, L., Lagnado, L., Perry, R. J., Robinson, D. W. and McNaughton, P. A., 1989, Extrusion of Calcium from rod outer segment is driven by both sodium and K gradient, *Nature*, 337:740-743.

Cervetto, L. and McNaughton, P.A., 1986, The effects of phosphodiesterase inhibitors and lanthanum ions on the light-sensitive current of toad retinal rods, *J. Physiol.*, 370:91-100.

Connor, J.A., 1986, Digital imaging of free calcium changes and of spatial gradients in growing processes in single, mammalian central nervous system cells,

Proc. Natl. Acad. Sci. U.S.A., 83:6179-6183.

Fesenko, E.E., Kolesnikov, S.S. and Lyubarsky, A.L., 1985, Induction by cyclic GMP of cationic conductance in plasma membrane of retinal rod outer segment, *Nature*, 313:310-313.

Fesenko, E.E., Kolesnikov, S.S. and Lyubarsky, A.L., 1986, Direct action of cGMP on the conductance of retinal rod plama membrane, *Biochim. Biophys. Acta*, 856:661-671.

Fleischman, D. and Denisevich, M., 1979, Guanylate cyclase of isolated bovine retinal rod axonemes, *Biochem.*, 18:5060-5066.

Forti, S., Menini, A., Rispoli, G. and Torre, V., 1989, Kinetics of phototransduction in retinal rods of the newt triturus Cristatus, *J. Physiol.*, 419:265-295.

George, J.S. and Hagins, W.A., 1983, Control of $[Ca^{2+}]$ in rod outer segment disks by light and cyclic GMP, *Nature*, 303:344-348.

Grynkiewicz, G.M., Poenie, M. and Tsien, R.Y., 1985, A new generation of $[Ca^{2+}]$ indicators with greatly improved fluorescence properties, *J. Biol. Chem.*, 260:3440-3450.

Hodgkin, A.L., McNaughton, P.A. and Nunn, B.J., 1985, The ionic selectivity and calcium dependence of the light-sensitive pathway in toad rods, *J. Physiol.*, 358:447-468.

Hodgkin, A.L., McNaughton, P.A. and Nunn, B.J., 1987, Measurement of sodium-calcium exchange in salamander rods, *J. Physiol.*, 391:347-370.

Koch, K.-W. and Kaupp, U.B., 1985, cGMP directly regulates a cation conductance n membranes of bovine rods by a cooperative mechanism, *J. Biol. Chem.*, 260:6788-6800.

Koch, K.-W. and Stryer, L., 1988, Highly cooperative feedback control of retinal rod guanylate cyclase by calcium ions, *Nature*, 334:64-66.

Korenbrot, J. I. and Miller, D.L., 1989, Cytoplasmic free calcium concentration in dark-adapted retinal rod outer segments, *Vision Res.*, In press.

Lagnado, L., Cervetto, L. and McNaughton, P.A., 1988, Ion transport by the Na-Ca exchange in isolated rod outer segments, *Proc. Natl. Acad. Sci. U.S.A.*, 85:4548-4552.

Lamb, T. D. 1986, Transduction in vertebrate photoreceptors: The roles of cyclic GMP and calcium, *Trends Neurosci.*, 9:224-228.

Lolley, R. N. and Racz, E., 1982, Calcium modulation of cyclic GMP synthesis in rat visual cells, *Vision Res.*, 22:1481-1486.

Matthews, G. 1986, Comparison of the light-sensitive and cyclic GMP-sensitive conductances in the rod photoreceptor: Noise characteristics, *J. Neurosci.*, 6:2521-2526.

Matthews, H.R., Murphy, R.L.W., Fain, G. and Lamb, T.D., 1988, Photoreceptor light adaptation is mediated by cytoplasmic calcium concentration, *Nature*, 334:67-69.

McNaughton, P.A., Cervetto, L. and Nunn, B.J., 1986, Measurement of intracellular free calcium concentration in salamander rods, *Nature*, 322:261-263.

Miller, D.L. and Korenbrot, J.I., 1986, Effects of the intracellular calcium buffer Quin-2 on the photocurrent responses of toad rods, *Biophys. J.*, 49:281a.

Miller, D.L. and Korenbrot, J.I., 1987, Kinetics of light-dependent Ca fluxes across the plasma membrane if rod outer segments, *J. Gen. Physiol.*, 90:397-425.

Nakatani, K. and Yau, K. -W., 1988, Calcium and light adaptation in retinal rods and cones, *Nature*, 334:69-71.

Pepe, I.M., Panfoli, I. and Cugnoli, C., 1986, Guanylate cyclase in rod outer segments of the toad retina, FEBS *Lett.*, 203:73.

Ratto, G.M., Payne, R., Owen, W.G. and Tsien, R.Y., 1988, The concentration f cytosolic free calcium in vertebrate rod outer segments measured with Fura-2, *J. Neu-*

rosci., 8(9):3240-3246.

Robinson, P. R., Kawamura, S., Abramson, B. and Bownds, D., 1980, Control of the cyclic GMP phosphodiesterase of frog photoreceptor membranes, *J. Gen. Physiol.*, 76:631-645.

Sickel, W., 1965, Respiratory and electrical responses to light stimulation in the retina of the frog, *Science*, 148:648-651.

Steinemann, A. and Stryer, L., 1973, *Biochemistry*, 12:1499-1502.

Torre, V., Matthews, H.R. and Lamb, T.D., 1986, The role of calcium in regulating the cyclic GMP cascade of phototransduction, *Proc. Natl. Acad. Sci. U.S.A.*, 83:7109-7113.

Yau, K.-W. and K. Nakatani, 1984, Electrogenic Na-Ca exchange in retinal rod outer segment, *Nature*, 311, 661:663.

Yau, K.-W. and Nakatani, K., 1985a, Light-induced reduction of cytoplasmic free calcium in retinal rod outer segment, *Nature*, 313:579-582.

Yau, K.-W. and Nakatani, K., 1985b, Light-suppressible, cGMP-sensitive conductance in the plasma membrane of a truncated rod outer segment, *Nature*, 317:252-255.

Yoshikami, S. and Hagins, W.A., 1971, Light, calcium, and the photocurrent of rods and cones, *Biophys. J.*, 11:47a.

CONTROL OF INTRACELLULAR CALCIUM IN VERTEBRATE PHOTORECEPTORS

P.A. McNaughton, L.Cervetto, L.Lagnado, R.J.Perry and D.W.Robinson

Physiological Laboratory, University of Cambridge, U.K.

Calcium seemed for several years to be the most likely candidate for the intracellular messenger which is responsible for linking the absorption of light by rhodopsin to the first electrical event in phototransduction, the suppression of current flowing across the outer segment membrane (Yoshikami and Hagins, 1971; Hagins, 1971). Evidence against this point of view has come from three main sources: the intracellular calcium concentration is now known to decline after a flash of light, and not to rise as was required by the original calcium hypothesis (McNaughton, Cervetto and Nunn, 1986; Ratto, Payne, Owen and Tsien, 1988); the pathway linking rhodopsin isomerization to the breakdown of cyclic GMP, which is now considered to be the internal messenger, has been characterized in detail and has been found to exhibit the properties required to generate the light response (see review by Stryer, 1986); and, finally, cyclic GMP has been shown to open light-sensitive channels in isolated patches of outer segment membrane, while calcium has little or no effect (Fesenko, Kolesnikov and Lyubarksy, 1985). The action of calcium seems instead to be through an indirect effect on the light-sensitive pathway, most probably through an inhibition of the guanylate cyclase responsible for synthesizing cyclic GMP from GTP (Lolley and Racz, 1982; Koch and Stryer, 1988).

Although calcium is not the primary transmitter in the light response, the level of internal calcium plays an important role both in the control of the light-sensitive current and in the process of light adaptation. Reducing external calcium causes a dramatic enhancement of the light-sensitive current (Hagins and Yoshikami, 1975; Yau, McNaughton and Hodgkin, 1981; Hodgkin, McNaughton, Nunn and Yau, 1984; Lamb and Matthews, 1988) probably because the consequent decline in internal calcium releases the guanylate cyclase from inhibition (Lolley and Racz, 1982; Hodgkin, McNaughton and Nunn, 1985; Koch and Stryer, 1988). If intracellular calcium is increased by inhibiting the Na:Ca,K exchange then the flash response becomes more sensitive, and the correlation between time to peak and amplitude is reminiscent of the desensitizing effects of background light (Yau, McNaughton and Hodgkin, 1981). When the internal calcium level is stabilized, either by incorporating a calcium buffer into the cell or by bathing the cell in a low-sodium, low-calcium medium to abolish calcium transport across the cell membrane, then the state of light-adaptation of the cell seems also to be stabilized (Torre, Matthews and Lamb, 1986; Matthews, Murphy, Fain and Lamb, 1988; Nakatani and Yau, 1988). These observations make it likely that calcium is an important modulator of the state of light or dark adaptation of the cell.

Sensory Transduction
Edited by A. Borsellino *et al.*
Plenum Press, New York, 1990

This review will be concerned less with what calcium actually does in a photoreceptor than with the mechanism of regulation of calcium in the outer segment. Calcium enters the outer segment through the light-sensitive channel, and is pumped out across the cell membrane by an ion exchange whose stoichiometry is $4Na^+:1Ca^{2+},1K^+$. Within the cell calcium is buffered in the short term by two mechanisms: a low-affinity buffer of large capacity, and a second system of much higher affinity but lower capacity. The interplay of these three factors - the influx, efflux and buffering of internal calcium - determines the rate of change of calcium in response to a perturbation such as the sudden onset of a bright light.

Calcium influx through the light-sensitive channel

In the usual picture of current flow presented in elementary textbooks (see Figure 1A) the light-sensitive current is represented as being entirely carried by sodium. This view was based on the observation (Sillman, Ito and Tomita, 1969; Hagins and Yoshikami, 1975; Yau, McNaughton and Hodgkin, 1981) that the light-sensitive current collapses when another monovalent ion replaces sodium. We now know that the light-sensitive channel exhibits only a low degree of selectivity for Na over other alkali metal ions, and that the reason the current collapses is that the Na:Ca,K exchange mechanism, which has an extremely high degree of selectivity for Na, is not sustained by other monovalent ions. The consequent accumulation of intracellular calcium is the main cause of the abolition of the light-sensitive current in the absence of external Na.

The light-sensitive channel also readily admits calcium ions, and, since these are a normal constituent of the external medium, a substantial fraction of the light-sensitive current is carried by calcium. The actual fraction carried by calcium is not easy to determine: in experiments in which Na was abruptly replaced by an impermeant ion about 20% of the light-sensitive current was sustained by calcium (Hodgkin, McNaughton and Nunn, 1985), but in normal external solution it seems that a somewhat smaller fraction (about 10%) is carried by calcium (Cervetto and McNaughton, 1986; Hodgkin, McNaughton and Nunn, 1987, Nakatani and Yau, 1988), probably because the presence of Na^+ ions in the external fluid tends to displace calcium ions from the channel. An amended diagram of the flow of light-sensitive current, including the calcium influx, is shown in Figure 1B.

There appear to be no significant sources of calcium influx into the rod outer segment apart from the substantial flow through the light-sensitive channel. This assertion is based on two observations: first, the rate of increase of $[Ca^{2+}]_i$, measured with aequorin, is undetectable when light-sensitive channels are closed, even with isotonic calcium in the external medium (McNaughton, Cervetto and Nunn, 1986); and second, the membrane resistance of an isolated outer segment is in excess of $40G\Omega$ when light-sensitive channels are closed and when the Na:Ca,K exchange is inactive (Lagnado, Cervetto and McNaughton, 1988). A resistance of $40G\Omega$ corresponds to a conductance of 25pS, or roughly the conductance of a typical single channel, and we conclude that the outer segment membrane does not contain more than at most a few ionic channels, other than the light-sensitive channels, and may contain none at all. There is clearly not much scope for calcium influx through light-insensitive channels in the outer segment; unlike other cell types, the photoreceptor outer segment does not contain the usual collection of voltage sensitive calcium channels or other channels, such as the Na channel, through which calcium can pass.

194

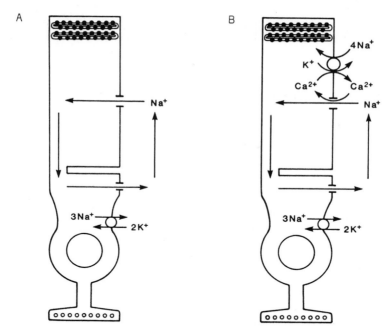

Fig. 1: Part A shows the picture of the dark current in a single rod as it is presented in many textbooks. The light-sensitive current, carried by Na$^+$ ions, flows in through the outer segment membrane and passes along the cell before leaving through the inner segment membrane. The principal current carrier at the inner segment is K$^+$, but in view of the complex nature of the current-carrying systems in the inner segment membrane no specific ion has been shown carrying the current. The ionic gradients of Na$^+$ and K$^+$ are maintained by an ATP-driven Na:K pump which is located only in the inner segment membrane. Two discs are shown schematically at the top: in reality, of course, the entire outer segment is filled with discs (about 2000 in number). Part B shows some necessary modifications to the scheme shown in A. The light-sensitive channel is imperfectly selective, and about 10% of the light-sensitive current is carried by Ca^{2+} ions. There is probably also an efflux of K$^+$ ions through the light-sensitive channel at the normal dark resting potential, as the channel selects poorly for Na$^+$ over K$^+$ and the resting potential in dark is far from E$_K$, but this has not yet been conclusively demonstrated. The calcium ions flowing in through the light-sensitive channel are pumped out by a 4Na$^+$:1Ca^{2+},1K$^+$ exchange which depends for its energy supply only on the transmembrane ion gradients of Na and K and not on cellular metabolic energy.

Calcium efflux carried by the Na:Ca,K exchange

With a typical light-sensitive current of 40pA in a salamander rod, of which 10% is carried by calcium, the total calcium concentration in the outer segment cytoplasmic volume of 1pl would be increasing at a rate of 40μM/sec in the absence of some means of extruding calcium from the cell. This influx must in the steady state be balanced by an equal efflux, which in the rod outer segment is carried by the Na:Ca,K exchange. In other cells, such as the squid axon (Baker and McNaughton, 1978), the major part of the calcium efflux is carried by an ATP-driven pump, but this mechanism seems to be absent (or at most of minor importance) in the rod. A possible reason for this difference is that the ATP- generating machinery is located in the inner segment, separated from the bulk of the outer segment membrane by the considerable diffusion barrier imposed by the stacked discs. It may be more feasible for a small ion like Na^+ to diffuse sufficiently rapidly from the outer segment, in order to maintain the gradient necessary to drive the Na:Ca,K exchange, than it would be to sustain the larger molecule ATP in sufficient quantity to drive a calcium pump. Another possible advantage is that a Na-linked system is self-regulating: an increase in $[Na^+]_i$ causes a decrease in Na:Ca,K exchange activity, an increase in intracellular $[Ca]$, and a consequent partial shutdown of the Na influx through light-sensitive channels.

The power of the Na:Ca,K exchange is shown in Figure 2, in which a rod outer segment, into which the calcium-sensitive photoprotein aequorin had been incorporated, was loaded with calcium by exposure to a solution containing isotonic calcium (plus the phosphodiesterase inhibitor IBMX, which serves to keep the light-sensitive channels open long enough to incorporate a sufficient calcium load). After the light-sensitive channels had been closed by a bright flash the internal free calcium concentration declined very slowly, if at all, until external Na was reintroduced, when calcium was rapidly pumped from the cell, with the free $[Ca^{2+}]_i$ declining at $30 \mu M/sec$. The rate of decline of the total $[Ca^{2+}]_i$ (i.e. free plus bound) was $600 \mu M/sec$.

Figure 2A also shows that the extrusion of calcium was accompanied by an inward current. From the ratio between the charge flowing when the cell is loaded with calcium by exposure to isotonic calcium to that flowing when the calcium is extruded by the exchange it has been concluded that one net charge flows into the cell for every calcium ion extruded (Yau and Nakatani, 1984; Hodgkin, McNaughton and Nunn, 1987; Lagnado, Cervetto and McNaughton, 1988). This stoichiometry is maintained over a wide range of external ionic conditions and transmembrane voltages (Lagnado, Cervetto and McNaughton, 1988; Lagnado and McNaughton, 1988), showing that the charge transported by the pump is tightly coupled to the inflow of calcium.

A stoichiometry of one charge countertransported for every Ca^{2+} extruded has always been assumed to correspond to a $3Na^+:1Ca^{2+}$ exchange. It was therefore something of a surprise to discover that the exchange depends on the presence of potassium on the same side of the membrane as calcium, and that potassium is cotransported by the exchange. This observation was made when we attempted to characterise the reversed exchange. We found that the exchange would only reverse when both potassium and calcium were present together in the external medium. Potassium could, of course, be merely a cofactor which (for instance) assists the binding of calcium without being cotransported, but this possibility was ruled out by showing that changes in external $[K^+]$ can perturb the equilibrium level of $[Ca^{2+}]_i$, and therefore that potassium must contribute energy to the exchange process by being transported across the membrane. The trace labelled "110Na" in Figure 3A shows that a reduction in $[K]_0$, with no change in $[Ca]_0$ or $[Na]_0$, causes a charge influx, corresponding to a movement of calcium from the cell, followed by an efflux of charge as the previous equilibrium level of $[Ca^{2+}]_i$ is restored.

The remaining traces in Figure 3A show the effects of simultaneous reductions in $[Na^+]_0$ and $[K^+]_0$. If n Na^+ ions exchange for one K^+ then the exchange will remain at equilibrium if $\{[Na^+]_1/[Na^+]_2\}^n = \{[K^+]_1/[K^+]_2\}$, where the suffixes refer to the concentrations in the solutions before and after the change. The solutions with $[Na^+]_0 = 51mM$ and $62mM$ were chosen to be near equilibrium for a tenfold change in $[K^+]_0$ and for $n = 3$ and 4 respectively, and it can be seen from Figure 3A that the equilibrium is maintained at $[Na^+]_0 = 62mM$, while $[Na^+]_0 = 51mM$ causes a net charge efflux, with a corresponding influx on return to $[Na^+]_0 = 110mM$. The charge movements on return to $[Na^+]_0 = 110mM$ are plotted as a function of $[Na^+]_0$ in Figure 3B. The intersection with the horizontal axis occurs near the value of $[Na^+]_0 = 61.9mM$ expected for a $4Na^+:1K^+$ exchange.

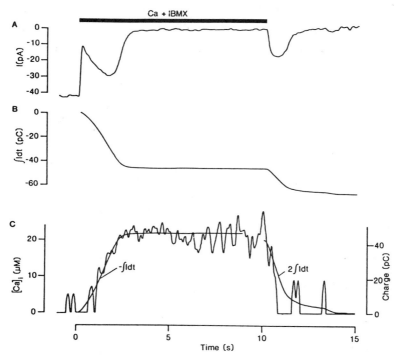

Fig. 2: An experiment investigating the calcium transport and buffering systems in a single salamander rod outer segment. Part A shows the outer segment membrane current, recorded with a suction pipette. The outer segment was exposed to isotonic calcium (black bar), with the PDE inhibitor IBMX included to prevent the closure of light-sensitive channels which would otherwise result from the inhibition of the guanylate cyclase by intracellular calcium. The calcium load introduced into the outer segment can be calculated from the integral of the light-sensitive current (shown in B). On reintroduction of the normal Na-containing solution the inward current is due to the action of the electrogenic Na:Ca,K exchange in transporting Ca from the cell. The ratio of 2:1 between the charge flowing through the light- sensitive channel and the charge transported by the exchange (see part B) shows that a single charge flows into the cell for every calcium ion transported out. Part C shows the intracellular free calcium ion concentration (noisy trace), measured from the light output from the calcium-sensitive photoprotein aequorin, and recorded at the same time as the trace in A. The rise in free $[Ca^{2+}]_i$ is proportional to the rise in total $[Ca^{2+}]_i$ (smooth trace, taken from the integral in B), showing that a constant fraction (about 5%) of the total $[Ca^{2+}]_i$ remains free. The rate of decline of free $[Ca^{2+}]_i$ with maximal Na:Ca,K exchange activation is $30\mu M.sec^{-1}$. (Figure modified from McNaughton, Cervetto and Nunn, 1986).

Figure 3C shows a similar equilibrium experiment in which $[K^+]_0$ and $[Ca^{2+}]_0$ were altered simultaneously. Here a tenfold increase in $[K^+]_0$ was balanced by a tenfold decrease in $[Ca^{2+}]_0$, showing that one K^+ is co-transported with one Ca^{2+}. In Figure 3D $[K^+]_0$ was reduced by a factor of 2.5 and the membrane potential was simultaneously depolarized until equilibrium was attained, as judged by the absence of net charge movement on return to control conditions. The reduction in $[K^+]_0$ was exactly balanced by a 23mV depolarization, as predicted if one net charge is countertransported for every K^+. The exchange stoichiometries obtained from these different measurements are therefore $4Na^+$ countertransported for one K^+ (Figs. 3A, 3B), one K^+ and one Ca^{2+} cotransported (Fig. 3C) and one K^+ countertransported for every positive charge (Fig. 3D). The only exchange stoichiometry consistent with all these measurements is $4Na^+:1Ca^{2+},1K^+$ where the colon indicates countertransport and the comma cotransport. This stoichiometry is, of course, also consistent with the known countertransport of one positive charge for one Ca^{2+} (see above).

A second surprising discovery about the Na:Ca,K transport in the rod outer segment concerned the location of the ion binding sites. Most models of membrane transport have assumed that the ion binding takes place at the membrane surface - whether external or internal - and that the exchange of ions occurs by some process of movement of the ion binding sites, with ions attached, through the full electric field which is expressed across the hydrophobic part of the membrane. According to this model ion binding should be independent of membrane potential, while the exchange process, which involves net charge transfer, would be expected to be sensitive to changes in membrane potential. This view appears to describe the binding of calcium at the internal membrane surface fairly well: changes in membrane potential do not affect the affinity of the internal binding site of the Na:Ca,K exchange for calcium (Lagnado, Cervetto and McNaughton, 1988). However, the same is not true for the binding of Na at the external face. The affinity for Na is increased by hyperpolarization, and when a concentration of Na sufficiently high to saturate the binding site is applied externally then the exchange rate becomes insensitive to membrane potential. The easiest way to accommodate this observation is to think of the Na binding site as being located within the membrane electric field, perhaps at the foot of a short well or channel intruding partway across the membrane. Hyperpolarization then increases the apparent affinity of the site for external Na, by increasing the local concentration of Na, but with a sufficiently high concentration of external Na the binding site will be saturated, irrespective of membrane potential.

A corollary of the finding that all of the sensitivity to membrane potential of the forward mode of the exchange is associated with the binding of external Na is that the current carried by the exchange must, when the external Na binding site has been saturated, be independent of membrane potential. This conclusion may seem contradictory when translocation of charge is involved, but it is exactly what is observed (Lagnado, Cervetto and McNaughton, 1988). There is in fact no fundamental difficulty, since the pump cycle will undoubtedly consist of a number of steps, and any one - not necessarily the step involving charge translocation across the membrane - could be rate-limiting in the exchange cycle.

Experiments on the Na:Ca,K exchange in the rod outer segment have provided some new insights into this much-studied and important exchange mechanism. Not only has the stoichiometry now been established for the first time with some degree of certainty - and turns out to involve transport of potassium rather than the simple exchange of Na for Ca which had always been assumed - but strong indications have emerged that the Na binding site must be sought, if and when detailed molecular models of the exchange become available, not at the external membrane surface, but located partway across the membrane.

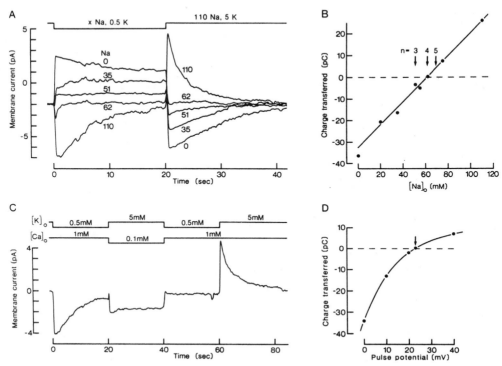

Fig. 3: Stoichiometry of the Na:Ca,K exchange. A: Currents observed during a tenfold reduction in external $[K^+]$, with a simultaneous reduction in $[Na^+]_0$ to various levels (shown alongside each trace) B: Relation between $[Na]_0$ during the exposures shown in A and charge flow on return to 110mM $[Na]_0$. Straight line, drawn by eye, crosses the axis at $[Na]_0 = 62mM$. Arrows show predicted crossing points for exchange stoichiometries of $nNa^+:1K^+$, where n = 3, 4 and 5. C: Similar experiment in which $[K]_0$ and $[Ca]_0$ were changed simultaneously. $[K]_0$ first reduced from 5mM to 0.5mM, which caused an inward current as Ca was extruded from the cell. At the next solution change $[Ca]_0$ was reduced tenfold from 1mM to 0.1mM and $[K]_0$ simultaneously elevated tenfold from 0.5mM to 5mM; the absence of a current change, apart from the steady offset which probably reflects a small junction potential, shows that the equilibrium was maintained. $[Ca]_0$ was then returned to 1mM and $[K]_0$ to 0.5mM, again with no time-dependent relaxation of current, but when $[K]_0$ was elevated to 5mM with $[Ca^{2+}]_0$ constant (final change) a substantial transient outward current was activated. $[Na]_0 = 110mM$ throughout. D: Effect of simultaneous change in membrane potential and $[K^+]_0$ on charge transfer. Rod was maintained in 110mM Na, 1mM Ca, 5mM K, and $[K^+]_0$ was reduced to 2mM for 40sec. Depolarizing pulse coincided with the solution change. Ordinate shows the charge transferred on return to control conditions, and abscissa the pulse amplitude (from a holding potential of -14mV). Arrow shows the expected change in V_m to null a 2.5-fold change in $[K^+]_0$ if one charge is countertransported for one K^+.

Intracellular calcium buffering

Investigations of the total quantity of $[Ca^{2+}]_i$ within the outer segment, using microanalytical methods on frozen or fixed tissue, have found concentrations ranging from 0.1mM (Somlyo and Walz, 1985) to 5mM (Fain and Schröder, 1985). This calcium pool seems to be for the most part firmly bound, as it exchanges on a time scale of hours with the external Ca^{2+}, although there are some indications that bright light or treatment with phosphodiesterase inhibitors can speed the exchange (Schröder and Fain, 1984; Fain and Schröder, 1987).

More important for the control of $[Ca^{2+}]_i$ on the time scale of the light response is the much smaller pool of calcium which is either free, or bound but rapidly exchangeable with the free calcium. The free calcium content of the rod outer segment is small; in the first measurements with aequorin the level was found to be below the limit of detection of about $0.6\mu M$ (McNaughton, Cervetto and Nunn, 1986). More recent measurements using a more sensitive version of the same technique have detected a resting free $[Ca^{2+}]_i$ in darkness of $0.4\mu M$. Experiments on intact retina using the calcium-sensitive fluorophore fura 2 have found $[Ca^{2+}]_i = 0.22\mu M$ (Ratto, Payne, Owen and Tsien, 1988).

In addition to the free $[Ca^{2+}]_i$ a small fraction of the bound calcium exchanges so rapidly with the free $[Ca^{2+}]_i$ that the buffering can be regarded as effectively instantaneous on the time scale of the light response. In Figure 2 this buffering can be seen from the discrepancy between the free internal $[Ca^{2+}]$, measured with aequorin, which amounts to a maximum level of about $22\mu M$, and the total calcium rise within the cell, which from integration of the light-sensitive current amounts to about $450\mu M$. As in many other cells, therefore, only a small fraction of the intracellular calcium is free.

Subsequent investigations (Hodgkin, McNaughton and Nunn, 1987; Cervetto, Lagnado and McNaughton, 1989) have elucidated further properties of the intracellular calcium buffering. The time constant of equilibration between the free $[Ca^{2+}]_i$ and the rapidly bound component of $[Ca^{2+}]_i$ is less than 10msec, and therefore any delays in the uptake and release of calcium from the rapidly bound component of calcium can be ignored on the time scale of events such as the response to a flash or a step of light. The rapid calcium buffer system consists of two components with quite distinct properties: a high-affinity component has a Michaelis constant in the sub-micromolar range and a limited capacity of around $200\mu M$, while a low-affinity component has a Michaelis constant so low that it behaves as a linear buffer which binds about 16 calcium ions for every one which is free. The behaviour of this low-affinity buffer can be seen in the experiment of Figure 2, where the total calcium is proportional to the free calcium, both while the cell is being calcium-loaded through the light-sensitive channels and during the phase of calcium extrusion by the Na:Ca,K exchange, but is larger by a factor of about 20.

Modulation of the light-sensitive pathway by intracellular calcium

The processes influencing the turnover of calcium within the rod outer segment and the interaction, as we at present understand it, of calcium with the light-sensitive pathway, are summarized in Figure 4. Two rate constants influence the level of cyclic GMP, and therefore the magnitude of the light-sensitive current: the rate constant β governing the synthesis of cGMP from GTP by the guanylate cyclase, and the rate constant a governing the breakdown of cGMP by the phosphodiesterase to inactive GMP. The second process is modulated by the complex series of events initiated by absorption

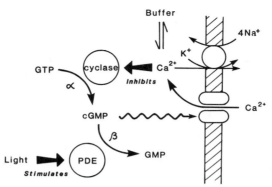

Fig. 4: Summary diagram showing mechanisms involved in the control of intracellular calcium in the rod outer segment, and the interaction of calcium with the light-sensitive pathway (see text).

of a photon of light by a molecule of rhodopsin. The activity of the guanylate cyclase, on the other hand, depends only, so far as we know, on the level of intracellular calcium. A direct dependence of guanylate cyclase activity on absorption of light has been proposed but not so far substantiated, and although this remains an interesting possibility, most of the main features of the light-sensitive pathway can be explained by the simpler model shown in Figure 4.

Calcium enters the rod through the light-sensitive channels, and is extruded by a $4Na^+:1Ca^{2+},1K^+$ exchange mechanism. Within the rod free calcium is in rapid equilibrium with calcium bound to a buffer system, which consists of a high-affinity and a low-affinity component. The time constant of turnover of internal free Ca^{2+} is of the order of 0.5sec. Free Ca interacts with the light-sensitive pathway by inhibiting the guanylate cyclase which produces cGMP from GTP. Thus the internal concentrations of both calcium and cGMP are under tight feedback control: an increase in $[Ca^{2+}]_i$ inhibits the cyclase, causing a decrease in the rate constant α, a reduction in [cGMP], a closure of light-sensitive channels and a consequent decline in $[Ca^{2+}]_i$; while a decline in [cGMP], caused for instance by an increase in the activity of the phosphodiesterase after a flash of light, will cause a closure of light-sensitive channels, a decline in $[Ca^{2+}]_i$ and a consequent dis-inhibition of the guanylate cyclase.

An important element in the control of the light-sensitive pathway is the time constant of turnover of internal Ca, τ_{Ca}. It can be shown that τ_{Ca} is approximated by the expression

$$\tau_{Ca} \approx \frac{K_{Ca}}{K_{buff}} \cdot \frac{C}{j_{sat}} \; FV$$

where K_{Ca} and K_{buff} are the Michaelis binding constants for internal calcium of the Na:Ca,K exchange and the high-affinity component of the calcium buffer, respectively; C is the capacity of the high-affinity buffer; j_{sat} is the maximal Na:Ca,K exchange current; V is the outer segment volume; and F is the Faraday.

Photoreceptors have, therefore, a number of ways in which the time constant of turnover of intracellular $[Ca^{2+}]$ could be regulated. The time constant is increased by increases in the binding constant K_{Ca} of the Na:Ca,K exchange, in the capacity C of the high-affinity buffer, or in the volume V of the outer segment; and it is decreased by increases in the binding constant K_{buff} of the high-affinity buffer or in the maximal Na:Ca,K exchange current, j_{sat}. Rods contain a substantial concentration (c. $250\mu M$) of a high-affinity calcium buffer, which will have the effect of slowing the relaxation in intracellular free $[Ca^{2+}]$ in response to a perturbation such as a suppression of the light-sensitive current, and therefore of slowing the onset of light adaptation which is thought to depend on the change in calcium. A possible reason for the smaller size of cones compared to rods in many species may be the need to reduce τ_{Ca}, in order to speed the process of light adaptation, by reducing the outer segment volume. Another interesting possibility, in view of the evidence for a regulation of the affinity of the Na:Ca,K exchange, is that K_{Ca} may be modulated in the process of light adaptation, thereby reducing the time constant of turnover of calcium in the light-adapted state.

REFERENCES

Baker, P. F. and McNaughton, P. A., 1978, The influence of extracellular calcium binding on the calcium efflux from squid axons, *J. Physiol.*, 276:127-150.

Cervetto, L., Lagnado, L. and McNaughton, P. A., Transport and buffering of calcium in salamander rod outer segments, *J. Physiol.*, submitted.

Cervetto, L., Lagnado, L., Perry, R. J., Robinson, D. W. and McNaughton, P. A., Extrusion of calcium from rod outer segments is driven by both sodium and potassium gradients, *Nature*, 337:740-743.

Cervetto, L., McNaughton, P. A.,1986, The effects of phosphodiesterase inhibitors and lanthanum ions on the light-sensitive current of toad retinal rods, *J. Physiol.*, 370:91-109.

Fain, G.L . and Schröder, W. H., 1985, Calcium content and calcium exchange in dark-adapted toad rods, *J. Physiol.*, 368:641-655.

Fain, G. L. and Schröder, W. H., 1987, Calcium in dark-adapted toad rods: evidence for pooling and cyclic guanosine-3', 5'-monophosphate-dependent release, *J. Physiol.*, 389:361-384.

Fesenko, E. E., Kolesnikov, S. S. and Lyubarsky, A. L., 1985, Induction by cyclic GMP of cationic conductance in plasma membrane of retinal rod outer segment, *Nature.*, 313:310-313.

Hagins, W. A., 1972, The visual process: excitatory mechanisms in the primary receptor cells, *Ann. Rev. Biophys. Bioeng.*, 1:131-158.

Hagins, W. A. and Yoshikami, S, 1975, Ionic mechanisms in excitation of photoreceptor, *Ann. N.Y. Acad. Sci.*, 264:314-325.

Hodgkin, A. L., McNaughton, P. A. and Nunn, B. J., 1985, The ionic selectivity and calcium dependence of the light-sensitive pathway in toad rods, *J. Physiol.*, 358:447-468.

Hodgkin, A. L., McNaughton, P. A. and Nunn, B. J., 1987, Measurement of sodium-calcium exchange in salamander rods, *J. Physiol.*, 391:347-370.

Hodgkin, A. L., McNaughton, P. A., Nunn, B. J. and Yau, K. -W., 1984, Effect of ions on retinal rods from Bufo marinus, *J. Physiol.*, 350:649-680.

Koch, K. -W. and Stryer, L., 1988, High degree of cooperativity in the inhibition of retinal rod guanylate cyclase by calcium *Nature.*, 334:64-66.

Lagnado, L., Cervetto, L., and McNaughton, P. A., 1988, Ion transport by the Na-Ca exchange in isolated rod outer segments, *P.N.A.S.*, 85:4548-4552.

Lagnado, L. and McNaughton, P. A., 1988, The stoichiometry of Na:Ca exchange in isolated salamander rod outer segments, *J. Physiol.*, 407:82P.

Lamb, T. D. and Matthews, H. R., 1988, External and internal actions of calcium in the response of salamander retinal rods to altered external calcium concentration, *J. Physiol.*, 403:473-494.

Lolley, R. N., and Racz, E., 1982, Calcium modulation of cyclic GMP synthesis in rat visual cells, *Vision Res.*, 22:1481-1486.

McNaughton, P. A., Cervetto, L., and Nunn, B. J., 1986, Measurement of the intracellular free calcium concentration in salamander rods, *Nature.*, 322:261-263.

Matthews, H. R., Fain, G. L., Murphy, R. L. W. and Lamb, T. D., 1988, Photoreceptor light adaptation is mediated by cytoplasmic calcium concentration, *Nature.*, 334:67-69.

Nakatani, K. and Yau, K. -W., 1988, Calcium and magnesium fluxes across the plasma membrane of the toad rod outer segment, *J. Physiol.*, 395:695-729.

Nakatani, K. and Yau, K. -W., 1988, Calcium and light adaptation in retinal rods and cones, *Nature.*, 334:69-71.

Ratto, G. M., Payne, R., Owen, W. G. and Tsien, R. Y, 1988, The concentration of cytosolic free calcium in vertebrate outer segments measured with fura-2, *J. Neurosci.*, 8:3240-3246.

Schröder, W. H. and Fain, G. L., 1984, Light-dependent calcium release from photoreceptors measured by laser micro-mass analysis, *Nature.*, 309:268-270.

Sillman, A. J., Ito, H. and Tomita, T., 1969, Studies on the mass receptor potential of the isolated frog retina. II. On the basis of the ionic mechanism, *Vision Res.*, 9:1443-51.

Somlyo, A. P. and Walz, B., 1985, Elemental distribution in Rana pipiens retinal rods: quantitative electron probe analysis, *J. Physiol.*, 358:183-195.

Stryer, L., 1986, Cyclic GMP cascade of vision, *Ann. Rev. Neurosci.*, 9:87-119.

Torre, V., Matthews, H. R. and Lamb, T. D., 1986, Role of calcium in regulating the cyclic GMP cascade of phototransduction in retinal rods, *Proc. Natl. Acad. Sci. U.S.A.*, 83:7109-7113.

Yau, K. -W., McNaughton, P. A. and Hodgkin, A. L., 1981, Effect of ions on the light-sensitive current in retinal rods, *Nature.*, 292:502-505.

Yau, K. -W. and Nakatani, K., 1984, Electrogenic Na-Ca exchange in retinal rod outer segment, *Nature.*, 311:661-663.

Yoshikami, S. and Hagins, W. A., 1971, Light, calcium, and the photocurrent of rods and cones, *Biophys. J.*, 11:47a.

LIGHT ADAPTATION IN RETINAL RODS OF THE NEWT

S. Forti[†], A. Menini[‡], G. Rispoli[*] , L. Spadavecchia [‡] and V. Torre

Dipartimento di Fisica, Universita' di Genova, Italy
[†] IRST, Povo, Trento, Italy
[‡] Istituto di Cibernetica e Biofisica, C.N.R., Genova, Italy

INTRODUCTION

It is known that guanosine 3',5'-cyclic monophosphate (cyclic GMP) is the internal transmitter of phototransduction (Caretta and Cavaggioni, 1983; Fesenko et al., 1985; Yau and Nakatani, 1985b; Haynes et al., 1986; Zimmermann and Baylor, 1986) and that Ca^{2+} is not a positive transmitter (Matthews et al., 1985; Lamb et al., 1986) as originally proposed (Hagins, 1972). Moreover, it is known that the light-sensitive channels present in the plasma membrane outer segments of rod are activated by cyclic GMP in a co-operative way (Fesenko et al., 1985; Haynes et al., 1986; Zimmermann and Baylor, 1986).

In this paper some aspects of phototransduction and light adaptation are reconsidered within the framework of this new evidence and a quantitative reconstruction of phototransduction is attempted. The relation between light intensity and amplitude of normalized photoresponse at different times, both in the dark and during light adaptation, is studied and the role of diffusion in phototransduction is analysed. It is shown, firstly that a parsimonious model of phototransduction can neglect effects because of diffusion of intracellular molecules, and, secondly that intracellular uniformity can be initially assumed. Changes of the concentration of intracellular Na^+ , Ca^{2+} and cyclic GMP are evaluated and their effects on light adaptation are discussed taking into consideration that changes of intracellular Ca^{2+} affect the cyclic GMP metabolism, thus controlling sensitivity and light adaptation (Cervetto et al., 1985; Torre et al., 1986; Matthews et al., 1988; Nakatani and Yau, 1988c). Finally, the reconstruction of the time course of photoresponses to both flashes and steps of light is attempted using the available knowledge of the cyclic GMP cascade.

METHODS

Recordings of suction-pipette current (Baylor et al., 1979a) were made from rods of the dark-adapted retina of the newt *Triturus cristatus* of northern Italy in order to extend the analysis to a new animal, which could be found locally.

---

* G. Rispoli's present address: Department of Physiology and Biophysics, University of Washington School of Medicine, Seattle, U.S.A.

Sensory Transduction
Edited by A. Borsellino *et al.*
Plenum Press, New York, 1990

The dissection of the retina and the apparatus for suction-electrode recording and optical stimulation of rods was similar to that described by Menini et al. (1988). Rods from the newt retina are rather similar to those from the tiger salamander retina (Lamb et al., 1986) but they are shorter and have a larger diameter than those from the toad retina (Cervetto et al., 1976). The length and diameter of the outer segment was 35 ± 3 μm, 9.5 ± 1 μm respectively while length and diameter of the inner segment was 32 ± 4 μm, 12 ± 1 μm respectively. It was rather difficult to obtain intact isolated rods from the newt retina with mechanical dissociation, which proved rather successful with the retina of the tiger salamander. The great majority of recordings were obtained from outer segments of rods from small pieces of retina. The outer segment of a rod was drawn into the suction pipette, and its light response tested. Provided the saturating response was sufficiently large (≥ 20 pA), and stable, the experiment was started. The largest photoresponse was 45 pA, while using the retina of the tiger salamander it was common to record photoresponses larger than 50 pA, using exactly the same experimental procedure. Experiments were performed at room temperature ($17\text{-}25°C$). Unpolarized light of wave-length 498 nm was used for all stimuli.

The Ringer solution contained (in mM) NaCl 110; KCl 2.5; $CaCl_2$ 1; $MgCl_2$ 1.6; HEPES 3; EDTA 0.1; glucose 5; buffered to pH 7.5 with TMAOH (tetramethylammonium hydroxide).

QUANTUM SENSITIVITY OF RODS OF THE NEWT

It was not possible to record single quantum events as in toad rods (Baylor et al., 1979b), probably because of their small size. A way of estimating the amplitude of single events is to look at the ratio between variance and mean response of a series of current recordings in response to very dim flashes of light. Fig. 1 shows the average response (A) and the variance (B) obtained from 75 responses to a flash of light, whose intensity was 0.5 photons/μm^2.

The change of variance caused by the light was 0.16 pA^2 and the mean amplitude of the photoresponse was 1.2 pA. From the ratio of these two quantities we can estimate the amplitude of the single event to be 0.13 pA. On average the amplitude from seven newt was about 0.3 pA. This single event, which we assume to be the response to a single absorbed photon, is much smaller than the photon response in toad rods (Baylor et al., 1979b). The effective collecting area was estimated to be 20 μm^2 as for rods of tiger salamander (Lamb et al., 1986).

THEORETICAL SECTION

Light intensity and amplitude of photoresponse

We now present the theoretical background for the model of phototransduction. The relation between light intensity I (assumed to be in arbitrary units) and the normalized amplitude of photoresponse $\frac{\Delta R}{R}$ has been described by a Michaelis-Menten relation:

$$\frac{\Delta R}{R} = \frac{I}{I + I_0} \tag{1}$$

where I_0 is the light intensity halving the photoresponse. Equation (1) was first proposed by Baylor and Fuortes (1970), who assumed the existence of a positive transmitter whose concentration inside the photoreceptor was uniform and which was able to block light-sensitive channels. It was later shown that the diffusion of excitation and therefore of the internal transmitter was rather restricted (Lamb et al., 1981), possibly just to a few disks. A more careful analysis of the relation between normalized suppressed photocurrent $\Delta J/J_S$ and I at fixed times showed that the experimental data were not fitted by eq.(1), but better by

$$\frac{\Delta J}{J_S} = 1 - e^{-I} \tag{2}$$

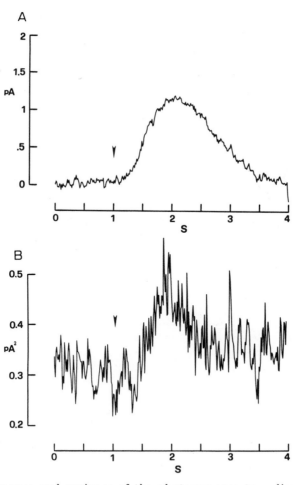

Fig. 1: Mean response and variance of the photoresponse to a dim flash response. A: average response to 75 brief flashes of monocromatic light (495 nm) of intensity 0.5 photons $/\mu m^2$. B: variance of 75 responses as in A.

Eq(2) can be obtained assuming that diffusion of excitation is restricted and that a photoisomerization completely blocked all channels within a narrow region (Lamb et al., 1981). When diffusion of excitation is restricted, in the presence of very dim lights, it is expected that the distribution of the internal transmitter along the outer segment to have many large minima and maxima. By increasing the light intensity the number of minima increases and minima eventually start to superimpose. It is evident that under these conditions the spatial profile of the distribution of the internal transmitter becomes more uniform and in the presence of a steady light exciting many rhodopsin molecules, or when light flashes producing thousands of photoisomerizations are used, the interior of the photoreceptor will appear as a well stirred compartment which has an almost uniform distribution of the internal transmitter. Consequently, during light adaptation there is no strong argument supporting eq.(2).

In a later section it will be shown that in dark adapted rods, eq.(2) fits the experimental data reasonably well but does not fit the data obtained during light adaptation. It is useful to consider theoretical relations between I and $\frac{\Delta J}{J_S}$ which could hold during light adaptation when the interior of the photoreceptor is well stirred.

It is now well established that the light-sensitive channel is activated by cyclic GMP and experiments with excised patches have shown that the cyclic GMP-activated current J, does not increase linearly with the cyclic GMP concentration g but as

$$\frac{J}{J_{max}} = \frac{g^3}{K^3 + g^3} \tag{3}$$

where J_{max} is the maximal cyclic GMP dependent current which can be recorded in an excised patch and K is the range between 10 and 50 μM (Haynes at el., 1986; Zimmerman and Baylor, 1986). The cube law implies a cooperativity of at least 3 cyclic GMP molecules. At a microscopic level eq.(3) could be explained by a scheme like

$$C_c + G \quad \overset{k_{12}}{\underset{k_{21}}{\rightleftarrows}} \quad C_1 + G \quad \overset{k_{23}}{\underset{k_{32}}{\rightleftarrows}} \quad .... \quad C_{n-1} + G \quad \overset{k_{n-1,n}}{\underset{k_{n,n-1}}{\rightleftarrows}} \quad C_{open} \tag{4}$$

where G is a cyclic GMP molecule, C_c is the closed channel, C_i represents the channel with i cyclic GMP molecules bound, C_{open} is the channel open by n molecules of cyclic GMP and $k_{i,j}$ the on and off rate constants. The original data of Haynes et al. (1986) and Zimmerman and Baylor (1986) can be fitted using equations derived from scheme (4) with n=3 or n=4 (Torre et al., 1987). Therefore it is not possible to decide whether 3 or 4 cyclic GMP molecules are required to open the channel.

It is now derived an equation describing the relation between light intensity I of a flash and amplitude of photoresponse which assumes the cooperativity between cyclic GMP molecules and a uniform distribution of cyclic GMP inside a rod. Let g_0 be the steady state level of cyclic GMP in some light adapted conditions. The simplest way to describe the decrease of cyclic GMP caused by the light I is an enzymatic cascade leading to a Michaelis-Menten relation:

$$g = \frac{g_0 I_0}{I + I_0} \tag{5}$$

where I_0 is the light intensity halving the steady level of cyclic GMP g_0. In darkness the resting level of free cyclic GMP is only a few micromolar (see Yau and Nakatani, 1985b; Stryer, 1986; Yau and Nakatani, 1988b) and therefore during light adaptation g_0 is much smaller than K. Consequently, during light adaptation or when $g << K$ from scheme (4) with n steps, the fraction of open channels in the presence of a concentration of cyclic GMP equal to g can be obtained, and is

$$\frac{C_{open}}{C_{tot}} \sim g^n \tag{6}$$

where C_{tot} is the total number of channels and n is equal to 3 or to 4. If C_0 is the number of conducting channels during a steady light for which the concentration of cyclic GMP is g_0, the fraction of channels closed by a superimposed light I is

$$\frac{C_0 - C_{open}}{C_0} = \frac{g_0^n - g^n}{g_0^n} \tag{7}$$

and using eq.(5), we finally obtain

$$\frac{C_0 - C_{open}}{C_0} = 1 - \left[\frac{I_0}{I + I_0} \right]^n \tag{8}$$

If it is assumed that the circulating photocurrent is proportional to the number of open channels, then the fraction of suppressed photocurrent $\Delta J / J_S$ is proportional to the fraction of channels closed by light and so eq.(8) also describes the dependence of $\Delta J / J_S$ on light intensity I.

Kinetics of photoresponse

In a variety of photoreceptors (Fuortes and Hodgkin, 1964; Baylor et al., 1974; Baylor et al., 1979a) the time course of the electrical response to a dim flash of light can be fitted by a Poisson equation:

$$t^n \, e^{-\alpha t} \tag{9}$$

which can be explained by a chain of n+1 slow stages:

$$h\nu \xrightarrow{\quad} y_1 \xrightarrow{\alpha} y_2 \xrightarrow{\alpha} \dots\dots y_{n+1} \xrightarrow{\alpha} \qquad \text{scheme (a)}$$

where y_i are intermediate photo products of the absorption of a photon $h\nu$ and y_{n+1} is a substance controlling the kinetics of phototransduction. Eq.(9) reproduces quite accurately, the time course of the photoresponse to a dim flash of light in several amphibian rods with n=3 and α around 3 sec^{-1}. As diffusion can play a relevant role in shaping the relation between amplitude of photoresponse and light intensity, it is therefore conceivable that longitudinal and radial acqueous diffusion can contribute to the shaping of the kinetics of the light response and that some of the slow stages implied by eq.(9) are simply caused by diffusion and do not represent independent chemical reactions. The number of slow stages can easily be determined by looking at the slope of the rising phase of photoresponses when displayed on a log- log plot (Fuortes and Hodgkin, 1964; Baylor et al., 1974; Baylor et al., 1979a). This kind of analysis shows that the two traces obtained in darkness and in the presence of a bright steady light have initially a very similar slope. Collected data from 11 cells show that in dark adapted conditions the value of n is 2.8 ± 0.3 and in the presence of bright steady lights is 2.6 ± 0.4. The difference in the value of n from 3 does not appear to be significant and it is concluded that diffusion can contribute to the shaping of the light response in darkness, but it is unlikely that the number of delays is primarily controlled by diffusion. Therefore it is more useful to try to understand the kinetics of phototransduction during light adaptation than those in dark adapted condition. Diffusion can be relevant in the latter, but not in the former case. The shape of the light response during light adaptation is still fitted by eq.(9) with n=3 and it is an obvious question to enquire into the possibility that the cube law could be the origin of the existence of the four slow stages. In this view the change in intracellular cyclic GMP does not have a time course described by eq.(9), which is brought into the electrical response by the cube law.

Let us suppose that the steady level of cyclic GMP is g_0 and the change of cyclic GMP is $\Delta g(t)$. Then using eq.(3) and because $g_0 << K$, we have

$$J(t) \approx \frac{J_{max}}{K^3} g^3(t) \tag{10}$$

for a small change of cyclic GMP, $\Delta g(t)$, a Taylor series expansion of eq.(10) gives

$$\Delta J(t) \approx \frac{J_{max}}{K^3} \left[3 g^2_0 \Delta g(t) + 3 g_0 \Delta g^2(t) + \Delta g^3(t) \right] \tag{11}$$

It is evident that since $\Delta g(t)$ is always smaller than g_0, the linear term in eq.(11) is always the dominant one, and that $\Delta J(t)$ will have a time course similar to $\Delta g(t)$. Therefore the four delays are not primarily caused by the cube law.

Photoresponses to steady lights

We now analyse the effect of prolonged illumination, which ultimately leads to light adaptation. We discuss changes of intracellular Na^+ and Ca^{2+} which occur when their influx through light-sensitive channels is decreased or blocked by the light. We analyse these changes assuming that the intracellular level of Ca^{2+} is involved in the control of the cyclic GMP metabolism and therefore contributing to light adaptation (Torre et al., 1986; Matthews et al., 1988; Sather et al., 1988). We also assume that the main site of action of intracellular Ca^{2+} is the guanyl cyclase (Lolley and Racz, 1982; Pepe et al., 1986; Koch and Stryer, 1988; Rispoli et al., 1988) and not on the phosphodiesterase as indicated by previous results (Robinson et al., 1980; Torre et al., 1986).

Changes of intracellular cyclic GMP, Ca^{2+} and Na^+

In the dark cyclic GMP is continuously produced by guanyl cyclase at a rate A and is hydrolised at a rate $\bar{V}$. The rate of cyclic GMP hydrolysis is increased by light in proportion to the activated phosphodiesterase PDE* (Liebman and Pugh, 1982). Therefore we can describe changes of free cyclic GMP g as

$$\dot{g} = A - g (\bar{V} + \sigma PDE^*) \tag{12}$$

where σ is a proportionality constant, assumed for simplicity to be $1 \ sec^{-1} \ \mu M^{-1}$. The concentration of free cyclic GMP in darkness is a few micromole (Yau and Nakatani, 1985b; Stryer, 1986; Nakatani and Yau, 1988b) and here it is assumed to be $2 \ \mu M$. The rate of dark activity of phosphodiesterase $\bar{V}$ has been estimated to be about $0.4 \ sec^{-1}$

by Hodgkin and Nunn (1988). The rate of synthesis of cyclic GMP A from eq.(12) at the steady state is equal to $g_0 \bar{V}$ or 0.8 μM sec^{-1}.

It now seems to be well established that the activity of the cyclase is controlled by intracellular Ca^{2+} in a cooperative way (Lolley and Racz, 1982; Pepe et al., 1986; Koch and Stryer, 1988; Rispoli et al., 1988) and the simplest way to describe this effect is to assume that

$$A = \frac{A_{max}}{1 + (\dfrac{c}{K_c})^m} \tag{13}$$

where A_{max} is the maximal activity of the cyclase and K_c is the intracellular Ca^{2+} producing half inhibition of the cyclase activity and m is the number of Ca^{2+} molecules necessary to inhibit a cyclase molecule. From Koch and Stryer (1988) m is close to 4 and K_c is close to 0.1 μM. Since the resting level of free intracellular Ca^{2+} in darkness is about 300 nM (Ratto et al., 1988), the modulating action of Ca^{2+} on the cyclase is expected to be high in darkness but when in the presence of steady bright lights, intracellular free Ca^{2+} has caused substantial fall in the activity of the cyclase, this modulating action is affected by intracellular Ca^{2+} to a lesser extent.

Intracellular Ca^{2+} , unlike Na$^+$, is buffered (McNaughton et al., 1986) and its changes $\dot{c}$ are controlled by the influx through light sensitive channels (Yau and Nakatani, 1984; Hodgkin et al., 1985; Nakatani and Yau, 1988a; Menini et al., 1988), by its extrusion by the Na$^+$/Ca^{2+} exchange (Yau and Nakatani, 1984; Hodgkin et al., 1987; Lagnado et al., 1988) and by binding and unbinding to internal buffers. The equations describing these mechanisms are

$$\dot{c} = \frac{J_{Ca}}{2Fv} - \gamma_{Ca}\, c - k_1\, (e_T - c_b)\, c + k_2\, c_b \tag{14a}$$

$$\dot{c}_b = k_1\, (e_T - c_b)\, c - k_2\, c_b \tag{14b}$$

where J_{Ca} is the total Ca^{2+} current, e_T is the total buffer concentration, c_b the Ca^{2+} concentration bound to the buffer, γ_{Ca} the rate of Ca^{2+} extrusion mediated by the Na$^+$/Ca^{2+} exchange, k_1 and k_2 the on and off rate for the binding of Ca^{2+} to the internal buffer. In tiger salamander rods the ratio between bound and free intracellular Ca^{2+} is at least 10 (McNaughton et al., 1986) and there is evidence to suggest the existence of low-affinity buffers which are not saturated under normal conditions ($e_T \gg c_b$) and high affinity buffers which are almost entirely saturated ($e_T \sim c_b$) with a dissociation constant $\frac{k_2}{k_1} \sim 30nM$ (Hodgkin et al., 1987). The steady state level of intracellular Ca^{2+} , c does not depend on the buffer concentration and is so obtained from equations (14a, b):

$$c = \frac{J_{Ca}}{2Fv\, \gamma_{Ca}} \tag{15}$$

which depends only on the Ca^{2+} influx and its rate of extrusion.

The total volume of the newt outer segment and inner segment is respectively 2.7 10^{-12} l and 3.6 10^{-12} l. Assuming that the free volume is half of the total volume, the total free volume of the rod can be taken as about 3.5 10^{-12} l. The resting level of free intracellular Ca^{2+} is about 300 nM (Ratto et al., 1988) and J_{Ca} is approximately 1/4 of the total photocurrent (Menini et al., 1988). Since in newt rods the largest photocurrent recorded with a suction electrode was 40 pA (giving a value of $J_{Ca}/2Fv$ in darkness of about 15 μM/sec), from eq.(15) γ_{Ca} is estimated to be about 50 sec^{-1}. The time constant of the electrogenic current carried by the Na^+/Ca^{2+} on the top of a bright flash response is about 1 sec^{-1}, a value significantly smaller than the rate constant of Ca^{2+} extrusion γ_{Ca} of the exchanger derived from eq.(15). As suggested by Professor Alan L. Hodgkin this apparent discrepancy can be accounted for by the presence of a high affinity Ca^{2+} buffer inside the rod which can prolong the time constant of the exchanger by as much as 100 times.

Equations (14a,b) describe the Ca^{2+} extrusion through the Na^+/Ca^{2+} exchange by a simple first order mechanism, neglecting the Ca^{2+} entry through the exchange itself, which will be relevant when $[Ca^{2+}]_i$ is very low. As pointed out by Blaustein and Hodgkin (1969), at equilibrium and when $[Ca^{2+}]_i$ is entirely controlled by the activity of the exchanger, the intracellular level of free Ca^{2+} c_0 is set by

$$c_0 = [Ca^{2+}]_i = [Ca^{2+}]_0 \frac{[Na^+]_i^3}{[Na^+]_0^3} e^{V_m F/RT} \tag{16}$$

taking $[Na^+]_i = 12mM$ (Torre, 1982) and $V_m = -60mV$ a value of c_0 equal to - 100 nM is obtained. Consequently eq.(14a) can be more appropriately rewritten as

$$\dot{c} = \frac{J_{Ca}}{2Fv} - \gamma_{Ca}(c - c_0) - k_1(e_T - c_b)c + k_2 c_b \tag{17}$$

The intracellular level of Na^+ is unlikely to be buffered and its changes $\frac{d}{dt}[Na^+]_i$ can be described by its influx and its extrusion, assumed to obey a first order kinetics:

$$\frac{d[Na^+]_i}{dt} = \frac{J_{Na}}{Fv} - \gamma_{Na}[Na^+]_i \tag{18}$$

where F is the Faraday constant, v the rod volume, J_{Na} the Na^+ current entering into the rod and $\gamma_{Na}[Na^+]_i$ is the Na^+ extrusion mediated by Na^+/K^+ pump. When the Na^+ influx is abolished intracellular Na^+ in toad rods is likely to decrease exponentially with a time constant of about 0.04 sec^{-1} (Torre, 1982). Assuming similar behaviour for newt rods, we have the value 0.04 sec^{-1} for γ_{Na}. Since a large component of the Na^+ influx is through the light-sensitive channels we can expect significant changes of $[Na^+]_i$ within a couple of minutes when the photocurrent is reduced by a steady light.

The reconstruction of the kinetics of photoresponses

Let us now attempt to provide a kinetic scheme able to reproduce several aspects of phototransduction. We will not discuss the very early events, occurring during the initial 50 mseconds, because they have already been thoroughly discussed by Cobbs and Pugh (1987). Our aim is not to obtain a perfect fitting of the experimental recordings, but to explain the major features of the kinetics of phototransduction using a parsimonious model which considers the known biochemistry of the cyclic GMP cascade and of changes in intracellular Ca^{2+} . It is now useful to summarize some conclusions so far obtained:

1- Acqueous diffusion of the internal transmitter is not responsible for the existence of the slow stages in the linear response to a flash of light. The main reason supporting this view is that four slow stages are still required to fit the linear response in the presence of a bright steady light. Under these conditions, given the high number of photoisomerizations, the cell interior is likely to be well stirred and the concentration of the internal transmitter can be expected to be homogeneous. In this view diffusion may shape the time course of the photoresponse in darkness, but it is rather unlikely to affect the kinetics during light adaptation.

2- We have already considered the possibility that the cooperativity between cyclic GMP molecules is responsible for an initial rising phase proportional to t^n , with n equal to 3. Since cyclic GMP is a negative transmitter it is argued that the time course of the photoresponse in the linear range will follow the time course of changes of cyclic GMP despite the cooperativity between molecules of cyclic GMP. Therefore, the slow stages are intrinsic to the cyclic GMP cascade. Furthermore, it has been shown (Karpen et al., 1988) that cyclic GMP opens channels in excised patches within very few msec. Therefore the binding of cyclic GMP to the channel is unlikely to contribute to the slow stages.

3- It is now well established that intracellular Ca^{2+} falls during light (Yau and Nakatani, 1985b; McNaughton et al., 1986; Ratto et al., 1988) and that changes of intracellular Ca^{2+} contributes to light adaptation (Torre et al., 1986) by modulating the cyclase (Pepe et al., 1986; Koch and Stryer, 1988). The time course of changes of intracellular Ca^{2+} are controlled by the many intracellular buffers, which differ according to their speed and affinity (McNaughton et al., 1986; Hodgkin et al., 1987). This parsimonious model assumes only the existence of a low affinity Ca^{2+} buffer inside the rod with a $K_D = \frac{k_2}{k_1} = 4 \ \mu M$ and a total concentration of 500 μM. These figures imply that the ratio between lightly bound Ca^{2+} and free Ca^{2+} is about 115, a value higher than the estimate of 10-20 of McNaughton et al. (1986). In order to reproduce the time constant of the reactivation of the photocurrent observed with steps of light (see Fig. 4 and 5) we have assumed $k_2 = 0.8 \ sec^{-1}$ and $k_1 = 0.2 \ sec^{-1} \ \mu M^{-1}$.

Origin of the slow stages

In their seminal paper on the kinetics of light responses in Limulus photoreceptors Fuortes and Hodgkin (1964) identified the slow stages as activations, but also as inactivations of the internal transmitter. Therefore a slow stage in phototransduction (Baylor et al., 1974; Cervetto et al., 1977; Baylor et al., 1979a; Lamb, 1986) of vertebrate rods could be the result of a slow activation in the cyclic GMP cascade but might also be caused by a slow inactivation. In scheme (b) we have reported the known biochemical events involved in the control of free intracellular cyclic GMP.

$$
\begin{array}{ccccccc}
 & & & & \alpha_1 & & & & \alpha_3 \\
h\nu & \longrightarrow & Rh & \longrightarrow & & Rh_i & \longrightarrow \\
 & & & \longleftarrow & & \\
 & & & \alpha_2 & \\
 & & & \epsilon \downarrow & & \beta_1 \\
 & & & T^* & & \longrightarrow \\
 & & \tau_1 & \downarrow \ \uparrow & \tau_2 \\
 & & & PDE^* \\
 & & & \downarrow \\
A & & & \\
\longrightarrow & cyclic \ GMP & \longrightarrow & GMP \\
 & & \bar{V}
\end{array}
$$

scheme (b)

213

where $h\nu$ is a photon, Rh and Rh_i are the activated and inactivated rhodopsin respectively. T^* and PDE^* are the active transducin and phosphodiesterase respectively.

The pathway for the activation of PDE^* is known in great detail, but the values of rate constants of different biochemical steps in the intact cell have yet to be established precisely. Light activates rhodopsin very quickly. Activated rhodopsin is inactivated by the encounter with the two proteins rhodopsin kinase and 48 KD protein (Applebury and Chabre, 1986; Stryer, 1986). It is assumed that the inactivation of photoexcited rhodopsin Rh is reversible, thus giving origin to the late response. The inactivation of Rh is a rather slow stage and is likely to be one of the slow stages in phototransduction.

It is well known that 1 activated rhodopsin is able to activate about 500 transducins per second (Bennett, Michel-Villaz and Kuhn, 1982; Vuong and Stryer, 1984) and 1 transducins activates 1 PDE within 100 msec (Liebman and Evanzuk, 1982), therefore indicating that the activation of PDE by transducins is another slow stage in photo-transduction. The only known pathway for PDE^* inactivation is through inactivation of transducin, which again could be another slow stage in phototransduction (Applebury and Chabre, 1986; Stryer, 1986). From scheme (b) we obtain the set of equations:

$$\dot{R}h = J_{h\nu}(t) - \alpha\, Rh + \alpha_2\, Rh_i \tag{19.1}$$
$$\dot{R}h_i = \alpha_1\, Rh - (\alpha_2 + \alpha_3)\, Rh_i \tag{19.2}$$
$$\dot{T}^* = \epsilon Rh\,(T_{Tot} - T^*) - \beta_1 T^* + \tau_2\, PDE^* \tag{19.3}$$
$$\dot{PDE}^* = \tau_1 T^*\,(PDE_{Tot} - PDE^*) - \tau_2\, PDE^* \tag{19.4}$$

where Rh, Rh_i are the photoexcited and inactive rhodopsin concentration, T_{Tot} and T^* are the total and activated trasducin concentration, PDE_{Tot} and PDE^* are the total and the active phosphodiesterase concentration and $J_{h\nu}(t)$ the flux of rhodopsin photoisomerizations. In equations (19) it is assumed that the concentrations of excitable rhodopsin are unlimited and that the maximal amount of excitable transducin T_{Tot} and phosphodiesterase PDE_{Tot} are 1000 μM and 100 μM respectively (see Stryer 1986 for a justification of these assumptions). The values of the rate constant $\alpha_1, \alpha_2, \alpha_3, \beta_1, \tau_1$ and τ_2 were deduced to be equal to 20 sec^{-1}, 0.0005 sec^{-1}, 0.05 sec^{-1}, 10.6 sec^{-1}, 0.1 sec^{-1} μM^{-1} and 10 sec^{-1} respectively. The values for α_2 and α_3 were selected so as to reproduce the time course and amplitude of the slow response. The values for $\alpha_1, \beta_1, \tau_1$ and τ_2 were primarily selected so as to account for the kinetics of dim flash responses.

In order to have a full reconstruction of the kinetics of phototransduction the four differential equations (19) describing the activation of phosphodiesterase and the following equations (20)-(23) were solved:

$$\dot{c} = b\,J - \gamma_{Ca}\,(c - c_0) - k_1\,(e_T - c_b)\,c + k_2\,c_b \tag{20}$$
$$\dot{c}_b = k_1(e_T - c_b)\,c - k_2\,c_b \tag{21}$$
$$\dot{g} = \frac{A_{max}}{1 + \left(\dfrac{c}{K_c}\right)^4} - g\,(\bar{V} + \sigma PDE^*) \tag{22}$$
$$J = J_{max}\,\frac{g^3}{g^3 + K^3} \tag{23}$$

The value of different parameters and the reason for their choice are given in Table 1.

Table 1: Value and reason of choice of parameters used in computer simulations.

Parameter	Value	Reason of choice
α_1	$20\ sec^{-1}$	To fit the time course of photoresponses to dim flash of light.
α_2	$0.0005\ sec^{-1}$	To obtain the appropriate amplitude of the late response.
α_3	$0.05\ sec^{-1}$	To fit the time course of the late response.
a	$0.5\ sec^{-1}\mu M^{-1}$	From Bennett et al. (1982)
T_{Tot}	$1000\ \mu M$	From Stryer (1986).
β_1	$10.6\ sec^{-1}$	To fit the time course of fast reactivation of the photocurrent at the cessation of a step of light.
τ_1	$0.1\ sec^{-1}\mu M^{-1}$	To fit the time course of photoresponse to dim flash of light.
τ_2	$10\ sec^{-1}$	
PDE_{Tot}	$100\ \mu M$	From Stryer (1986).
σ	$1\ sec^{-1}\mu M^{-1}$	Unitary proportionality constant.
b	$0.625\ \mu M\ sec^{-1}pA^{-1}$	From eq.(17) and the value of γ_{Ca}, c_o and of resting free Ca^{2+}.
γ_{Ca}	$50\ sec^{-1}$	From eq(15) and the value of resting free Ca^{2+} of 300 nM (Ratto et al., 1988).
c_o	$100\ nM$	From equation (16).
k_1	$0.2\ sec^{-1}\mu M^{-1}$	To reproduce the kinetics of the reactivation
k_2	$0.8\ sec^{-1}$	of the photocurrent with steps of light and
e_T	$500\ \mu M$	to have an affinity buffer with K_D close to 10 μM (Hodgkin et al., 1987).
$\bar{V}$	$0.4\ sec^{-1}$	From Hodgkin and Nunn (1988)
A_{max}	$65.6\ \mu M\ sec^{-1}$	To have 2 μM for the resting level of free cyclic GMP.
K_c	$100\ nM$	From Koch and Stryer (1988)
K^3	$1000\ \mu M^3$	From Zimmerman and Baylor (1986)
J_{max}	$5040\ pA$	From eq(3) and to have 40 pA of resting current and a resting level of 2 μM of free cyclic GMP

The seven differential equations were numerically integrated using the routine RKF45BC kindly provided by Dr. Greg Bernstein, based on the method of Shampine, 1977, with the initial conditions:

$$g(0)\ =\ 2\ \mu M \qquad c(0)\ =\ 300\ nM \qquad c_b(0)\ =\ \frac{500}{1+\frac{4}{0.3}}\ \mu M\ =\ 34.9\ \mu M$$

$$Rh^*(0)=0 \qquad Rh_i(0)=0 \qquad T^*(0)=0 \qquad PDE^*(0)=0$$

EXPERIMENTAL SECTION

Responses to brief flashes

Fig.2 illustrates two families of photoresponses to brief flashes of light in darkness (A) and in the presence of a steady light equivalent to 1300 Rh*/sec (B). The circulating photocurrent was 30 pA in darkness and was reduced to 15 pA by the steady light. The flash sensitivity in darkness was about 0.28 pA/Rh* and decreases to 5 fA/Rh* in the presence of the background of light. In darkness the time to peak of the dim flash response was 950 msec and decreases to 650 msec by increasing the light intensity by 100 times. When flashes are superimposed on backgrounds of steady light the time to peak of dim flash responses shortens to 330 msec and the acceleration of the time to peak with brighter flashes is not observed with very bright steady lights.

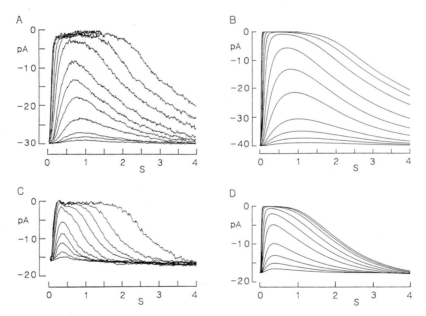

Fig. 2: Families of responses to brief flashes in darkness (A) and superimposed on a steady light equivalent to 1300 Rh*/sec (C). Flash intensity in A were: 3, 6, 12, 25, 58, 115, 230, 600, 1150, 2350, Rh*. In C: 58, 115, 230, 600, 1150, 2350, 6100, 12500, 24000, 50000 Rh*. Response of amplitude less than 5 pA were obtained as the average of at least 8 responses. Theoretical curves in panels B and D were obtained from equations (19) - (23). B: theoretical photocurrents obtained with pulses of $J_{h\nu}(t)$ of duration 10 msec equal to 0.2, 0.5, 1, 2, 5, 10, 20, 50, 100, 500 Rh_t^*/sec. D: theoretical photocurrents obtained with pulses of $J_{h\nu}(t)$ of duration 10 msec, equal to 2, 5, 10, 20, 50, 100, 1000, 2000 Rh_t^*/sec, superimposed on a steady light equivalent to 0.1 Rh_t^*/sec. Pulses of $J_{h\nu}(t)$ were given 40 seconds after the onset of the steady light.

A family of traces obtained by solving the set of 7 differential equations (19)-(23), in which the flux of photoisomerization $J_{h\nu}(t)$ had a duration of 10 msec, is illustrated in Fig. 2C. The theoretical curves have a time course rather similar to the experimental traces shown in Fig. 2A. The time to peak of theoretical traces, which is about 1080 msec for dim lights, shortens to 690 msec for bright lights, similar to the experimental behaviour. The time to peak shortens because as intracellular Ca^{2+} drops, the cyclase is stimulated, thus more efficiently counteracting the light induced phosphodiesterase activation.

The theoretical families of photocurrents, obtained by simulating a pulse superimposed to a steady flux of photoisomerizations, are shown in Fig. 2D. In agreement with the experimental traces, the time-to-peak of the dim flash response shortens to 450 msec, and does not accelerate with brighter flashes. In this case the simulation shows that the cGMP drops to only about 0.2 μM, although this causes a complete suppression of the photocurrent. The cooperative action of cyclic GMP on the channel (see eq.(23)) is responsible for the almost complete suppression of the photocurrent even though cyclic GMP is only reduced by 10 times. This feature is likely to be a remarkable property of the phototransduction machinery. The calculated changes of intracellular Ca^{2+} are smaller and slower.

The experimental traces obtained in response to bright flashes of light have a duration longer than the theoretical traces. This defect of the model is more severe in darkness than in the presence of a steady light.

In Fig.3 some theoretical relations between normalized photoresponse $\Delta J/J_S$ and light intensity I are reproduced and the experimental data obtained in darkness (A) and in the presence of a steady light equivalent to 110.000 Rh*/sec (B) are shown. In Fig. 3A we have drawn the Michaelis-Menten relation (eq.(1), thin continuous line), the exponential relation (thick continuous line) and the relation described by eq.(8) with n equal to 2, 3 and 4 (broken lines, see figure legends). These theoretical curves are rather similar to those already proposed by Baylor et al., (1974) and Lamb et al. (1981). As shown in Fig. 3A and 3B the experimental points at different times appear slightly shifted and the same relation fits the experimental data obtained at different times. In darkness (see Fig. 3A) the experimental points lie between the exponential eq.(2) and eq.(8) with n=4. In the presence of a bright steady light equivalent to 110.000 Rh*/sec the experimental points are better fitted by the Michaelis-Menten eq.(1) or by eq.(8) with n=2.

These results show that eq.(2) fits the experimental data obtained in darkness reasonably well, but not those obtained in the presence of bright steady lights. Consequently, we do not expect diffusion to shape the photoresponse during light adaptation, where different mechanisms, like those described by eq.(8), are likely to control the relation between amplitude of photoresponse and light intensity. However, the exact value of n cannot be decided, because our data were not collected under voltage clamp conditions and voltage dependent currents may have slightly affected the traces of most bright photoresponses (Baylor and Nunn, 1986; Cobbs and Pugh, 1987).

Photoresponses to lights of different durations

When a light is impinging onto a rod for a time longer than a second or so, the rod changes its responsiveness to light and initiates light adaptation. Fig. 4A reproduces

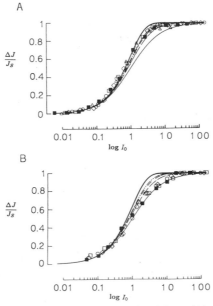

A

B

Fig. 3: Thick continuous line has been obtained with eq(2), thin continuous line wih eq.(1), the broken lines have been obtained with eq(8) with n=2 (— - — -), n=3 (- - - - -) and n=4 (- - — - -). Theoretical curves have been shifted so as to have same rising phase for dim light intensities. I_0 is the light intensity causing a photoresponse of 0.5 with relation (1). Comparison of the theoretical curves with the experimental data obtained in darkness at the following times (sec): 0.7 ($\triangle$), 0.8 ($\blacksquare$), 0.9 ($\square$), 1 ($\diamond$) and 1.1 ($\bigcirc$) (A) and in the presence of a steady light equivalent to 110000 Rh*/sec at the times (sec): 0.275 ($\blacksquare$), 0.3 ($\square$), 0.325 ($\diamond$) and 0.35 ($\bigcirc$) (B). The experimental points have been shifted so as to lie on the same curve. The data were obtained from traces shown in Fig. 2A-C.

photoresponses to a steady light equivalent to 4500 Rh*/sec of different durations (1, 2, 5, 10, 20, 30 and 40 seconds) . The photocurrent, which is initially fully suppressed, partially reactivates within 10 or 20 seconds. When the light is turned off, after longer exposures, the photocurrent recovers with a complex time course. The time course of the reactivation of the photocurrent after a steady light of 5, 10 and 20 seconds is shown in greater detail in the inset of Fig. 4A.

The photocurrent, following a flash of 5 seconds, remains saturated for an additional 2.5 seconds, while after longer illumination promptly recovers when the light is turned off. The two traces, however, cross because, with time, another slow component of the photoresponse is turned on. This component, only appears clearly following strong lights.

A full speeding up of the reactivation of the photocurrent, when the light is turned off, is obtained with illuminations exceeding 10 seconds. When the same experiment is repeated in the presence of a steady light equivalent to 1050 Rh*/sec the slow component of the photoresponse is reduced or even abolished as shown in Fig. 4C. In this case, when the test flash is terminated, the experimental traces do not cross (see inset) as in the absence of the steady light.

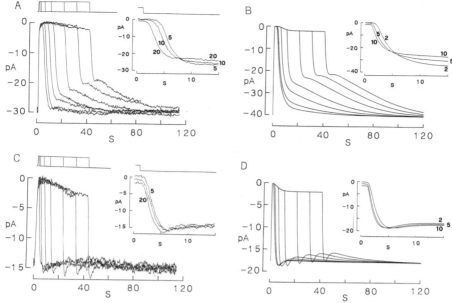

Fig. 4: Photoresponses to flashes of different duration. A: photoresponses to steps of light equivalent to 4500 Rh*/sec of 1, 2, 5, 10, 20, 30 and 40 seconds of duration. Inset: the reactivation of the photocurrent following a step of light of 5, 10 and 20 sec. Time 1 second coincides with the extinction of the light. C: photoresponses to steps of light equivalent to 4500 Rh*/sec of 1, 2, 5, 10, 20, 30 and 40 seconds of duration superimposed to a steady light equivalent to 1050 Rh*/sec . Inset: the reactivation of the photocurrent following a step of light of 5, 10 and 20 sec. Time 1 second coincides with the extinction of the light. B: Theoretical curves obtained with pulses of $J_{h\nu}(t)$ of different duration, pulses of 1, 2, 5, 10, 20, 30, 40 seconds of $J_{h\nu}(t)$ equal to 10 Rh_t^*/sec. Inset: the reactivation of the photocurrent following steps of 2, 5, 10 seconds of duration. Time 1 second coincides with the extinction of the light. D: Theoretical curves obtained with pulses of 1, 2, 5, 10, 20, 30 and 40 seconds of $J_{h\nu}(t)$ equal to 10 Rh_t^*/sec, superimposed on a steady light equivalent to 0.1 Rh_t^*/sec. Inset: the reactivation of the photocurrent following steps of light of 2, 5 and 10 seconds of duration. Time 1 second coincides with the extinction of the light.

It is of some importance to see whether or not the proposed model is able to account for these aspects of phototransduction. Fig. 4B reproduces theoretical curves obtained by simulating photoresponses of different durations. In the inset the theoretical curves at the cessation of the light are reproduced on an enlarged time scale to illustrate the time course of the reactivation of the photocurrent. Fig. 4D and the inset reproduce similar theoretical curves obtained in the presence of a constant flux of photoisomerizations.

As shown in Fig. 4, most of the experimental features are well reproduced by the model: the shortening of the latency of the reactivation of the photocurrent with longer lights and the progressive appearence of the slow response; the crossing of the repolarization phase after a brief and a long step of light; the reduction or suppression of the slow response; and the absence of the crossing of the repolarization phase in the presence of a steady light.

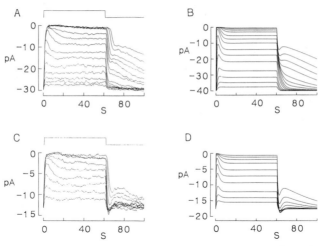

Fig. 5: A: photoresponses to steps of light of 60 seconds of duration. The intensity
of steady lights was equivalent to 5, 11, 20, 45, 100, 210, 480, 1050, 2200, 4500,
11500, 22000, 45000, 110000 and 220000 Rh*/sec. C: photoresponses to steps of
light of 60 seconds in the presence of a steady light equivalent to 1050 Rh*/sec.
The intensity of steps of light was estimated to be 480, 1050, 2200, 4500, 11500,
22000, 45000, 110000 and 220000 Rh*/sec. Each trace is the average of at least
two responses. Responses to dim lights (amplitude < 5 pA) were obtained as the
average of at least 5 responses. B: Theoretical curves obtained as in Fig. 4 but
with steps of $J_{h\nu}(t)$ of 60 seconds of duration. The intensity of $J_{h\nu}(t)$ was 0.002,
0.005, 0.01, 0.02, 0.05, 0.1, 0.2, 0.5, 1, 2, 5, 10, 20, 50 Rh_t^*/sec. D: Theoretical
curves obtained with steps of $J_{h\nu}(t)$ of 60 seconds of duration superimposed on a
steady light equivalent to 0.1 Rh_t^*/sec. The intensity of $J_{h\nu}(t)$ was 0.05, 0.1, 0.2,
0.5, 1, 2, 5, 10, 20 and 50 Rh_t^*/sec.

Responses to steps of light

Photoresponses to steps of light of 60 seconds of increasing intensity are shown in
Fig.5A. Responses to weak steps of light showed large fluctuations and many responses
were averaged, while responses to bright steps of light revealed only the usual instrumental noise, and fewer responses were averaged. The low frequency noise present with weak
light is likely to be caused by quantum fluctuation as in toad rods (Baylor et al., 1979b).
Photoresponses to intermediate light intensities reach a peak within one or two seconds
and at later times present a partial reactivation of the photocurrent. This delayed reactivation of the photocurrent is the major mechanism underlying light adaptation. When
the step of light is switched off the photocurrent reactivates with a complex time course.
Following light intensities not fully suppressing the photocurrent, the current reactivates
with a delay of less than 100 mseconds and with a time constant of about 0.8 sec^{-1}.
In the presence of brighter lights the photocurrent remains fully suppressed for a few
seconds before commencing to reactivate. After a partial fast recovery of the photocurrent a long lasting component of the photoresponse is observed. This component of the
photoresponse, which is observed in darkness and can be named the late or the slow
response, extinguishes at a rate constant of about 0.02 sec^{-1}. Following Lamb (1981)
we attribute the origin of the late response to the existence of a reversible step in the

inactivation of photoexcited rhodopsin or to a partial ability of inactive rhodopsin to activate transducin.

Photoresponses to steps of light of 60 seconds of increasing intensity, in the presence of a steady light equivalent to 1050 Rh*/sec are shown in Fig. 5C. The effect of the steady light on the kinetics of photoresponses is most pronounced at the cessation of the steps of light where an overshoot can be observed and the kinetics of the late response is slightly accelerated. The time course of the reactivation of the photocurrent, observed with steps of light initially blocking a large fraction of the circulating current, is not accelerated. Similarly the time course of reactivation of the photocurrent at the termination of non-saturating steps of light is hardly affected. A family of theoretical traces, simulating a flux of photoisomerizations of 60 seconds is shown in Fig. 5B.

In agreement with the experimental traces the photocurrent, after a partial or complete suppression, reactivates with a delay. This reactivation is produced by the second slow fall of intracellular Ca^{2+}, caused by the Ca^{2+} uptake by the internal buffer. A noticeable feature of the proposed model is evident by the analysis of changes of cyclic GMP where its free level during a steady light does not fall below the level set by the maximal activity of phosphodiesterase and cyclase. From eq.(20) intracellular Ca^{2+} cannot fall below 100 nM (i.e. the value of c_0) and from eq.(22) at the steady state the lower level of cyclic GMP is

$$
\frac{A_{max}}{\left(1 + \left(\frac{c_0}{k_c}\right)^4\right)\left(\bar{V} + \sigma PDE_{TOT}\right)} \tag{24}
$$

which from the value of Table 1 is 0.33 μM. Consequently, at the steady state the level of free cyclic GMP decreases by a factor of only 6, while acting as an internal transmitter able to transduce light intensity over 4 log units.

At the termination of the light the theoretical curves show almost the same behaviour as the experimental traces, where the photocurrent reactivates with a fast and a slow component. The fast reactivation of the photocurrent is due to the substantial depletion of intracellular Ca^{2+}, which activates the cyclase (Hodgkin and Nunn, 1988). This enzyme remains activated even during the 2 or 3 seconds of the fast reactivation, because the influx of Ca^{2+}, through light sensitive channels, is absorbed by the intracellular Ca^{2+} buffer. The slow component is caused by the reversible inactivation of photoactivated rhodopsin.

A defect of the model evident in the traces shown in Fig. 5 is the almost instantaneous reactivation of the photocurrent at the cessation of the step of light. A more complex scheme for the enzymatic cascade is likely to remove this defect.

Fig. 5D reproduces theoretical traces simulating a step of light of 60 seconds but in the presence of a steady light. The theoretical traces reproduce the time course of the experimental recordings shown in Fig. 5C quite well. The presence of a steady light in the model does not accelerate the time course of the reactivation of the photocurrent, in agreement with the experimental observation. In a similar manner to the experimental traces, a rebound of the photocurrent is observed at the cessation of the step of light. In the model the rebound is accounted for by a faster deactivation of transducin and phosphodiesterase in the presence of a steady light and by the presence of the Ca^{2+} buffer.

DISCUSSION

In this paper a quantitative reconstruction of the kinetics of phototransduction has been attempted, obtaining a satisfactory agreement between theoretical and experimental curves. A noticeable feature of the proposed model is the relatively small modulation of internal messengers, while preserving the ability of the photoreceptor to transduce light intensity over a 4 log unit range. During steady bright lights, the intracellular cyclic GMP decreases from 2 μM to about 0.3 μM and free Ca^{2+} from 300 nM to 100 nM. This ability to modulate the response over a rather wide range of light intensity, with small changes in the concentration of internal transmitters, is due to the cooperative action of cyclic GMP on the channel and of Ca^{2+} on the cyclase activity. These observations may suggest that any transduction mechanism is unlikely to require large changes of internal second messengers, such as a rise or a fall of free Ca^{2+} by several log units, and so, consequently, the obvious design to increase the operating range of the transduction process would be to use highly cooperative mechanisms.

It can be seen that the model is able to account for many features of the kinetics of phototransduction, but not all the details, as discussed in the experimental section.

In the equations used to obtain the theoretical curves, the concentration of different substances has been assumed to be uniform, thus highly simplifying the model. It is well known, however, that internal diffusion inside a photoreceptor is restricted (Lamb et al., 1981) and the concentration of the internal transmitter is not expected to be uniform when just a few photons are absorbed in a second. The main reason for making this simplifying assumption is the observation that in the presence of steady bright lights, producing several tens of thousands of photoisomerizations per second, the kinetics of phototransduction is accelerated, but the number of slow stages required to fit the time course of the rising phase of photoresponse in the linear range is not changed. Since, under these conditions, given the high number of photoisomerization, the cell interior is expected to be almost uniform, it is concluded that internal diffusion could shape the time course of the photoresponse in dark adapted conditions, but not in the presence of bright steady lights. The understanding of photoresponses to dim or moderate flashes in dark adapted conditions can be refined by introducing longitudinal diffusion into the model.

Ca^{2+} permeates through light sensitive channels (Hodgkin et al., 1985; Yau and Nakatani, 1985b; Nakatani and Yau, 1988a; Menini et al., 1988) and controls light adaptation (Torre et al., 1986; Matthews et al., 1988) by inhibiting the activity of the cyclase (Lolley and Racz, 1982; Pepe et al., 1986; Koch and Stryer, 1988). Internal Ca^{2+} is extruded by Na^+ / Ca^{2+} exchange and can be bound by internal buffers (Hodgkin et al., 1987). If the role of internal Ca^{2+} now seems to be rather well understood, there are still some open questions, relevant to a quantitative analysis of its role in phototransduction. In the proposed model we have assumed that the internal low affinity buffer was slow, that is the on rate was well below the limit set by diffusion of 100 sec^{-1} μM^{-1}. This assumption is based on computer simulation in which the presence of fast Ca^{2+} buffers, such as BAPTA, caused the appearance of rebounds as those observed by trapping BAPTA inside a rod (Torre et al., 1986). Therefore it is concluded that the existence of large amounts of fast Ca^{2+} buffers inside a rod give rise to an unusual time course of photoresponses. In this model only one kind of internal Ca^{2+} buffer has been considered. Other experimental evidence also indicates the existence of a high affinity buffer (Hodgkin et al., 1987). The agreement with the experimental data can be expected to improve by including a second Ca^{2+} buffer in the model.

The proposed model assumes a fast action of Ca^{2+} on the cyclase and an almost instantaneous equilibrium between active and inhibited cyclase. The effect of Ca^{2+} on the cyclase does not seem to be direct, but appears to require a soluble component (Koch and Stryer, 1988), which is not considered in the model. The argument in favour of a fast action of Ca^{2+} on the cyclase is the rapid blockage of the photocurrent when intracellular Ca^{2+} rises (Hodgkin et al., 1985) and the rapid activation when intracellular Ca^{2+} falls (Lamb and Matthews, 1988).

The model also assumes that the inactivation of rhodopsin is slightly reversible, thus giving origin to the late response (Lamb, 1986). It is also possible to account for the late response by supposing that inactive rhodopsin is still able to activate transducin with a much lower efficacy than activated rhodopsin. The large buffering of cyclic GMP has not been included in the model, and this could be relevant and play a role in light adaptation. The reason for not including the cyclic GMP buffering is essentially one of parsimony and uncertainty of its kinetic properties. The model is not very critical with respect to the values of the rate constants, but is sensitive to their order of magnitude. Biochemical measurements of these rate constants, possibly in situ, will provide good tests for this model.

ACKNOWLEDGEMENTS

We acknowledge Dr. L. Cervetto for many useful discussions and Dr. Franco Gambale for his continuous encouragement and helpful advice. Dr. D. Bertrand gave us his computer program DATAC. Mr. E. Gaggero, G. Franzone, G. Gaggero, D. Magliozzi and P.G. Gagna built many parts of the apparatus. Ms. C. Rosati kindly typed the manuscript and prepared the illustrations. Clive Prestt checked the English.

REFERENCES

Applebury, M. L. and Chabre, M., 1986, Interaction of photoactivated rhodopsin with photoreceptors proteins: the cyclic GMP cascade, *In*: "The Molecular Mechanism of Photoreception", ed. Stieve H., pp. 51-66, Dahlem Konferenzen, Berlin, Springer.

Baylor, D. A. and Fuortes, M. G. F., 1970, Electrical responses of single cones in the retina of the turtle, *J. Physiol.*, 207:77-92.

Baylor, D. A. and Hodgkin, A. L., 1974, Changes in time scale and sensitivity in turtle photoreceptors, *J. Physiol.*, 242:729-758.

Baylor, D. A., Hodgkin, A. L. and Lamb, T. D., 1974, The electrical response of turtle cones to flashes and steps of light, *J. Physiol.*, 242:685-727.

Baylor, D. A., Lamb, T. D. and Yau, K. W., 1979a, The membrane current of single rod outer segments, *J. Physiol.*, 354:203-223.

Baylor, D. A., Lamb, T. D. and Yau, K. W., 1979b, Responses of retinal rods to single photons, *J. Physiol.*, 288:613-634.

Baylor, D. A. and Nunn, B. J., 1986, Electrical properties of the light-sensitive conductance of salamander rods, *J. Physiol.*, 371:115-145.

Bennett, N., Michel-Villaz, M. and Kuhn, H., 1982, Light induced interaction between rhodopsin and the GTP-binding protein: metarhodopsin II is the major photoproduct involved, *Eur. J. Biochem.*, 127:97-103.

Blaunstein, M. P. and Hodgkin, A. L., 1969, The effect of cyanide on the efflux of calcium from squid axons, *J. Physiol.*, 200:497-527.

Caretta, A. and Cavaggioni, A., 1983, Fast ionic flux activated by cyclic GMP in the membrane of cattle rod outer segments, *Eur. J. Biochem.*, 132:1-8.

Cervetto, L., Pasino, E. and Torre, V., 1977, Electrical responses of rods in the retina of Bufo Marinus, *J. Physiol.*, 267:17-51.

Cervetto, L., Torre, V., Rispoli, G. and Marroni, P., 1985, Mechanisms of light adaptation in toad rods, *Exp. Biol.*, 44:147-157.

Cervetto, L., Menini, A., Rispoli, G. and Torre, V., 1988, Modulation of the ionic selectivity of the light-sensitive current in isolated rods of the tiger salamander, *J. Physiol.*, 406:181-198.

Cobbs, W. H. and Pugh, E. H., 1987, Kinetics and components of the flash photocurrent of isolated retinal rods of the larval salamander, Ambystoma Tigrinum, *J. Physiol.*, 394:529-572.

Fesenko, E. E., Kolesnikov, S. S. and Lyubarsky, A. L., 1985, Induction by cyclic GMP of cationic conductance in plasma membrane of retinal rod outer segment, *Nature*, 313:310-313.

Fuortes, M. G. F. and Hodgkin, A. L., 1964, Changes in time scale and sensitivity in the ommatidia of Limulus, *J. Physiol.*, 172:239-263.

Hagins, W. A., 1972, The visual process: excitatory mechanisms in the primary receptor cells, *Ann. Rev. Biophys. Bioeng.*, 1:131-158.

Haynes, L. W., Kay, A. R. and Yau, K. W., 1986, Single cyclic GMP-activated channel activity in excised patches of rod outer segment membrane, *Nature*, 321:66-70.

Hodgkin, A. L., McNaughton, P. A. and Nunn, B. J., 1985, The ionic selectivity and calcium dependence of the light-sensitive pathway in toad rods, *J. Physiol.*, 358:447-468.

Hodgkin, A. L., McNaughton, P. and Nunn, B. J., 1987, Measurement of sodium calcium exchange in salamander rods, *J. Physiol.*, 391:347-370.

Hodgkin, A. L. and Nunn, B. J., 1988, Control of light-sensitive current in salamander rods, *J. Physiol.*, 403:473-494.

Karpen, J. W., Zimmerman, A. L., Stryer, L. and Baylor, D. A., 1988, Gating kinetics of the cyclic-GMP activated channel of retinal rods: flash photolysis and voltage-jump studies, *Proc. Natl. Acad. Sci. USA*, 85:1287-1291.

Koch, K. W. and Stryer, L., 1988, Highly cooperative feedback control of retinal rod guanylate cyclase by calcium ions, *Nature*, 334:64-66.

Lagnado, L., Cervetto, L. and McNaughton, P. A., 1988, Ion transport by the Na-Ca exchange in isolated rod outer segments, Proc. Natl. Acad. Sci., in press.

Lamb, T. D., 1981, Spontaneous quantal events induced in toad rods by pigment bleaching, *Nature*, 287:349-351.

Lamb, T. D., McNaughton, P. A. and Yau, K. W., 1981, Spatial spread of activation and background desensitization in toad rod outer segments, *J. Physiol.*, 272:463-496.

Lamb, T. D., 1986, "Photoreceptor adaptation - vertebrates in the molecular mechanism of photoreception," Ed. H. Stieve, 267-286, Dahlem Konferenzen 1986, Berlin Springer-Verlag.

Lamb, T. D., Matthews, H. R. and Torre, V., 1986, Incorporation of calcium buffers into salamander retinal rods: a rejection of the calcium hypothesis of phototransduction, *J. Physiol.*, 372:315-349.

Lamb. T. D. and Matthews, H. R., 1988, External and internal actions in the response of salamander retinal rods to altered external calcium concentration, *J. Physiol.*, 403:473-494.

Liebman, P. A. and Evanzuk, A. T., 1982, Real time array of rod disk membrane cyclic GMP phosphodiesterase and its controller enzymes, *Methods in Enzymol-*

ogy, 81:532-542.

Liebman, P. A. and Pugh, E. N., 1982, Gain, speed and sensitivity of GTP binding vs. PDE activation in visual excitation, *Vision Res.*, 22:1475-1480.

Lolley, R. H. and Racz, E., 1982, Calcium modulation of cyclic GMP synthesis in rat visual cells, *Vision Res.*, 22:1481-1486.

Matthews, H. R., Torre, V. and Lamb, T. D., 1985, Effects on photoresponse of calcium buffers and cyclic GMP incorporated into the cytoplasm of retinal rods, *Nature*, 313:582-585.

Matthews, H. R., Murphy, R. I., Fain, G. I. and Lamb, T. D., 1988, Photoreceptor light adaptation is mediated in isolated rods of the tiger salamander, *Nature*, 334:67-69.

McNaughton, P. A., Cervetto, L. and Nunn, B. J., 1986, Measurement of the intracellular free calcium concentration in salamander rods, *Nature*, 322:261-263.

Menini, A., Rispoli, G. and Torre, V., 1988, The ionic selectivity of the light-sensitive current in isolated rods of the tiger salamander, *J. Physiol.*, 402:279-300.

Nakatani, K. and Yau, K. -W., 1988a, Calcium and magnesium fluxes across the plasma membrane of the toad rod outer segment, *J. Physiol.*, 395:695-730.

Nakatani, K. and Yau, K. -W., 1988b, Guanosyne 3', 5' - cyclic monophosphate - activated conductance studied in a truncated rod outer segment of the toad, *J. Physiol.*, 395:731-754.

Nakatani, K. and Yau, K. -W., 1988c, Calcium and light adaptation in retinal rods and cones, *Nature*, 334:69-71.

Nunn, B. J. and Baylor, D. A., 1982, Visual transduction in retinal rods of the monkey Macaca Fascicularis, *Nature*, 299:726-728.

Pepe, I. M., Panfoli, I. and Cugnoli, C., 1986, Guanylate cyclase in rod outer segments of the toad retina, *FEBS Letters*, 203:73-76.

Ratto, G. M., Payne, R., Owen, W. G. and Tsien, R. Y., 1988, The concentration of cytosolic free Ca^{2+} in vertebrate rod outer segments measured with Fura 2, *J. Neurosci.*, in the press.

Rispoli, G., Sather, W. A. and Detwiler, P. B., 1988, Effect of triphosphate nucleotides on the response of detached rod outer segments to low external calcium, *Biophys. J.*, 53:388a.

Robinson, P. R., Kawamura, S., Abramson, B. and Brownds, M. D. , 1980, Control of the cyclic GMP phosphodiesterase of frog photoreceptor membranes, *J. Gen. Physiol.*, 76:631-645.

Sather, W. A., Rispoli, G. and Detwiler, P. B., 1988, Effect of calcium on light adaptation in detached gecko rod outer segments, *Biophys. J.*, 53:390a.

Shampine, L. F., 1977, Stiffness and nonstiff differential equation solvers, II: Detecting stiffness with Runge-Kutta methods, *ACM Transactions on Mathematical Software*, Vol.3, N.1, 44:53.

Stryer, L., 1986, Cyclic GMP cascade of vision, *Ann. Rev. Neurosci.*, 9:87-119.

Torre, V., 1982, The contribution of the Electrogenic Na - K Pump to the Electrical Activity of Toad Rods, *J. Physiol.*, 333:315-341.

Torre, V., Matthews, H. R. and Lamb, T. D., 1986, Role of calcium in regulating the cyclic GMP cascade of phototransduction in retinal rods, *Proc. Natl. Acad. Sci. U.S.A.*, 83:7109-7113.

Torre, V., Rispoli, G., Menini, A. and Cervetto, L., 1987, Ionic selectivity, blockage and control of light-sensitive channels, *Neurosci. Res.*, suppl.6: S25:S44.

Vuong, T. M. and Stryer, L., 1984, Millisecond activation of transducin in the cyclic nucleotide cascade of vision, *Nature*, 311:659-661.

Yau, K. W. and Nakatani, K., 1984, Electrogenic Na-Ca exchange in retinal rod outer segment, *Nature*, 311:661-663.

Yau, K. W. and Nakatani, K., 1985a, Light-induced reduction of cytoplasmic free calcium in retinal rod outer segment, *Nature*, 313:579-582.

Yau, K. W. and Nakatani, K., 1985b, Light-suppressible, cyclic GMP-sensitive conductance in the plasma membrane of a truncated rod outer segment, *Nature*, 317:252-255.

Zimmerman, A. L. and Baylor, D. A., 1986, Cyclic-GMP sensitive conductance of retinal rods consists of aqueous pores, *Nature*, 321:70-72.

DIFFERENCES IN RESPONSE KINETICS AND ABSOLUTE SENSITIVITY

BETWEEN RED-, BLUE- AND ULTRAVIOLET-SENSITIVE CONES OF

THE TIGER SALAMANDER

R. J. Perry, A. J. Craig and P. A. McNaughton

Physiological Laboratory
University of Cambridge, Cambridge, U.K.

ABSTRACT

Three cone types have been identified in the retina of the tiger salamander, using the suction electrode technique. Their spectral sensitivity functions peak at 605nm ("red-sensitive" cone), 440nm ("blue-sensitive" cone) and below 400nm ("ultraviolet-sensitive" cone). The estimated absolute sensitivity of red-sensitive cones is 3.7 fA photons^{-1} μm^2 and of blue-sensitive cones is 41.9 fA photons^{-1} μm^2. The time-to-peak and the overall duration of the current response to a dim flash are longer in blue-sensitive cones than in red- or ultraviolet-sensitive cones.

INTRODUCTION

Attwell, Werblin and Wilson (1982) have recorded intracellularly from cones of the larval tiger salamander, and have shown the spectral sensitivity curve of a single cone type, peaking at about 620nm. This confirms Liebman's (1972) claim, based on microspectrophotometry, that larval salamander single cones only contain one type of pigment, showing peak absorption at 620nm. However, Attwell et al. mention that one batch of salamanders gave some cones showing their peak sensitivity at a wavelength shorter than 450nm, so it cannot be assumed that there is only one type of cone in the larval tiger salamander. A more thorough study of the spectral sensitivity of the cones in this species has therefore been undertaken. A preliminary report of this work has appeared (Craig and Perry, 1988).

METHODS

The methods for isolating, optically stimulating and recording current from salamander cones were essentially identical to those applied to toad rods by Baylor, Lamb and Yau (1979) with the modifications of Lamb, McNaughton and Yau (1981) and Hodgkin, McNaughton, Nunn and Yau (1984). The suction electrodes were of internal diameter 7-8 μm. The spectral sensitivity curve peaking at about 605nm was obtained with the outer segment in a Ringer-filled suction pipette as Baylor et al. originally described.

The rest of the data, however, was collected with the inner segment in the pipette as described for rods by Yau, McNaughton and Hodgkin (1981), except that the pipette was, in this case, filled with 3mM glucose Ringer. This was found to increase the stability and longevity of the cones, and also facilitated further experiments involving rapid changes in the solution bathing the outer segment.

"Monochromatic" stimuli were produced using 19 interference filters, which were calibrated for peak wavelength transmitted and for half band-width (c. 20nm) using a Varian DMS 90 UV/Visible spectrophotometer. The duration of test flashes used was always the same within each cell, but varied between 10ms and 100ms between cells according to the minimum duration of an unattenuated flash required to elicit at least a just-saturating response. The spectral sensitivity of each cone type, relative to the cone's own peak wavelength, was determined by a method similar to that of Naka and Rushton (1966; see also Baylor and Hodgkin, 1973; Baylor, Nunn and Schnapf, 1984).

RESULTS

Spectral and absolute sensitivities

Three cone types were defined, with spectral sensitivity functions peaking at 605nm ("red-sensitive" cone), 440nm ("blue-sensitive") and below 400nm ("ultraviolet-sensitive"). The sensitivity of the ultraviolet-sensitive cone was relatively low at all wavelengths tested, but rose sharply as the wavelength approached 400nm, the shortest wavelength stimulus that we were able to deliver. This suggests that the peak sensitivity of this cone occurs well within the ultraviolet range. The part of the ultraviolet-sensitive cone curve shown has of necessity been arbitrarily aligned on the axis of log. relative sensitivity, since the sensitivity at the peak wavelength is not available. The spectral sensitivity obtained for the red-sensitive cone is compatible, within the range of experimental error, with that published by Attwell et al. (1982).

Since the absolute sensitivities of isolated photoreceptors are rather variable, only tentative conclusions can be drawn in view of the rather small sample of cones. Absolute sensitivity seems to depend critically on the state of health of the cone, and therefore cones giving unusually small or noisy light responses have been excluded from our estimates. The absolute sensitivity (mean $\pm$ SEM) of 8 red-sensitive cones was 3.7 ± 0.5 fA photons$^{-1}\mu$m^2 and of 3 blue-sensitive cones was 41.9 ± 10.1 fA photons$^{-1}\mu$m^2. We were unable to measure the absolute sensitivity of ultraviolet-sensitive cones, since the wavelength at which these cells are maximally sensitive falls outside the range of stimuli available.

Response characteristics

Figure 1 illustrates the very different response properties of these three cone types. The time-to-peak of a dim flash response was shortest in red-sensitive cones (A, mean $\pm$ SEM = 141 ± 7ms, n=12), slightly longer in ultraviolet-sensitive cones (C, 187 ± 12ms, n=3) and longest in blue-sensitive cones (B, 544 ± 114ms, n=6). The difference in the time-course of the recovery of the light-sensitive current after a flash is not so easily quantified, but the responses shown in Figure 1 are typical. After a flash which shuts off about 3/4 of the light-sensitive current, the red-sensitive cone (A) has recovered all this current within half a second and the ultraviolet-sensitive cone (C) has recovered within about a second. The equivalent response in the blue-sensitive cone (B), however, lasts

228

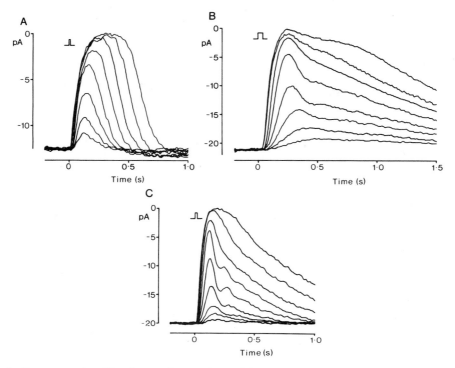

Fig. 1: Response families from: A. a red-sensitive cone, B. a blue-sensitive cone and C. an ultraviolet-sensitive cone. For each cone, the flash intensity was incremented by a factor of approximately 2 between successive traces. Each trace is the average of 5-10 identical flash responses. Flash durations as indicated. Reproduced with permission from Craig and Perry, 1988.

for well over two seconds. The spectral sensitivity of a cone could therefore be predicted reliably from the time-course of its flash responses.

The form of the flash responses of each cone type was independent of the wavelength of stimulating light, the only difference being in the photon density required to elicit a given response. The relationship between log. photon density and response amplitude was therefore consistent in shape at different wavelengths. These observations are in accordance with the "principle of univariance" (Naka and Rushton, 1966).

The time-to-peak of the responses of red-sensitive cones to all but the dimmest flashes increased with increasing flash intensity. This may seem a little surprising, since in rods the time-to-peak is invariably reduced as the flash intensity increases, as is also observed in the blue- and ultraviolet-sensitive cone traces shown. However, the increased time-to-peak at brighter flash intensities is not unique to red-sensitive cones; this phenomenon has also been observed in a minority of blue- and ultraviolet-sensitive cones (traces not shown).

At intermediate flash intensities, the ultraviolet-sensitive cones consistently showed marked current oscillations superimposed on the recovery of light-sensitive current after flashes of intermediate intensity. A hint of such behaviour can also usually be seen on the recovery of blue-sensitive cones. The underswing which is frequently observed after the red-sensitive cone response may be a related phenomenon.

DISCUSSION

The chief finding of the present study was that the tiger salamander possesses at least three types of cone, as defined by spectral sensitivity. Their peak wavelengths were at 605nm (red-sensitive), at 440nm (blue-sensitive) and well below 400nm (ultraviolet-sensitive). Blue-sensitive cones were found to be about ten times more sensitive than red-sensitive cones. We have also shown that the response of the photocurrent to flashes differs in the three cone types. Most notably: (i) the time-to- peak of the flash response increases with increasing intensity in the red-sensitive cone, but usually decreases in the other two populations of cones; (ii) the blue-sensitive cone has an unusually long time course and (iii) the ultraviolet-sensitive cone shows prominent oscillations superimposed on the recovery of light-sensitive current from flashes of intermediate intensity.

Spectral sensitivity

The 165nm separation between the respective red- and blue-sensitive cone peak wavelengths is perhaps surprisingly wide: the separation between "blue-sensitive" and "green-sensitive" cone sensitivity peaks in the human is only about 100nm, while the "red"-"green" separation is only about 30nm. We cannot rule out the possibility that a further cone type exists in the salamander retina, but such a cone would need to be extremely rare or fragile to have escaped detection. Not only have the spectral sensitivities of 20 cones been thoroughly established in the present study, but more than 50 other cones have been quickly checked for spectral sensitivity during the course of more recent experiments, and all of these conformed to one or other of the spectral sensitivities reported here.

The salamander may not need a green-sensitive cone. Salamander rods are unlike those of primates in that they can adapt to light, thereby considerably extending their effective dynamic range. It is therefore conceivable that the prevailing illumination in the murky world of the salamander is such that this usually operates in its mesopic range. Since the peak sensitivities of the "green" and "red" rods are 438 and 527nm respectively (see Liebman, 1972), these would fill the gap between the "blue-" ($\lambda_{max} = 440$nm) and "red-sensitive" ($\lambda_{max} = 605$nm) cones.

Sensitivity to ultraviolet light has long been known to exist in ants and bees, and more recently has been demonstrated in flies and spiders. Pigeons, humming-birds and chickens are also credited with sensitivity to ultraviolet light (see Stark and Tan, 1982, for a review). Ultraviolet-sensitive photoreceptors have been identified by microspectrophotometry in two species of cyprinid fish: roach (Avery, Bowmaker, Djamgoz and Downing, 1983) and Japanese dace (Harosi and Hashimoto, 1983). However, we here report the first electrophysiological recordings from ultraviolet-sensitive photoreceptors, and thereby confirm that absorption in the ultraviolet range is indeed physiologically relevant, rather than, for example, representing absorption by the 11-cis and all-trans isomers of free retinal$_2$ ($\lambda_{max} = 344$nm and 350nm, respectively) or retinol$_2$ ($\lambda_{max} = 393$nm and 401nm, see Morton, 1972).

Red cone time-to-peak

Previous work on amphibian photoreceptors has shown that the time-to-peak of the current trace always decreases with increasing flash intensity, but in the present study we have shown an increase in the time-to-peak of the red cone flash response as the stimulus intensity is increased (Figure 1A). However, this does not necessarily mean that the closure of light-sensitive channels takes longer to peak in response to brighter flashes. If the rising phase of the bright flash response of the red cone is limited by the time constant of the cell membrane, as is suggested by the dramatic speeding of this rising phase when the cone is voltage-clamped (Cobbs, Barkdoll and Pugh, 1985), then an increased time constant at the peak of the bright flash response could be responsible for the anomalous time-to-peak behaviour of the red cone.

Slow time course of blue-sensitive cone response

It is difficult to relate current responses, as recorded here, with the signal which the cell is transmitting at its synaptic pedicle. However, in view of the relatively slow time-course of the current response of the salamander's blue-sensitive cone, it seems plausible that the red cone may be able to transmit a higher temporal frequency than the blue-sensitive cone. It is therefore interesting that Wisowaty and Boynton (1979) have demonstrated psychophysically that the flicker fusion frequency is about 3x higher for the red-sensitive system than for the blue-sensitive system in human vision (see also Brindley, du Groz and Rushton, 1965). If a relatively slow time course is a feature of blue photoreceptors across many species, then the difference in flicker fusion frequency between the human red- and blue-sensitive channels may arise, at least in part, at the level of the photoreceptors.

Baylor, Nunn and Schnapf (1987) do not report any difference in kinetics between the three cone types of the cynomolgus monkey, which is believed to have cones indistinguishable form those of the human (DeValois, Morgan, Polson, Mead and Hull, 1974; Bowmaker, Dartnall and Mollon, 1980; Dartnall, Bowmaker and Mollon, 1983). However, Abraham, Alpern and Kirk (1985) have evidence from electroretinograms obtained from humans that the kinetics of the three cone mechanisms may differ.

Current oscillations during recovery of flash response

Baylor and his colleagues have shown that the current response of cones from the cynomolgus monkey (Baylor et al., 1987) and man (Schnapf, Kraft and Baylor, 1987) show a considerable underswing (ie. rebound increase in photocurrent from the dark level) after the transient arrest of the photocurrent by a light flash. A smaller, but longer-lived underswing was observed after the red cone response in the salamander. This phenomenon and the current oscillations in the other two cone types are characteristic of a delayed negative feedback. This feedback may be intrinsic to the outer segment. Recent evidence would seem to suggest that the fall in free calcium as a result of the suppression of the photocurrent in response to light (Yau and Nakatani, 1985; McNaughton, Cervetto and Nunn, 1986; Ratto, Payne, Owen and Tsien, 1988) may disinhibit guanylate cyclase (Koch and Stryer, 1988), leading to an increase in the level of 3',5'-cyclic guanosine monophosphate (cGMP) and a corresponding increase in photocurrent.

Supported by the MRC.

REFERENCES

Abraham, F.A., Alpern, M. and Kirk, D.B., 1985, Electroretinograms evoked by sinusoidal excitation of human cones, *J. Physiol.*, 363:135-150.

Attwell,D., Werblin, F.S. and Wilson, M., 1982, The properties of single cones isolated from the tiger salamander retina, *J. Physiol.*, 328:259-283.

Avery, J.A., Bowmaker, J.K., Djamgoz, M.B.A. and Downing, J.E.G., 1983, Ultra-violet sensitive receptors in a freshwater fish, *J. Physiol.*, 334:23P-24P.

Baylor, D.A. and Hodgkin, A.L., 1973, Detection and resolution of visual stimuli by turtle photoreceptors, *J. Physiol.*, 234:163-198.

Baylor, D.A., Lamb, T.D. and Yau, K.-W., 1979, The membrane current of single rod outer segments, *J. Physiol.*, 288:589-611.

Baylor, D.A., Nunn, B.J. and Schnapf, J.L., 1984, The photocurrent, noise and spectral sensitivity of the monkey Macaca fascicularis, *J. Physiol.*, 357:575-607.

Baylor, D.A., Nunn, B.J. and Schnapf, J.L., 1987, Spectral sensitivity of the cones of the monkey Macaca fascicularis. *J. Physiol.*, 390:145-160.

Bowmaker, J.K., Dartnall, H.J.A. and Mollon, J.D., 1980, Microspectrophotometric demonstration of four classes of photoreceptor in an old world primate Macaca fascicularis, *J. Physiol.*, 298:131-143.

Brindley, J.D., du Groz, J.J. and Rushton, W.A.H., 1965, The flicker fusion frequency of the blue sensitive mechanism of colour vision, *J. Physiol.*, 183:497-500.

Cobbs, W.H., Barkdoll III, A.E. and Pugh Jr, E.N., 1985, Cyclic GMP increases photocurrent and light-sensitivity of retinal cones, *Nature*, 317:64-66.

Craig, A.J. and Perry, R.J., 1988, Differences in absolute sensitivity and response kinetics between red-, blue- and ultraviolet-sensitive isolated salamander cones. *J. Physiol*, 407:83P.

Dartnall, H.J.A., Bowmaker, J.K. and Mollon, J.D., 1983, Human visual pigments: microspectrophotometric results from the eyes of seven persons, *Proc. R. Soc.*, B 220:115-130.

DeValois, R.L., Morgan, H.C., Polson, M.C., Mead, W.R. and Hull, E.M., 1974, Psychophysical studies of monkey vision, I. Macaque luminosity and color vision tests, *Vision. Res.*, 14:53-67.

Harosi, F.I. and Hashimoto, Y., 1983, Ultraviolet visual pigment in a vertebrate: a tetrachromatic cone system in the dace, *Science*, 222:1021-1023.

Hodgkin, A.L., McNaughton, P.A., Nunn, B.J. and Yau, K.-W., 1984, The effect of ions on retinal rods from Bufo marinus, *J. Physiol.*, 350:649-680.

Koch, K.-W. and Stryer, L., 1988, Highly cooperative feedback control of retinal rod guanylate cyclase by calcium ions, *Nature*, 334:64-66.

Lamb, T.D., McNaughton, P.A. and Yau, K.-W., 1981, Longitudinal spread of activation and background desensitisation in toad rod outer segments, *J. Physiol*, 319:463-496.

Liebman, P.A., 1972, Microspectrophotometry of photoreceptors. *In*: "Handbook of Sensory Physiology," vol. VII/1, ed., Dartnall, H.J.A., pp.481:528, New York, Springer-Verlag.

McNaughton, P.A., Cervetto, L. and Nunn, B.J., 1986, Measurement of the intracellular free calcium concentration in salamander rods, *Nature*, 322:261-263.

Morton, R.A., 1972, The chemistry of the Visual Pigments, *In*: "Handbook of Sensory Physiology," vol.VII/1, ed., Dartnall, H.J.A., pp.33:68, New York, Springer-Verlag.

Naka, K.I. and Rushton, W.A.H., 1966, S-potentials from colour units in the retina of fish (Cyprinidae), *J. Physiol.*, 185:536-555.

Ratto, G.M., Payne, R., Owen, W.G. and Tsien, R.Y., 1988, The concentration of Cytosolic free calcium in vertebrate rod outer segments measured with Fura-2, *J. Neurosci.*, 8(9):3240-3246.

Schnapf J.L., Kraft T.W. and Baylor, D.A., 1987, Spectral sensitivity of human cone photoreceptors, *Nature*, 325:439-441.

Stark, W.S. and Tan, K.E.W.P., 1982, Ultraviolet light: photosensitivity and other effects on the visual system, *Photochem. & Photobiol.*, 36:371-380.

Wisowaty, J.J. and Boynton, R.M., 1979, Temporal modulation sensitivity of the blue mechanism: measurements made without chromatic adaptation, *Vision Res.*, 20:895-909.

Yau, K.-W. and Nakatani, K., 1985, Light-induced reduction of cytoplasmic free calcium in retinal rod outer segment, *Nature*, 313:579-582.

Yau, K.-W., McNaughton, P.A. and Hodgkin, A.L., 1981, Effects of ions on the light-sensitive current of retinal rods, *Nature*, 292:502-505

A PRESYNAPTIC ACTION OF GLUTAMATE ON CONE

PHOTORECEPTORS

K. Everett, M. Sarantis and D. Attwell

Department of Physiology, University College London, England

INTRODUCTION

The photoreceptors of lower vertebrates are thought to use glutamate as a neuro-transmitter (Miller and Schwartz, 1983). The neurotransmitter actions of glutamate in the outer retina are well documented. Glutamate gates channels in postsynaptic bipolar (Attwell et al., 1987; Miller and Slaughter, 1986; Nawy and Copenhagen, 1987), and horizontal cells (Slaughter and Miller, 1983; Ishida et al., 1984). These neurotransmitter actions of glutamate may be terminated in part by electrogenic transport of glutamate into retinal glial (Müller) cells (Brew and Attwell, 1987).

The present study sought to investigate whether the photoreceptor transmitter, glutamate, had any effect on the photoreceptors themselves. Previous reports have suggested that glutamate has no effect on rods and cones (Murakami et al., 1972), or that it hyperpolarizes photoreceptors (Cervetto and MacNichol, 1972). However there is ultrastructural evidence (Lasansky, 1973) for a chemical synapse from rods to cones in the tiger salamander retina. Since glutamate can act as a transmitter from photoreceptors to second order neurones, it may also mediate a signal from rods to cones via this synapse. In addition, radiotracing experiments (White and Neal, 1976; Miller and Schwartz, 1983) suggest that glutamate is taken up into photoreceptors, to assist in terminating its transmitter action and for efficient recycling of the released transmitter. Such uptake might be electrogenic, as in Müller cells (Brew and Attwell, 1987), and thus polarize the cone when glutamate is applied.

We investigated the actions of glutamate on cone photoreceptors from the tiger salamander (*Ambystoma tigrinum*) retina. Glutamate was found to gate an unusual anion channel at the cone synaptic terminal, activating a positive feedback mechanism which may increase the gain of cone phototransduction.

METHODS

Experiments were carried out on cone photoreceptors isolated from the tiger salamander retina by papain dissociation (Bader et al., 1979). Glutamate responses (n=170) were obtained from single cones (n=50) and from both principle and accessory members of double cones (n=120).

SOLUTIONS

The normal external Ringer's solution contained: 104.5mM NaCl; 2.5mM KCl; 3mM $CaCl_2$; 0.5mM $MgCl_2$; 15mM glucose; 5mM HEPES; with the pH adjusted to 7.25 with NaOH. Barium chloride (6mM) was often added to this solution, as it has been shown to block most of the cell's resting potassium conductance in Müller cells (Newman, 1985), reducing the current noise and increasing the range of voltages over which glutamate-induced currents can be studied (Brew and Attwell, 1987). Barium also decreased the cone conductance and current noise. Barium had no effect on the magnitude of the glutamate-evoked current in cones. To change the chloride concentration in the external medium, NaCl was replaced by sodium gluconate.

Patch pipettes used for whole-cell recording contained the following solutions. For 101mM chloride in the pipette: 80mM KCl; 15mM K-acetate; 5mM NaCl; 5mM HEPES; 7mM $MgCl_2$; 5mM Na_2ATP; 1mM $CaCl_2$; 5mM K_2EGTA; with the pH adjusted to 7.0 using KOH. For 30mM pipette chloride, 71 of the 80mM KCl was replaced by potassium acetate. For 9mM pipette chloride, all of the KCl was replaced by potassium acetate, and 6mM $MgCl_2$ was replaced by $Mg(acetate)_2$. The junction potentials at the pipette tips were measured (Fenwick et al., 1982), and added to the apparent membrane potential.

DRUG APPLICATION

L-glutamate was applied by iontophoresis or by bath perfusion. The glutamate analogues kainate, quisqualate and N-methyl-D-aspartate were applied by bath perfusion, dissolved in normal Ringer's solution. For iontophoresis of glutamate, micro-electrodes filled with 1M sodium L-glutamate (pH 8.0, resistance 30MΩ when measured in Ringer's) were used. A negative current (-40nA) was used to eject glutamate, and a positive backing current (+30nA) was used to minimize loss of the drug from the electrode when no ejection current was applied. Results obtained with bath perfusion of glutamate were similar to those obtained with iontophoresis.

RESULTS

Glutamate produces a current in isolated cone photoreceptors

Fig. 1 shows the currents produced by iontophoretically applied glutamate in a cone voltage-clamped at the potentials shown alongside each trace. With 101mM chloride in the patch pipette, glutamate produces an inward current at potentials below 0mV, which becomes larger at more negative potentials. The glutamate-induced current reverses around 0mV, and becomes outward at more positive potentials. The current-voltage relation shows inward rectification around the reversal potential.

The glutamate receptors are kainate-type receptors

Various analogues of glutamate were applied to isolated cones by bath perfusion. The currents induced in an isolated double cone (voltage-clamped to -40mV) by 30μM glutamate, kainate, quisqualate and N-methyl-D-aspartate (NMDA) are shown in Fig. 2. Kainate produced a current which was larger than that produced by glutamate.

236

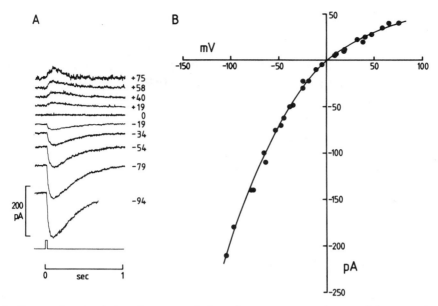

Fig. 1: Dependence of the glutamate-induced current on voltage with a patch pipette
chloride concentration of 101mM. The extracellular solution was Ringer's containing 6mM barium chloride. A. Currents induced by iontophoretic application of
glutamate to an isolated cone, held in voltage-clamp at different membrane potentials shown alongside each trace. Timing of iontophoresis is shown by the square
trace below the data. B. Current-voltage relation for the currents shown in A.
Peak glutamate-induced currents are plotted as a function of membrane potential.

Quisqualate and NMDA had no appreciable effect, even at $100\mu M$. The currents evoked
at -40mV by $30\mu M$ glutamate, kainate, quisqualate and NMDA had relative magnitudes
0.45:1:0:0 (mean data from 2 cells). This is a similar finding to that of Kaneko and
Tachibana (1987), who report an action of glutamate on turtle cones with some properties
similar to those described here.

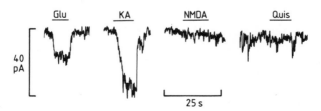

Fig. 2: Currents induced by $30\mu M$ concentrations of glutamate (Glut) and its analogues,
kainate (KA), N-methyl-D-aspartate (NMDA) and quisqualate (Quis), in Ringer's
containing 6mM barium chloride, in an isolated cone held in voltage-clamp at -
50mV. Timing of drug application is shown by the bars.

Noise changes associated with the glutamate-induced current

The change in current produced by glutamate or kainate was associated with an increase in membrane current fluctuations (noise), consistent with at least some of the glutamate-induced current resulting from the opening of ion channels. The increase in current noise produced when $10\mu M$ kainate was applied to an isolated cone voltage-clamped at -56mV is shown in Fig. 3A. The current-voltage relation for the kainate-induced current (with 101mM chloride in the patch pipette: Fig. 3B), is similar in shape to that obtained with glutamate in Fig. 1.

The glutamate-induced current is carried largely by chloride ions

Most glutamate-gated ion channels are relatively non-specific cation channels permeant to both sodium and potassium ions (Mayer and Westbrook, 1985).

We investigated the possibility that sodium passes through the glutamate-gated channel in cones. Lowering the external sodium concentration reduced the current produced by glutamate. However there was no obvious change in reversal potential of the glutamate response, which remained at 0mV (with 101mM chloride in the patch pipette), implying that sodium does not pass through the channel (Sarantis et al., 1988).

A possible explanation for the sodium-dependence of the glutamate response is that removal of external sodium ions may inhibit sodium-calcium exchange in the membrane, resulting in accumulation of calcium inside the cell. This increase in intracellular calcium may close the glutamate-gated channel, for example via a calcium-dependent phosphorylation mechanism.

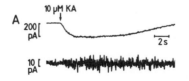

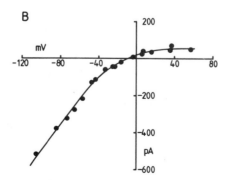

Fig. 3: A. Current induced by $10\mu M$ kainate in an isolated cone held at -56mV. Lower trace shows the noise increase during kainate application (2Hz high pass filtered; 1000Hz low pass filtered). B. Current-voltage relation for $10\mu M$ kainate-induced currents of the cone in A. External solution was Ringer's containing 6mM barium chloride. Patch pipette contained 101mM chloride.

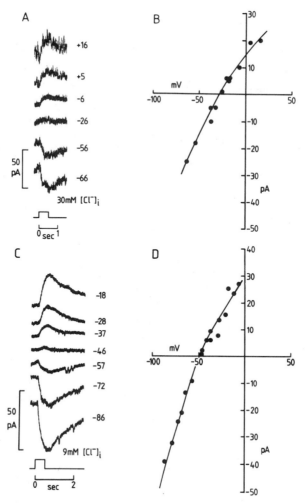

Fig. 4: Dependence of the glutamate-induced current on internal chloride concentration. A and C. Currents induced by iontophoretically applied glutamate in two isolated cones at the potentials shown beside each trace. Patch pipette chloride concentration was 30mM in A, 9mM in C. The external solution was barium Ringer's. Bottom trace shows timing of iontophoresis. B and D. Peak glutamate-induced currents as a function of membrane potential for the experiments in A and C respectively.

We found however, that removal of external sodium still abolished the glutamate response in the absence of external calcium (i.e. with 0.5mM EGTA and 0.1mM added calcium, giving a free calcium concentration of 3×10^{-8} M: conditions which should avoid an increase in internal calcium when external sodium is removed). Thus, a rise in intracellular calcium in the presence of low external sodium cannot be the cause of the reduction in the glutamate-evoked current. Therefore, it may be that sodium is required externally in some way, for glutamate to bind to receptors in the membrane or for the channel to open.

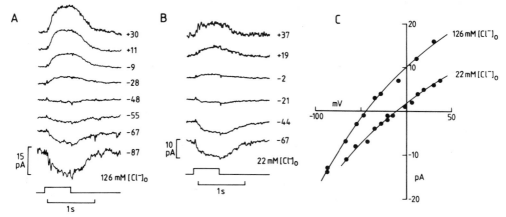

Fig. 5: Dependence of the glutamate-induced current on external chloride concentration.
A and B. Currents induced by glutamate (applied by iontophoresis) in the same
isolated cone, voltage-clamped to potentials shown. Bottom trace shows timing of
iontophoresis. External chloride concentration was 126mM (barium Ringer's) in
A, 22mM in B. Patch pipette chloride concentration was 9mM. C. Current-voltage
relation showing peak glutamate-induced currents as a function of membrane potential for each value of external chloride.

We did not try changing the external potassium concentration, but it has been
reported that a similar glutamate-induced current in turtle photoreceptors is not affected
by external potassium (Tachibana and Kaneko, 1988).

An invertebrate glutamate-gated chloride conductance has been reported by Cull-
Candy (1976), so we investigated the effect of changing the patch pipette chloride concentration. The internal chloride concentration of undisturbed isolated cones (in the
turtle) has been estimated to be between 12mM and 24mM (Kaneko and Tachibana,
1986). In Fig. 1, glutamate produced a current with a reversal potential of 0mV with
101mM chloride in the patch pipette. Lowering the patch pipette chloride concentration to 30mM or 9mM (replaced by acetate, Fig. 4), made the reversal potential of
the glutamate-induced current more negative. Similarly, lowering the external chloride
concentration (4 eells, chloride replaced by gluconate) made the reversal potential more
positive (Fig. 5). We conclude that chloride ions pass through the glutamate-gated ion
channel.

The reversal potentials for different internal chloride concentrations are shown in
Fig. 6. The observed reversal potentials do not fit the Nernst prediction, $E_{Cl} = (RT/F)$
ln $\{[Cl]_0/[Cl]_i\}$, with $[Cl]_0 = 126mM$ (barium Ringer's), for a chloride-specific channel
(dotted line), so the glutamate-gated channel is not chloride-specific.For each value of
internal chloride, the reversal potential is positive to E_{Cl}. Similarly for the data in
Fig. 5, the Nernst potential for chloride is -67mV for an external chloride concentration
of 126mM and -23mV for an external chloride of 22mM, i.e. more negative than the
observed reversal potentials of -44mV and -10mV respectively. There are two ways in
which this situation may be achieved:

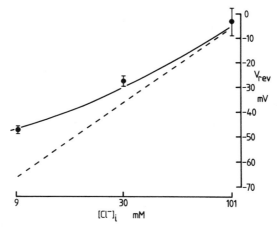

Fig. 6: Reversal potential of the current induced by glutamate as a function of chloride concentration in the patch pipette. The dotted line shows the Nernst prediction for a chloride-specific channel. Points show mean and standard deviation of the reversal potential in 4, 6 and 10 cells for 9, 30 and 101mM internal chloride respectively. The smooth curve is the Goldman-Hodgkin-Katz equation for a channel with a permeability ratio $P_{acetate}/P_{Cl} = 0.1$.

1. Positive ions, for example sodium ions, flowing into the cell through the glutamate-gated channels, would shift the reversal potential of the response positive to E_{Cl}. However we have already shown that sodium does not pass through the glutamate-gated channel in cones.

2. Another negative ion, for example acetate (the other main anion in our internal media), may flow out through the glutamate-gated channels.

On the basis of the most plausible suggestion above (2), assuming that acetate passes through the channel, the data in Fig. 6 were fitted with the Goldman-Hodgkin-Katz equation (solid curve) for a channel with a permeability ratio $P_{acetate}/P_{Cl} = 0.1$ (although we have yet to test whether acetate is the other current-carrying ion). The same permeability ratio also gave a rough fit to the dependence of the reversal potential on external chloride (assuming gluconate to be impermeant).

Dependence of the glutamate-induced currents on external glutamate concentration

Fig. 7A shows the dependence of the glutamate-induced current on various concentrations of glutamate applied sequentially by bath perfusion to an isolated cone, held in voltage-clamp at -41mV. A significant inward current (-28pA) was induced by 1μM glutamate, and the currents became larger (up to -70pA) as the external glutamate concentration was increased to 100μM.

The normalized dose-response curve obtained from the data in Fig. 7A is shown in Fig. 7B, in addition to that for similar experiments on the same cell at -62mV and +28mV. The maximum currents evoked at these potentials by 100μM glutamate were -93 and +24pA respectively. The shape of the dose-response curve is almost independent of voltage. The linear dependence of glutamate-induced current on glutamate concen-

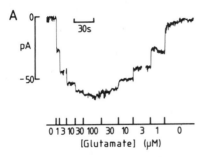 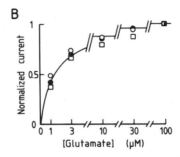

Fig. 7: Dose response curve for the glutamate-induced currents. A. Currents induced by different concentrations of locally perfused glutamate in barium Ringer's, in a single cone at -42mV. B. Normalized dose-response curve from the data in A (●), and from similar experiments on the same cell at -62mV (○) and at +28mV (□). The curve through the points is a Michaelis-Menten relation with a K_m of $1.4\mu M$, obtained from a Lineweaver-Burke plot of the data for -42mV.

tration at low doses of glutamate suggests that activation of the current depends on the binding of only one glutamate ion. The line drawn through the points is a curve describing first order Michaelis-Menten kinetics, having the form

$$I/I_{max} = [\text{glutamate}]/\{[\text{glutamate}] + K_m\}$$

for which the apparent K_m is $1.4\mu M$. This low value of K_m indicates that glutamate binds with high affinity to its receptor. In two other cones, Lineweaver-Burke plots gave K_m values of 2.6 and $5.5\mu M$ at -40mV.

The channels opened by glutamate are restricted to the synaptic terminal of the cone

We investigated the possibility that the receptors for glutamate were spatially localized in the cone cell membrane. Glutamate was iontophoresed at various positions around the cone. The glutamate-induced current was largest and fastest in onset (10-50msec), when glutamate was applied at the synaptic terminal. The current became smaller and slower in onset as the iontophoretic electrode was moved to other parts of the cell (e.g. near the cell body, near the outer segment). Presumably this is because of the time taken for glutamate to diffuse from these more distant sites to the synaptic terminal. These data suggest that the response is localized to the cone synaptic terminal (Sarantis et al., 1988, Fig. 3). Iontophoresis of glutamate at the end of the outer segment (approx. $50\mu m$ from the terminal) produced a response which eventually (after 1.2 sec) reached two-thirds of the glutamate response produced by iontophoresis near the synaptic terminal. This relatively large (if slow) response presumably reflects the low K_m for glutamate's action - only a little glutamate needs to diffuse from the iontophoretic electrode to the terminal to activate most of the current.

DISCUSSION

Glutamate alters the membrane current of isolated cone photoreceptors, by acting on receptors located at the cone synaptic terminal. The fact that the glutamate-evoked current shows a clear reversal potential (Figs. 1, 4 and 5), and is associated with an increase in current noise, implies that at least part of the glutamate-evoked current is produced by glutamate opening ion channels. In vertebrates, glutamate usually gates relatively non-specific cation channels which allow both sodium and potassium ions to pass (Mayer and Westbrook, 1985), with a dependence of current on voltage which is ohmic or outwardly rectifying. The cone glutamate-induced current is carried largely by chloride ions (see below), and rectifies inwardly in a manner similar to that of the glutamate-activated transport mechanism in retinal Müller cells (Brew and Attwell, 1987). Glutamate uptake into cones would serve to terminate the transmitter action of glutamate at the cone to bipolar and horizontal cell synapses, and would be a useful mechanism for returning the released glutamate to the photoreceptors for re-use. However, we can rule out a major contribution from glutamate uptake to the glutamate-evoked current in cones, because changing the external or internal chloride concentration has no effect on the glutamate-evoked current in Müller cells (Barbour et al. 1988), but has a large effect on the magnitude and reversal potential of the glutamate-evoked current in cones.

Changing the internal and external chloride concentrations revealed that chloride was the principle ion carrying the current induced by glutamate. However, the reversal potential for the response did not indicate a chloride-specific conductance. The reversal potentials for different values of internal and external chloride could be fitted by the Goldman-Hodgkin-Katz equation for a channel with a permeability ratio $P_{acetate}/P_{Cl} = 0.1$. With a physiological internal chloride concentration of between 12 and 24mM (in turtle: Kaneko and Tachibana, 1986), the glutamate-evoked current would reverse in the range -34 to -43mV, close to the cone dark potential of -40mV (Attwell et al., 1982).

The glutamate-induced current in cones was unaffected by bicuculline (100μM) or strychnine (10μM), at doses which reduced or abolished chloride currents evoked in isolated retinal bipolar or ganglion cells by iontophoresis of gamma-aminobutyric acid (GABA) or glycine. Thus the glutamate-gated anion conductance in cones is quite different from the GABA-gated chloride-specific conductance observed by Kaneko and Tachibana (1986) in turtle cones, and from glycine-gated chloride conductances seen in other cell types.

The dose-response data for glutamate show that glutamate binds with a high affinity ($K_m = 1$-6μM) to receptors at the synaptic terminal of the cone. By applying glutamate analogues (Fig. 2), we found that the glutamate receptors are of the kainate type. Second order retinal neurones, bipolar and horizontal cells, also have kainate-type glutamate receptors (Shiells et al., 1981; Slaughter and Miller, 1983). Our finding in cones is consistent with that of Kaneko and Tachibana (1987) who show that, in the turtle, both rods and cones have kainate-type glutamate receptors. However, our results differ markedly from theirs in the nature of the ions carrying the glutamate-evoked current: in salamander we have demonstrated that much of the current reflects chloride movement, while the data of Kaneko and Tachibana (1988) suggest that the glutamate-gated channel in turtle cones is a cation channel. We have so far been unable to demonstrate a glutamate-gated conductance in salamander rods.

The localization of the glutamate receptors in the cone synaptic terminal suggests that glutamate released from a cone may have a feedback action on its own terminal. The glutamate receptors on cones are presynaptic (with respect to transmission of information from photoreceptors to second order neurones). Presynaptic glutamate receptors have previously been suggested to exist in invertebrates (Thieffrey and Bruner, 1978; Thieffrey et al., 1979), and in vertebrate hippocampus (Errington et al., 1987) and olfactory cortex (Collins et al., 1982). In general, "autoreceptors" mediate a negative feedback action (with the exception of the receptors at the terminals of noradrenergic neurones) on the release of neurotransmitter (Starke, 1981). We suggest that the presynaptic glutamate receptors that we have characterized could initiate a positive feedback mechanism, serving to increase the gain of cone phototransduction, as follows.

In the intact retina, maximum glutamate release from retinal photoreceptors occurs in the dark. When a cone is brightly illuminated and polarized to (say) -65mV, so there is no glutamate release, a slight dimming of the light will depolarize the cone and initiate release of glutamate. The reversal potential for the glutamate-induced current in cones, -34 to -43mV for an estimated physiological internal chloride of 12-24mM (in turtle: Kaneko and Tachibana, 1986), is near the cone dark potential (assuming the intracellular acetate (or other permeant anion) concentration is the same *in vivo* as is present inside our patch pipettes). Thus, glutamate released will produce an inward current in the synaptic terminal of the cone it is released from, depolarizing it further and causing further release of glutamate. Consequently, for a certain change in light intensity, there will be a greater change in cone voltage, and in the amount of glutamate released, than would occur in the absence of the glutamate autoreceptors (Sarantis et al., 1988). There will therefore be an increase in the gain for the conversion of changes in light intensity into changes of glutamate release. The quantitative significance of this proposed feedback loop remains to be determined.

Supported by the Wellcome Trust, M.R.C., Royal Society and Wolfson Foundation.

REFERENCES

Attwell, D., Mobbs, P., Tessier-Lavigne, M., and Wilson, M., 1987, Neurotransmitter-induced currents in retinal bipolar cells of the axolotl, *Ambystoma mexicanum*, *J.Physiol.*, 387:125.

Attwell, D., Werblin, F. S., and Wilson, M., 1982, The properties of single cones isolated from the tiger salamander retina, *J. Physiol.*, 328:259.

Bader, C. R., Macleish, P. R., and Schwartz, E. A., 1979, A voltage-clamp study of the light response in solitary rods of the turtle, *J. Physiol.*, 296:1.

Barbour, B., Brew, H., and Attwell, D., 1988, Electrogenic glutamate uptake in glial cells is activated by intracellular potassium, *Nature*, 335:433.

Brew, H., and Attwell, D., 1987, Electrogenic glutamate uptake is a major current carrier in the membrane of axolotl retinal glial cells, *Nature*, 327:707.

Cervetto, L., and Macnichol, E. F., 1972, Inactivation of horizontal cells in turtle retina by glutamate and aspartate, *Science*, 178:767.

Collins, G. G. S., Anson, J., and Surtees, L., 1982, Presynaptic kainate and N-methyl-D-aspartate receptors regulate excitatory amino acid release in the olfactory cortex, *Brain Res.*, 265:157.

Cull-Candy, S. G., 1976, Two types of extrajunctional L-glutamate receptors in locust muscle fibres, *J. Physiol.*, 255:449.

Errington, M. L., Lynch, M. A., and Bliss, T. V. P., 1987, Long-term potentiation in the dentate gyrus: Induction and increased glutamate release are blocked by D(-)aminophosphonovalerate, *Neuroscience*, 20:279.

Fenwick, E. M., Marty, A., and Neher, E., 1982, A patch-clamp study of bovine chromaffin cells and of their sensitivity to acetylcholine, *J. Physiol.*, 331:577.

Ishida, A. T., Kaneko, A., and Tachibana, M., 1984, Responses of solitary retinal horizontal cells from *Carassius auratus* to L-glutamate and related amino acids, *J. Physiol.*, 348:255.

Kaneko, A., and Tachibana, M., 1986, Effects of gamma-aminobutyric acid on isolated cone photoreceptors of the turtle retina, *J. Physiol.*, 373:443.

Kaneko, A., and Tachibana, M., 1987, Effects of L-glutamate on isolated turtle photoreceptors, ARVO Abstracts, *Invest. Opthalmol. Vis. Sci.*(Suppl)., 28:50.

Lasansky, A., 1973, Organization of the outer synaptic layer in the retina of the larval tiger salamander, *Phil. Trans. Roy. Soc. Lond.*, B, 265:471.

Mayer, M., and Westbrook, G. L., 1985, The action of N-methyl-D-aspartic acid on mouse spinal neurones in culture, *J. Physiol.*, 361:65.

Miller, A. M., and Schwartz, E. A., 1983, Evidence for the identification of synaptic transmitters released by photoreceptors of the toad retina, *J.Physiol.*, 334:325.

Miller, R. F., and Slaughter, M. M., 1986, Excitatory amino acid receptors of the retina: diversity of subtypes and conductance mechanisms, *Trends Neurosci.*, 9:211.

Murakami, M., Ohtsu, K., and Ohtsuka, T., 1972, Effects of chemicals on receptors and horizontal cells in the retina, *J.Physiol.*, 227:899.

Nawy, S., and Copenhagen, D., 1987, Multiple classes of glutamate receptors on depolarizing bipolar cells in retina, *Nature*, 325:56.

Newman, E., 1985, Voltage-dependent calcium and potassium channels in retinal glial cells, *Nature*, 317:809.

Sarantis, M., Everett, K., and Attwell, D., 1988, A presynaptic action of glutamate at the cone output synapse, *Nature*, 332:451.

Shiells, R. A., Falk, G., and Naghshineh, S., 1981, Action of glutamate and aspartate analogues on rod horizontal and bipolar cells, *Nature*, 294:592.

Slaughter, M. M., and Miller, R. F., 1983, The role of excitatory amino acid transmitters in the mudpuppy retina: An analysis with kainic acid and N-methyl aspartate, *J. Neurosci.*, 3:1701.

Starke, K., 1981, Presynaptic receptors, *Ann. Rev. Pharmacol. Toxicol.*, 21:7.

Tachibana, M., and Kaneko, A., 1988, L-glutamate-induced depolarization in solitary photoreceptors: A process that may contribute to the interaction between photoreceptors *in situ*, *Proc. Natl. Acad. Sci. USA*, 85:5315.

Thieffrey, M., and Bruner, J., 1978, Direct evidence for a presynaptic action of glutamate at a crayfish neuromuscular junction, *Brain Res.*, 156:402.

Thieffrey, M., Bruner, J., and Personne, P., 1979, The presynaptic action of L-glutamate at the crayfish neuromuscular junction, *J. Physiol.*, Paris, 75:635.

White, R. D., and Neal, M. J., 1976, The uptake of L-glutamate by the retina, *Brain Res.*, 111:79.

Wu, S. M., 1986, Effects of gamma-aminobutyric acid on cones and bipolar cells of the tiger salamander retina, *Brain Res.*, 365:70.

PROPERTIES OF ELECTROGENIC GLUTAMATE UPTAKE

IN RETINAL GLIA

B. Barbour, H. Brew and D. Attwell

Department of Physiology, University College London, England

INTRODUCTION

In the dark photoreceptors have a membrane potential of about -40mV. At this potential they continuously release neurotransmitter molecules onto bipolar and horizontal cells. When a photoreceptor absorbs light the membrane hyperpolarises (to a maximum of about -70mV) and the release of neurotransmitter is suppressed. The transmitter released by photoreceptors is thought to be glutamate (Marc and Lam, 1981; Brandon and Lam, 1983; Miller and Schwartz, 1983; Ayoub et al., 1988). The light-induced suppression of glutamate release from photoreceptors results in a voltage change in postsynaptic bipolar and horizontal cells. The speed of onset of the postsynaptic response is determined (at least partly) by how quickly the extracellular concentration of glutamate falls when release from photoreceptors is suppressed.

After release of acetylcholine at a cholinergic synapse (e.g. the neuromuscular junction) the neurotransmitter is hydrolysed extracellularly by acetylcholinesterase. A similar mechanism cannot terminate the synaptic action of glutamate as there are no extracellular enzymes which catalyse the breakdown of glutamate. Instead, glutamate is inactivated by uptake into neighbouring cells and by diffusion away from the synapse. Glutamate uptake may occur into neurones and/or glia. Glutamate uptake has been demonstrated into photoreceptors (Marc and Lam, 1981; Brandon and Lam, 1981), and into Müller cells, the main type of retinal glial cell (White and Neal, 1976). Müller cells may extend fine processes sufficiently close to glutamatergic synapses to contribute directly to removal of glutamate from the synaptic cleft. Even if this turns out not to be the case, the avid glutamate uptake into glia is certainly responsible for maintaining low extracellular concentrations of glutamate, facilitating diffusion away from the synapse and preventing the extracellular glutamate concentration from reaching neurotoxic levels (Rothman and Olney, 1987). Glutamate uptake into glia and photoreceptors is thus important in determining the properties of synaptic transmission at the photoreceptor output synapse.

An understanding of glutamate uptake may be important, not only for understanding how glutamate is removed from the synaptic cleft, but also for understanding how glutamate is released from photoreceptors in the first place. Miller and Schwartz (1983) and Schwartz (1986) have suggested that a significant fraction of glutamate release may be due to a Ca^{2+}-independent, presumably non-vesicular, release mechanism. One possibility is that reversal of glutamate uptake mediates Ca^{2+}-independent glutamate release. Knowledge of the stoichiometry and voltage-dependence of glutamate uptake would enable an assessment of the physiological significance of such a mechanism.

Glutamate uptake has been studied extensively using radioactive tracer techniques in a number of preparations: brain slices, whole retinae, synaptosomal vesicles, renal brush border vesicles and a variety of cultured cell lines. This work has shown that glutamate uptake is accompanied by an influx of sodium and possibly by an efflux of potassium ions. A major problem with these methods is that they do not allow control of the membrane potential which, as we show below, is a major determinant of the rate of glutamate uptake.

We tried a different approach: whole-cell patch-clamping of isolated Müller cells enables simultaneous control of intracellular and extracellular environments under voltage-clamp conditions. Uptake of glutamate proved to be electrogenic and could thus be studied by monitoring the inward current evoked by glutamate under voltage-clamp conditions. We have shown that this glutamate-evoked current is due to glutamate uptake and not to a glutamate-activated channel. We summarise here the research to date from our laboratory concerning the properties of electrogenic glutamate uptake in Müller cells (Brew and Attwell, 1987; Barbour et al., 1988).

METHODS

The methods are described in detail in Brew and Attwell (1987) and Barbour et al. (1988). Briefly, enzymatically-isolated Müller cells were whole-cell patch-clamped with pipettes containing a pseudo- internal solution and bathed in a modified Ringer's solution (important modifications of these solutions are stated in the text, e.g. replacement of external sodium with choline). In most experiments the external bathing solution contained 6mM barium to block the cell's potassium conductance. In the presence of barium the resting potential of a Müller cell was about -40mV. Unless otherwise stated, the experiments were performed at a holding potential of -40mV, 6mM $[Ba^{2+}]_0$ was present in the bathing solution and glutamate was bath applied at a concentration of $30\,\mu M$.

RESULTS

The application of glutamate to whole-cell clamped Müller cells evoked an inward current (fig. 1A, glutamate applied iontophoretically). This inward current was large at negative holding potentials and became progressively smaller as the holding potential was made more positive. Above approximately +50mV no glutamate-evoked current could be detected. Fig. 1B displays this graphically as a current-voltage plot. The data used for fig. 1B come from the cell in fig. 1A (filled circles) which was bathed with barium-containing external solution and from another cell (open circles) which was bathed in an external solution containing no barium. Addition of barium to the bathing medium had no effect on the magnitude or voltage-dependence of the glutamate-evoked current. The magnitude of the glutamate-evoked current depends on the concentration of glutamate applied (fig. 1C). Fig. 1D shows the mean magnitude of the glutamate-evoked current, from experiments such as in fig. 1C, plotted as a function of the concentration of applied glutamate. The data conform approximately to Michaelis-Menten kinetics: the smooth

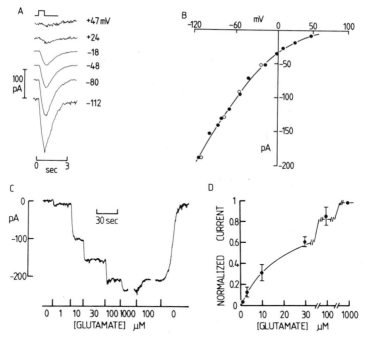

Fig. 1: Inward current evoked by glutamate in voltage-clamped Müller cells. A. Effect of varying holding potential on the inward current (plotted downwards) evoked by the iontophoretic application of glutamate (duration indicated by square pulse at the top). B. Data from two different cells (see text) plotted as a current-voltage relation. C. Currents evoked by different concentrations of bath-applied glutamate. D. Mean currents ($\pm$ S.D.) from experiments such as in C plotted against glutamate concentration. Data from at least 3 cells for each point.

curve is a Michaelis-Menten relation with an apparent K_m for glutamate of $22 \mu M$. Between holding potentials of -100mV and -20mV the dependence of the current on glutamate was independent of voltage. The application of glutamate was not accompanied by any detectable increase in noise of the membrane current between 0.5 and 500Hz, suggesting that the glutamate-evoked current is not caused by a glutamate-gated channel.

The current evoked by a number of compounds related to glutamate was compared to the current evoked by glutamate itself. Examples of raw data from such experiments are shown in fig. 2. None of the glutamate analogues used to classify glutamate-activated channels (kainate, N-methyl-D-aspartate and quisqualate; Watkins and Evans, 1981) evoked currents comparable to the response to glutamate. Of the stereo-isomers of glutamate and aspartate, D-glutamate was without effect, but both D-, and L-aspartate evoked responses. The relative magnitude of the responses to these compounds (all at $30 \mu M$, $V_H = -40mV$) was L-glu: L-asp: D-asp: NMDA: kainate: quisqualate: D-glu = 1: 0.43: 0.25: 0.06: 0.005: 0: 0 (mean data from three cells).

The actions of drugs known to block glutamate uptake in other preparations (Balcar and Johnston, 1972) were also investigated. Threo-3-hydroxy-DL-aspartate ($30 \mu M$) reduced the current evoked by $30 \mu M$ glutamate at -40mV by approximately 80%. Another blocker of glutamate uptake, p-chloro-mercuriphenylsulphonate ($100 \mu M$) reduced the current evoked by $30 \mu M$ glutamate at -40mV by approximately 30%.

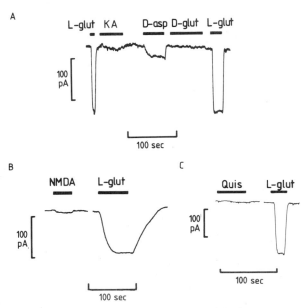

Fig. 2: Comparison of the glutamate-evoked current with the currents evoked by various
analogues of glutamate. A. L-glutamate, kainate, D-aspartate, D-glutamate and
L-glutamate. B. N-methyl-D-aspartate and L-glutamate. C. Quisqualate and
L-glutamate.

In accordance with the fact that glutamate uptake is accompanied by the co-
transport of sodium ions (Baetge et al., 1979), the current evoked by glutamate was
found to be sodium-dependent. Reduction of external sodium from 107mM to 2mM
(replaced by choline) abolished the response to glutamate. The mean current evoked by
30μM glutamate at different concentrations of external sodium is plotted against the

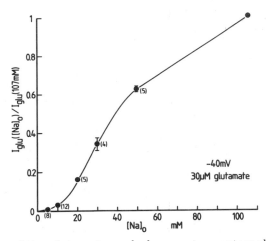

Fig. 3: Dependence of the glutamate-evoked current on external sodium concentration.
Graph of mean normalised currents ($\pm$ S.E.M.) plotted against sodium concentration. Small figure by each point indicates the number of cells tested at that
sodium concentration.

external concentration of sodium in fig. 3. The form of the relation is sigmoid, suggesting that more than one sodium ion is co-transported with each glutamate molecule. At low external sodium ($<$40mM) the glutamate-evoked current was approximately proportional to $[Na^+]_0$ $^{2.1}$ at a holding potential of -40mV.

In Fig. 4A the experimental traces show the effect of replacing internal potassium with choline. This was accomplished by whole-cell patch-clamping cells with different solutions in the patch pipette and allowing the filling solution to equilibrate with the cell interior. With lowered concentrations of potassium in the pipette solution the current evoked by glutamate was reduced. With zero potassium in the pipette solution the glutamate-evoked current was almost completely abolished. To compare quantitatively responses from different cells, the magnitude of the current from each cell was normalised to the capacitance of the cell. This procedure compensates for variation of current magnitude due to variation of cell surface area (and presumably, therefore, variation in the number of carrier molecules). The experiments were carried out in a bathing solution which contained no potassium. This was found to be necessary to prevent potassium leaking into the cell at the holding potential of -40mV and making the intracellular potassium concentration higher than that in the pipette. Fig. 4B is a graph of the mean currents from experiments like those in fig. 4A plotted against the potassium concentration in the pipette filling solution. The smooth curve is a Michaelis-Menten relation with a K_m of 15mM.

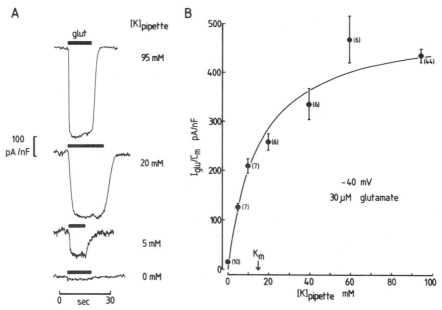

Fig. 4: The effect of removing internal potassium on the glutamate-evoked current. A. Specimen traces from 4 cells in which different proportions of internal potassium have been replaced with choline. B. Mean responses ($\pm$ S.E.M.), from experiments like in A, plotted against internal potassium concentration. The small figure by each point indicates the number of cells tested. The external solution contained no potassium for this experiment.

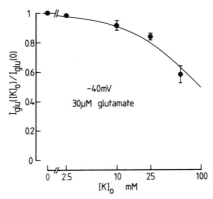

Fig. 5: The effect of raising external potassium on the glutamate-evoked current. The
graph shows the mean normalised current ($\pm$ S.E.M.) plotted against the external
concentration of potassium. Data from at least 5 cells for each point.

These data show that the glutamate uptake current is activated by internal potassium, the first order kinetics suggesting that binding of one potassium causes the effect.
If potassium is transported out of the cell by the uptake carrier then potassium ions
might also be expected to have an effect externally on the rate of uptake. This is demonstrated in fig. 5. Raising the concentration of external potassium (replacing choline,
$[Na^+]_0 = 52mM$) reduces the size of the glutamate-evoked current. The smooth curve is
of the form $K_m / \{[K^+]_0 + K_m\}$ with $K_m = 100mM$.

We have also tested the possible involvement in electrogenic glutamate uptake of
other ions in our solutions. None seem to influence the glutamate-evoked current. Varying the concentration of external calcium between about 100nM and 10mM had no ob-
servable effect on the magnitude or voltage dependence of the glutamate-evoked current.
Nor did complete removal of external magnesium or the barium used to block potassium
channels have any effect. Complete replacement of either internal chloride by the much
larger gluconate anion also had no effect on the magnitude or voltage-dependence of
the current evoked by glutamate. It thus seems unlikely that these ions are involved in
electrogenic glutamate uptake.

DISCUSSION

Channel vs uptake

The pharmacological profile and ion-dependence of the glutamate-evoked current in
voltage-clamped glia are unlike those previously reported for glutamate-activated channels (Watkins and Evans, 1981), but are consistent with the reported properties of glutamate uptake (e.g. Balcar and Johnston, 1972; Kanner and Sharon, 1978; Stallcup et
al., 1979). We can exclude the possibility that the current is due to a novel glutamate-
activated channel conducting Ca^{2+}, Mg^{2+}, Ba^{2+} or Cl^-, because dramatic variations of
the concentrations of these ions have no effect on the magnitude or voltage-dependence
of the glutamate- evoked current. The absence of a detectable increase of current noise
when glutamate is applied to Müller cells makes it even less likely that the glutamate-
evoked current represents a glutamate-gated channel. Furthermore, preliminary results

(Attwell et al., 1988) show that inclusion of glutamate in the pipette solution reduces the size of the glutamate-evoked current; a result difficult to explain except in terms of a carrier transporting glutamate.

Stoichiometry

In voltage-clamped Müller cells, the first order dependence of the glutamate-evoked current on glutamate concentration suggests that the uptake carrier transports one glutamate molecule per 'cycle'. This result agrees with numerous tracer studies (Hertz, 1979; Lerner, 1987) which indicate that high affinity glutamate uptake conforms to Michaelis-Menten kinetics with an apparent K_m for glutamate of the order of $10\mu M$ (we found values between 6 and $22\mu M$). Glutamate has three ionisable groups. At physiological pH over 99% of the glutamate is in the form with with all three groups charged, giving a net one negative charge. The widespread assumption is that this is the form which is transported.

Many studies agree that high affinity glutamate uptake is sodium-dependent and Stallcup et al. (1979) and Baetge et al. (1979) show that glutamate uptake is linked to sodium co-transport. High affinity glutamate uptake can accumulate glutamate against large glu_i/glu_o ratios (Erecinska et al., 1983; Lerner, 1987). Linking glutamate uptake to the electrochemical gradient for sodium provides the energy required to drive this process. Suggestions as to the ratio of glu:Na co-transported range from 1:1 to 1:3 (White and Neal, 1976; Schousboe et al., 1977; Baetge et al., 1979; Barbour et al., 1988). In the only studies to monitor uptake of labelled glutamate and sodium simultaneously (Baetge et al., 1979; Stallcup et al., 1979) a simple ratio of 1:2 was observed in cultured glia and in cultured neurones. Electrogenic uptake in glia displays a sigmoid dependence on external sodium suggesting that more than one sodium per glutamate is transported. At low sodium concentrations the relationship approximates to a dependence on $[Na^+]_0{}^2$.

A number of studies have suggested that glutamate uptake is also accompanied by the counter-transport of potassium (Kanner and Sharon, 1978; Kanner and Marva, 1982; Schneider and Sacktor, 1980; Burckhardt et al., 1980). The general finding is that a potassium gradient, with $[K^+]_i > [K^+]_o$, can drive glutamate accumulation into membrane vesicles in the absence of other ion gradients. However, as such membrane vesicles are substantially permeable to potassium, the potassium gradient will generate an inside negative potential which would stimulate electrogenic uptake. Kanner and Sharon (1978) and Burckhardt et al. (1980) showed that the effect of potassium is not due *solely* to changes of potential, but their results cannot distinguish between transport of potassium by the carrier and an allosteric modification of the carrier, following the binding of potassium, which catalyses glutamate uptake.

The voltage-clamped glial cell preparation enabled us to exclude all effects of potassium due to changes of potential. Internal potassium activates, and external potassium inhibits, electrogenic glutamate uptake. Although this is consistent with potassium counter-transport, it is also possible that the binding of potassium (inside and outside) allosterically alters the properties of the uptake carrier. We were, however, able to exclude some putative mechanisms for such allosteric effects: the reduction of uptake observed when $[K^+]_o$ was raised or $[K^+]_i$ was lowered could not be accounted for by changes of affinity for external glutamate or sodium (Barbour et al., 1988). Direct evidence of potassium counter-transport is still lacking, but it seems the simplest explanation of the data.

If potassium counter-transport is accepted, an obvious suggestion for the stoichiometry of glutamate uptake in glia is: co-transport of one glutamate anion with two sodium ions linked to the counter-transport of one potassium ion. However, this stoichiometry is electroneutral whereas uptake in glial cells is, at least in part, electrogenic (although whole-cell patch-clamping allows us to monitor electrogenic glutamate uptake as a membrane current, it gives no information concerning possible electroneutral uptake). It is therefore necessary to postulate the transport of an extra positive ion into the cell or a negatively charged ion out of the cell. We have eliminated most of the ions in our solutions from possible involvement in electrogenic uptake and (ignoring ATP, HEPES and EGTA) the two remaining possibilities for the extra transported ion are sodium or a proton. Thus two possible stoichiometries for electrogenic glutamate uptake into glia are: glu:2Na:H/K and glu:3Na/K. The study of Baetge et al. (1979) indicates that only two sodium ions per glutamate are transported into cultured glia and neurones. There is convincing evidence in kidney vesicles that glutamate uptake is accompanied by proton co-transport (Nelson et al., 1983) and a report that this also occurs in synaptosomes (Erecinska et al., 1983), but there is no conclusive evidence concerning glia.

Functional significance

The stoichiometries we suggest are both capable of accumulating glutamate against a very steep concentration gradient (see Barbour et al., 1988 for calculation). The magnitude of the glutamate-evoked current (up to 800pA with 30μM glutamate at the Müller cell resting potential of -90mV) shows that the rate of glutamate uptake into Müller cells would be fast enough to clear glutamate rapidly from the surrounding fluid. Based on the size of this current, an estimate of how quickly a Müller cell could clear the extracellular space around it of glutamate suggests that this could occur on a timescale comparable to that on which retinal synapses operate (for calculation see Brew and Attwell, 1987).

Radioactively labelled glutamate applied to the whole retina is accumulated preferentially into Müller cells (White and Neal, 1976), suggesting that glial uptake is responsible for maintaining low extracellular concentrations of glutamate. This result, however, does not prove that endogenous glutamate released at the synapse is taken up by glial cells rather than by neurones. As mentioned in the introduction, glial cell processes containing the uptake carrier would have to extend very close to the synapse in order to contribute directly to the clearing of glutamate from the synaptic cleft. Anatomical evidence on this point is not available.

The ability of glutamate uptake reversal to mediate Ca^{2+}-independent glutamate release depends critically on the composition of the solution at the synapse, about which very little is known. However, the proposed stoichiometries render reversal of glutamate uptake unlikely as a mechanism for non-vesicular transmitter release. The reversal potential for the carrier is given by

$$E_{rev} = 2E_{Na} - E_K - E_{glu} + (E_{Na} \ or \ E_H) \qquad (1)$$

where E_x is the Nernst potential for ion x, and the final term is E_{Na} if $3Na^+$ are transported and E_H if $2Na^+$ and H^+ are carried. An estimate for E_{glu} is +200mV: Hamberger et al. (1983) report that extracellular free glutamate is 3μM, while the total glutamate content of brain is of the order of 10 μmole/g (Johnson, 1978) so the free intracellular glutamate concentration is probably 10mM or less. Substituting this value into eqn. (1) with E_{Na} = +50mV, E_K = -95mV and E_H = -17mV, gives E_{rev}

= -22mV and +45mV for 2Na:H and 3Na stoichiometries respectively. The value for E_{glu} is obviously a very rough estimate so any conclusions must be tentative. If the reversal potential for glutamate uptake is +45mV then reversal of the carrier is unlikely to contribute to glutamate efflux under any physiological conditions. With a reversal potential of -20mV the carrier would reverse during action potentials in neurones which spike, but not during the graded changes of potential which occur in photoreceptors.

Supported by the Wellcome Trust, MRC, SERC, Royal Society and Wolfson Foundation.

REFERENCES

Attwell, D., Barbour, B. and Brew, H., 1988, The effect of internal ions on electrogenic glutamate uptake in glial cells isolated from the tiger salamander retina, *J. Physiol. Lond.*, 401:92P.

Ayoub, G. S., Korenbrot, J. I. and Copenhagen D. R., 1988, Glutamate is released from individual photoreceptors, *Investig. Ophthalmol. Suppl.*, 29:273.

Baetge, E. E., Bulloch, K. and Stallcup, W. B., 1979, A comparison of glutamate transport in cloned cell lines from the central nervous system, *Brain. Res.*, 167:210.

Balcar, V. J. and Johnston, G. A. R., 1972, The structural specificity of the high affinity uptake of L-glutamate and L-aspartate by rat brain slices, *J. Neurochem.*, 19:2657.

Barbour, B., Brew, H. and Attwell, D., 1988, Electrogenic glutamate uptake in glia is activated by intracellular potassium, *Nature*, 335:433.

Brandon, C. and Lam, D. M. K., 1983, L-glutamic acid: a neurotransmitter candidate for cone photoreceptors in human and rat retinas, *Proc. Natl. Acad. Sci. USA*, 80:5117.

Brew, H. and Attwell, D., 1987, Electrogenic glutamate uptake is a major current carrier in the membrane of axolotl retinal glial cells, *Nature*, 327:707.

Burckhardt, G., Kinne, R., Stange, G. and Murer, H., 1980, The effects of potassium and membrane potential on sodium-dependent glutamic acid uptake, *Biochim. Biophys. Acta.*, 599:191.

Erecinska, M., Wantorsky, D. and Wilson, D. F., 1983, Aspartate transport in synaptosomes from rat brain, *J. Biol. Chem.*, 258:9069.

Hamberger, A., Berthold, C., Karlsson, B., Lehmann, A. and Nystrom, B., 1983, Extracellular GABA, Glutamate and Glutamine *in vivo-* perfusion dialysis of the rabbit hippocampus, *in*: "Glutamine, Glutamate and GABA in the Central Nervous System", Hertz, L., Kvamme, E., McGeer, E. G. and Schousboe, A., ed., Alan R. Liss, NY.

Hertz, L., 1979, Functional interactions between neurones and astrocytes. 1. Turnover and metabolism of putative amino acid transmitters, *Prog. Neurobiol.*, 13:277.

Johnson, J. L., 1978, The excitant amino acids glutamic and aspartic acid as neurotransmitter candidates in the vertebrate central nervous system, *Prog. Neurobiol.*, 10:155.

Kanner, B. I. and Marva, E., 1982, Efflux of L-glutamate by synaptic plasma membrane vesicles isolated from rat brain, *Biochemistry*, 21:3143.

Kanner, B. I. and Sharon, I., 1978, Active transport of L-glutamate by membrane vesicles isolated from rat brain, *Biochemistry*, 17:3949.

Lerner, J., 1987, Acidic amino acid transport in animal cells and tissues, *Comp. Biochem. Physiol.*, 87B:443.

Marc, R. and Lam, D. M. K., 1981, Uptake of aspartic and glutamic acid by photoreceptors in goldfish retina, *Proc. Natl. Acad. Sci.*, 78:7185.

Miller, A. M. and Schwartz, E. A., 1983, Evidence for the identification of synaptic transmitters released by photoreceptors of the toad retina, *J. Physiol. Lond.*, 334:325.

Nelson, P. J., Dean, G. E., Aronson, P. S. and Rudnick, G., 1983, Hydrogen ion cotransport by the renal brush border glutamate transporter, *Biochemistry*, 22:5459.

Rothman, S. M. and Olney, J. W., 1987, Excitotoxicity and the NMDA receptor, *Trends Neurosci.*, 10:299.

Schneider, E. G. and Sacktor, B., 1980, Sodium gradient-dependent L-glutamate transport in renal brush border membrane vesicles, *J. Biol. Chem.*, 255:7645.

Schousboe, A., Svenneby, G. and Hertz, L., 1977, Uptake and metabolism of glutamate in astrocytes cultured from dissociated mouse brain hemispheres, *J. Neurochem.*, 29:999.

Schwartz, E. A., 1986, synaptic transmission in amphibian retinae during conditions unfavourable for calcium entry into presynaptic terminals, *J. Physiol. Lond.*, 376:411.

Stallcup, W. B., Bulloch, K. and Baetge, E. E., 1979, Coupled transport of glutamate and sodium in a cerebellar nerve cell line, *J. Neurochem.*, 32:57.

Watkins, J. C. and Evans, R. H., 1981, Excitatory amino acid transmitters, *Ann. Rev. Phamacol. Toxicol.*, 21:165.

White, R. D. and Neal, M. J., 1976, The uptake of L-glutamate by the retina, *Brain.*, 111:79.

THE PHOTORESPONSES OF THE TROUT PINEAL CELLS

C. Kusmic, P.L. Marchiafava and E. Strettoi*

Dipartimento di Fisiologia e Biochimica, Universita' di Pisa, Italy
*Istituto di Neurofisiologia del CNR, Pisa, Italy

The pineal body of the trout is a strongly vascularized rounded structure (about 1 mm diameter), lying upon the dorso- lateral portion of the diencephalon and attached to it by means of a peduncle about 2 mm long. The pineal body is essentially a photoneuroendocrine organ since several lines of evidences indicate its light regulated influence on locomotion, gonadotropic activity as well as on the production of melatonin, an hormon involved in body pigmentation (Oksche, 1986).

The pineal organ shares a common embriological origin with the retina, both evaginating from the diencephalon. During development the retina acquires structural and functional properties that make it an ideal organ for form and contrast discrimination. The pineal organ, on the other hand, being covered by a multilayered opaque tissue, can only performe luminance detection.

The light dependent activity of the pineal organ in lower vertebrates depends upon photoreceptors which are located within the organ itself (Holmgren, 1959; Hartwig and Oksche, 1982). The photoresponse is transmitted to the brain through second order pineal ganglion cells (Dodt and Meissl, 1982), which in higher vertebrates, however receive their inputs from photoreceptors located in the retina. Thus, the pineal body of fish subserve two important functions: 1. phototransduction, and 2. activation of photodependent neuroendocrine cells either within the organ itself or at more distant brain sites.

Although the electrical activity of the pineal organ has attracted interest by almost 30 years (Heerd and Dodt, 1961), the great majority of recordings are limited to the extracellular activity of the cell axons running within the epiphyseal peduncle (see Meissl, 1986). From these works it appears that pineal spiking cells produce on or off-type responses and show a maximal sensitivity at about 380 nm or 525 nm respectively. More recently, intracellular recordings from pineal photoreceptors have been obtained showing hyperpolarizing photoresponses (Hanyu, Niwa and Tamura, 1969; Meissl and Ekstrom , 1988; Kusmic, Marchiafava and Strettoi, 1988) , which under several aspects resembled those recorded from the lateral eye of other teleosts (Marchiafava, 1985).

There is no available information so far about the neural pineal circuitry and the structural organization underlying the signal transmission between pineal photoreceptors and ganglion cells. The aim of the present research was to analyze these aspects of

Sensory Transduction
Edited by A. Borsellino *et al.*
Plenum Press, New York, 1990

the epiphysis by recording intracellularly photoresponses from pineal cells which were subsequently identified by their structural characteristics. This was made possible by systematically injecting fluorescent or electron- opaque substances into the recorded cells.

The preparation consisted of the whole pineal organ of the trout, Salmo irideus, isolated after decapitation under dim light illumination. The organ was perfused with oxigenated teleostean saline solution containing (in mM) NaCl 110, KCl 2.6, MgCl$_2$ 2, CaCl$_2$ 2, Hepes 14, glucose 5 at pH 7.4. The preparation lied upon the stage of an inverted microscope equipped with an infrared TV monitor to direct and follow the advancement of the microelectrode tip to any desired portion of the isolated tissue. Light from an iodine 100W quartz lamp reached the preparation by epillumination producing a diffused spot of about 1 mm diameter. The light intensity at the source could be attenuated to any desired value by a series of neutral density and/or narrow band interference Baltzer B-40 filters. For intracellular recording ultrafine glass microelettrodes filled with KCl 1M or LiCl 1M solution, with the addition of staining substances were used. The electrode resistance ranged from 150 to 300 MΩ.

Figure 1 shows a series of superimposed responses to 50 msec white flashes of increasing intensity. The time course of these responses resembled those of retinal cone cells of lower vertebrates (Baylor and Fuortes, 1970). The time to peak ranged from 1 second in the smallest detectable photoresponses, to less than 100 millisecond in the largest responses.

Subsequent morfological identification at both light and electron microscope of the cells producing these photoresponses revealed the presence of a typical outer segment filled with disk membranes and of an inner segment of about 5 microns diameter ending with a large peduncle.

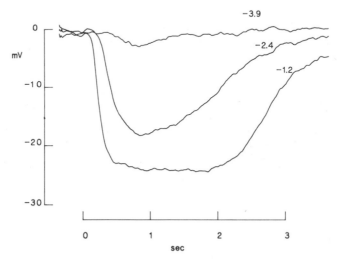

Fig. 1: Photoresponses from pineal photoreceptors. Superimposed recordings of three receptor responses to 50 msec monochromatic (521 nm) flash of increasing intensity. Numbers above each record indicate logarithmic units of attenuation. Zero in the ordinate indicates membrane potential in the dark.

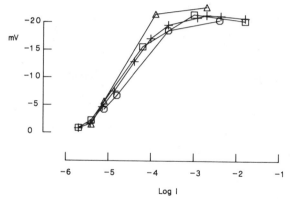

Fig. 2: Amplitude - intensity relationship in pineal receptor cells. Data obtained from three cells, indicated by various symbols. The average function is marked by the crossed line. Zero in the ordinate indicate membrane potential in the dark.

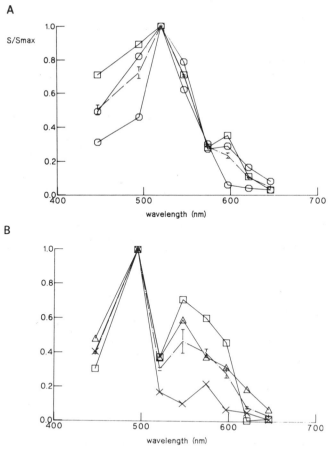

Fig. 3: Action spectra obtained from pineal photoreceptors. In A and B the interrupted line indicates the mean function and s.d. For explanation, see text.

Fig. 2 shows the average V/Log I function obtained from three photoreceptors in a dark adapted state. The cells responded over about 3.5 log units of light intensity, a value similar to that observed in retinal cones. The absolute sensitivity of receptor cells, measured with monocromatic flashes of just suprathreshold intensity producing linear responses was about 1200 μV ph-1 μm^2, at 521 nm. This values is also close to that measured in retinal cones.

Fig. 3 shows the action spectra obtained from two groups of receptor cells which could be distinguished by their sensitivity peaks. A first group of photoreceptors (Fig. 3A) shows a single peak at about 521 nm, with a rapid decay toward the longer wavelenghts. In these cells the responses to different wavelenghts at various intensities could be matched by their amplitude and time course, indicating their dependence on the number of photons captured indipendently of the wavelenght. The second group of photoreceptors (Fig. 3B) produces a more complex spectral function, characterized by two distinct peaks at 495 nm and 547 nm respectively. At present there are no indications whether the second group of cells contains distinct pigment types, or some neuronal interaction, within the pineal body, may contribute to the distribution of their sensitivity peaks.

During the course of these experiments, there were also recorded slow, graded depolarizing photoresponses which had not been previously described in the pineal organ of any animal species. Upon histological examination these cells showed an elongated shape reaching about 10-13 microns, a cell body located at one extremity and a few, short terminals departing from the other end.

A typical series of depolarizing responses to illumination of increasing intensity is shown in fig. 4. These were characterized by a slow rising phase at low light intensity and their time course underwent changes with light intensity very similar to the receptors

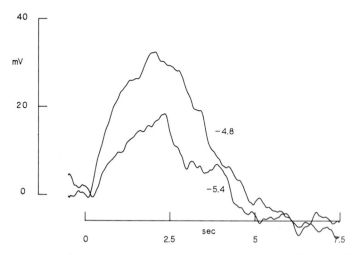

Fig. 4: Depolarizing photoresponses recorded from an histologically identified pineal cell. The light stimulus was a 50 msec white flash occurring at zero time. Numbers to the right indicate logarithmic unit of attenuation of the light source. Zero in the ordinate indicates membrane potential in the dark.

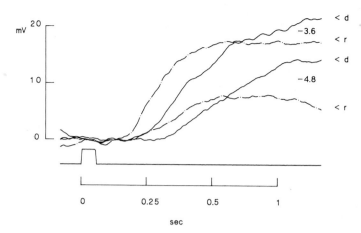

Fig. 5: Comparison between the latency of receptor and depolarizing cells responses to 50 msec flashes of white light of two different intensities, as indicated by the numbers to the right. pineal cells. r points to the receptor responses (dashed traces), d points to the depolarizing cells responses. For the sake of comparison, receptor responses are shown at inverted polarity.

cells, with the exception of the reversed polarity. A characteristic feature of depolarizing responses was a longer latency to all illumination intensities compared to photoreceptors under the same stimulus conditions, as shown in fig. 5. At low light intensity, the delay of depolarizing responses was about 80 msec longer than that of receptor cells, and it slightly decreased with increasing the light intensity.

Figure 6 shows the average V/Log I curve derived from four depolarizing cells. It appeared that the slope of the curve was steeper than photoreceptors (cf. Fig. 2) and it

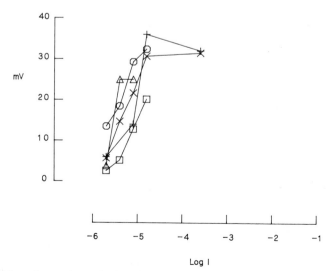

Fig. 6: Amplitude - Intensity relationship of depolaring cells. The line marked by an x refers to the average function. Zero in the ordinate indicate membrane potential in the dark.

was shifted to the left, indicating a greater light sensitivity. The absolute sensitivity of depolarizing cells at 521 nm was calculated from photoresponses in the linear range and the value obtained was about 7000 μV.Ph-1.μm^2, i.e. by about .8 log units more sensitive than photoreceptors.

The longer response latency observed in depolarizing cells, their steeper V/Log i function and the greater sensitivity with respect of photoreceptors are suggestive elements supporting the postsynaptic nature of depolarizing cells, which therefore receive a significant convergence of photoreceptor inputs.

We have no element at present to define the nature of the depolarizing cell, since there are no reports in the literature recalling either their morphology or their electrical activity. We are tempted to consider these new cell type as postsynaptic to photoreceptors, whose signals are instrumental to elicit there a voltage-dependent neurosecretion. A local release of substances from depolarizing cells within the pineal organ is supported by the dense vascularization present within the organ itself.

REFERENCES

Baylor, D. A. and Fuortes, M. G. F. , 1970, *J. Physiol.*, 207:77-92.

Dodt, E. and Meissl, H., 1982, *Experientia*, 38:996-1000.

Hanyu, I., Niwa, H. and Tamura, T., 1969, *Vision Res.*, 9:621.

Hartwig, H. G. and Oksche, A., 1982, *Experientia*, 38:991-996.

Heerd, E. and Dodt, E., 1961, *Pflüger Arch.*, 274:33.

Holmgren, U., 1959, "Goteborg Vetenskaps," Vitterhetssamh. Handl., B8, 1-66.

Kusmic, C., Marchiafava, P. L., and Strettoi, E., 1988, *J. Physiol.*, in press.

Marchiafava, P. L., 1985, *Proc. R. Soc. Lond. B*, 226:211-215.

Meissl, H., 1986, *In*: "Pineal and retinal relationships," O'Brien, P. J. and Klein, ed., D. C., Academic Press, pp 33-45.

Meissl, H. and Ekstrom, P., 1988, *Vision. Res.*, 28:49-56.

Oksche, A., 1986, *In*: "Pineal and retinal relationships," O'Brien, P. J. and Klein, D. C., ed., Academic Press, pp 1-14.

CONTRIBUTORS

INTERNATIONAL SCHOOL OF BIOPHYSICS ON "SENSORY TRANSDUCTION"

ERICE 9 - 19 JUNE 1988

LIST OF LECTURERS

Borsellino A. (Director of the School), SISSA, Strada Costiera 11 - 34014 Trieste, Italy

Torre V. (Director of the Course), Dipartimento di Fisica, Via Dodecaneso 33 - 16146 Genova, Italy

Cervetto L. (Director of the Course), Istituto di Neurofisiologia CNR - Via S. Zeno 51 - 56100 Pisa, Italy

Ashmore J., Department of Physiology, Medical School, University Walk, Bristol BS8 1TD , England

Baylor D., Department of Neurobiology Stanford University - Stanford, CA 94305, U.S.A.

Cavaggioni A., Istituto di Fisiologia Umana, Via A. Gramsci 14 - 43100 Parma, Italy

Chabre M., CNRS, Institut de Pharmacology, Route des Lucioles, Sophia Antipolis, 06560 Valbonne, France

Detwiler P., Department of Physiology and Biophysics SJ-40 - University of Washington School of Medicine, Seattle, WA 98195 U.S.A.

Fornili S.L., Dipartimento di Fisica, Via Archirafi 36 - 90123 Palermo, Italy

Gold G., Monell Chemical Senses Center, 3500 Market Street, Philadelphia, PA 3308 U.S.A.

Hamm H., University of Illinois, Department of Physiology and Biophysics, 835 S. Wolcott, Chicago 60608, U.S.A.

Horn R., Roche Institute of Molecular Biology, Department of Neurosciences, Nutley NJ 07110 U.S.A.

Kaissling K.E., Max-Planck-Institut fur Verhaltensphysiologie, Abteilung Schneider Post Stannberg, 8130 Seewiesen, West Germany

Kaupp B., Fachbereich Biologie/Chemie KFA. IBI. Postfach 1913, 5170 Julich, West Germany

Kros C., School of Biological Sciences, University of Sussex - Falmes, Brighton BN1 9QG, England

McNaughton P., Physiological Laboratory, University of Cambridge, Cambridge CB2 3EG, England

Stieve H., Institut fur Biologie II - Zoologie RWTH Aachen, Kopernikus Strasse 16, 5100 Aachen, West Germany

LIST OF PARTICIPANTS

Barbour, B.J., Department of Physiology, University College London, Gower Street, London WC1E 6BT, England

Bhaskaran, R., Department of Physics, BharathidasanUniversity, Tamilnadu, India

Benke, T., Dept. of Electrical and Computer Enginnering, Rice University, Houston TX 77251-1892, U.S.A.

Bizzarri, A.R., SISSA, Strada Costiera 11 - 34014 Trieste, Italy

Bornacin, F., CNRS, Institut de Pharmacology, Route des Lucioles, Sophia Antipolis, 06560 Valbonne, France

Cantatore, G., Istituto di Biofisica, Via San Lorenzo 26 - 56100 Pisa, Italy

Celebi, G., Ege University School of Medicine, Division of Basic Medical Sciences, Bornova, Izmir, Turkey

Colombaioni, L., Istituto di Neurofisiologia, Via S. Zeno 51 - 56100 Pisa, Italy

D'Amato, E., Dipartimento di Fisica, 38050 Povo, Trento, Italy

D'Arrigo, C.M., Istituto di Chimica Agraria, Via Valdisavoia 1 - Catania, Italy

Di Maio, V., Istituto di Cibernetica C.N.R., Via Toiano 6 - 80076 Arco Felice, Napoli, Italy

Distasi, C., Dipartimento di Biologia Animale, Laboratorio di Fisiologia Generale, Corso Raffaello 30 - 10126 Torino, Italy

Everett, K., Department of Physiology, University College London, Gower Street, London WC1E 6BT, England

Forti, L.C., Dipartimento di Fisiologia e Biochimica Generali, Via Celoria 26 - 20133 Milano, Italy

Forti, S., I.R.S.T., 38050 Povo, Trento, Italy

Grolli, S., Istituto di Biologia Molecolare, Via del taglio - 43100 Pisa, Italy

Hare, W., Department of Biophisics, 213 Donner Laboratory, University of California, Berkeley CA 94720, U.S.A.

Iappolo, A., Istituto di Chimica Agraria, Via Valdisavoia 1 - Catania, Italy

Karpen, J.W., Department of Cell Biology, Stanford University School of Medicine, Stanford CA 94305, U.S.A.

Kusmic, C., Dipartimento di Fisiologia e Biochimica, Via San Zeno 29/31 - 56100 Pisa, Italy

Lagnado, L., Physiological Laboratory, University of Cambridge, Cambridge CB2 3EG, England

Lazard, D., Weizmann Institute for Science, 76100 Rehovot, Israel

Lin Fan, SISSA, Strada Costiera 11 - 34014 Trieste, Italy

Lucia, S., Istituto di Biofisica, Via San Lorenzo 26 - 56100 Pisa, Italy

Lühring, H., Fachbereich Biologie/Chemie, Universitat Osnabruck, 4500 Osnabruck, West Germany

Mammano, F., SISSA, Strada Costiera 11 - 34014 Trieste, Italy

Masetto, S., Istituto di Fisiologia Generale, Via Forlanini 6 - 27100 Pavia, Italy

Mazzoni, M.R., Dept. of Physiology and Biophysics, University of Illinois at Chicago, College of Medicine, 835 S. Wolcott, Chicago IL 60612, U.S.A.

Menini, A., Istituto di Cibernetica e Biofisica, CNR, Via Dodecaneso 33 - 16146 Genova, Italy

Nassivera, E., Dipartimento di Fisica, 38050 Povo, Trento, Italy

Perry, R., Physiological Laboratory, University of Cambridge, Cambridge CB2 3EG, England

Piskin, K.A., Hacettepe University, Faculty of Medicine, Department of Medical Biology, Sihhiye, Ankara, Turkey

Puia, G., SISSA, Strada Costiera 11 - 34014 Trieste, Italy

Ratto, G.M., Istituto di Neurofisiologia CNR, Via San Zeno 51 - Pisa, Italy

Raynauld, J.P., Department of Physiology, Universite' de Montreal, Montreal, Quebec, Canada

Rennie, K., Dept. of Physiology, Medical School, University Walk, Bristol BS8 1TD, England

Rispoli, G., University of Washington School of Medicine, Dept. of Physiology & Biophysics, Seattle, WA 98195, U.S.A.

Schneeweis, D.M., University of Michigan, 1103 East Huron, Ann Arbor, Michigan 48104, U.S.A.

Sciancalepore, M., SISSA, Strada Costiera 11 - 34014 Trieste, Italy

Sericano, M., Dipartimento di Fisica, Via Dodecaneso 33 - 16146 Genova, Italy

Trimarchi, C., Istituto di Neurofisiologia, Via S. Zeno 51 - 56100 Pisa, Italy

Usai, C., Istituto di Cibernetica e Biofisica, CNR, Via Dodecaneso 33 - 16146 Genova, Italy

Xu Yaozhong, Dipartimento di Fisica, 38050 Povo, Trento, Italy

INDEX